石油高等院校特色规划教材

Drilling Engineering
钻井工程

（英汉对照·富媒体）

刘志坤　张　冰　主编

李　琪　主审

石油工业出版社

内容提要

本书以英汉对照结合富媒体展示的方式系统阐述钻井工程的基本理论和技术，内容包括旋转钻机的基本组成、钻井工程地质条件、钻柱、钻头、钻井液、钻井水力学、定向钻井、井控、固井及钻井事故与复杂井况等。

本书可作为石油工程专业高年级本科学生和留学生的教材，也可供油气井工程研究生和从事石油钻井的工程技术人员参考。

图书在版编目（CIP）数据

钻井工程：富媒体：英汉对照 / 刘志坤，张冰主编 . —北京：石油工业出版社，2022.8

石油高等院校特色规划教材

ISBN 978-7-5183-5505-1

Ⅰ . ①钻… Ⅱ . ①刘…②张… Ⅲ . ①钻井工程–高等学校–教材–英、汉 Ⅳ . ① TE2

中国版本图书馆 CIP 数据核字 (2022) 第 137912 号

出版发行：石油工业出版社
（北京市朝阳区安华里2区1号楼　100011）
网　　址：www.petropub.com
编辑部：（010）64523733
图书营销中心：（010）64523633
经　　销：全国新华书店
排　　版：三河市燕郊三山科普发展有限公司
印　　刷：北京中石油彩色印刷有限责任公司

2022年8月第1版　2022年8月第1次印刷
787毫米×1092毫米　开本：1/16　印张：25.5
字数：636千字

定价：64.00元
（如发现印装质量问题，我社图书营销中心负责调换）
版权所有，翻印必究

Preface/ 前言

With the continuous strengthening of China's foreign cooperation in the oil and gas field, the number of professional technicians working abroad and international students coming to China from the Belt and Road is increasing. This puts forward new requirements for the talent training of Petroleum Engineering Specialty under the background of international cooperation, which needs to be supported by corresponding bilingual textbooks. This textbook is a professional course textbook for undergraduate, graduate and international students majoring in petroleum engineering. Based on the principle of combining theory with practice, comparing Chinese and English, wide coverage and reflecting the technological frontier as much as possible, the textbook systematically expounds the basic theories and methods involved in drilling engineering from the basic composition of rotary drilling rig, drilling engineering geological conditions, drilling tools, drilling fluid, directional drilling technology, oil and gas well pressure control technology, cementing technology and other aspects. The content arrangement of the textbook basically conforms to the principle of step-by-step, which is conducive to classroom explanation and students' self-study. The content of the textbook is suitable for classroom teaching of about 50 class hours.

The features of this textbook are as follows: firstly, it is for different process links of drilling engineering, using Chinese-English bilingual method is adopted to highlight the basic theory, basic calculation, engineering design and basic process. Secondly, this textbook adopt Chinese and English, which is suitable for bilingual teaching. Thirdly, it insert a large number of pictures and videos to avoid single text introduction and help readers deepen their understanding.

随着我国油气领域对外合作的不断加强，石油钻井涉外专业技术人员和一带一路沿线国家来华留学生的数量均在不断增加。这对国际合作背景下石油工程专业的人才培养提出了新的要求，需要有与之相适应的双语教材作为支撑。本教材是根据这一实际需求，面向石油工程专业本科生、研究生和留学生编写的专业课教材。全书本着理论与实际相结合、中英对照、覆盖面广、尽量反映技术前沿的原则，从旋转钻机的基本组成、钻井工程地质条件、钻进工具、钻井液、定向钻井技术、油气井压力控制技术、固井技术等多方面，系统阐述了钻井工程所涉及的基本理论和方法。教材在内容编排上符合循序渐进的原则，有利于课堂讲解和学生自学。教材内容适合50学时左右的课堂讲授。

本书的特点是：首先，针对钻井工程不同的工艺环节，采用中英对照的方式，突出基本理论、基本计算、工程设计和基本工艺过程；其次，采用中英文本对照，适合双语教学；最后，插入大量的图片和视频，在避免文字介绍单一的同时有利于读者加深理解。

This textbook is compiled by Liu Zhikun、Zhang Bing、Wang Liupeng、Cao Jie from the oil and gas well engineering teaching and Research Office of Xi'an Shiyou University. The third, sixth and ninth chapters are compiled by Liu Zhikun, the first and second chapters are compiled by Zhang Bing, the fifth and seventh chapters are compiled by Wang Liupeng, and the fourth and eighth chapters are compiled by Cao Jie. The audio of this textbook was recorded by Zhong Zebo, Zhang Luxiaohe, and Zhang Fan. The textbook was edited by Liu Zhikun and Zhang Bing, and the final draft was uniformly revised. After the completion of the book, Professor Li Qi conducted a comprehensive review.

In particular, it should be noted that graduate students Liu Chenglu, Zhang Wei, Zhang Yinghui, Zhang Chaoqun, Li Ke, Li Aixin, Guo Shuai, Luo Rongtao and Zhao Yifan participated in the collection and arrangement of some materials. We appreciate them for their hard work and dedication.

In the process of editing this book, Ni Weijun, Gao Xiaorong, AI Erxin have made their contributions. This textbook is supported by the Excellent Academic Publication Fund of Xi'an Shiyou University, the National Natural Science Foundation of China (52004214) and the Oil and Gas Well Theory and Technology Innovation Team of Xi'an Shiyou University.

This book quotes a large number of previous research results (including pictures, photos and videos), and we would like to express our thanks. Some citations are not given directly in the text, but listed in the references at the end of the textbook.

Due to the limitation of the editor's level, there are some mistakes. Please use this textbook for criticism and correction.

本教材由西安石油大学油气井工程教研室的刘志坤、张冰、王六鹏、曹杰编写而成，其中第三章、第六章、第九章由刘志坤编写，第一章、第二章由张冰编写，第五章、第七章由王六鹏编写，第四章、第八章由曹杰编写。本书音频由钟泽波、张露小荷、张帆录制。本书由刘志坤、张冰主编并就最后定稿进行了统一修改。李琪教授对本书进行了全面审查。

特别需要指出的是，研究生刘成路、张薇、张莹辉、张超群、李轲、李瑷昕、郭帅、罗荣涛、赵一凡参与了资料的收集整理与文字整理等工作，对他们的辛勤付出在此表示真诚的感谢。

在本书编写过程中得到了西安石油大学倪维军、高晓荣、艾二鑫等老师的支持和帮助，得到了西安石油大学优秀学术著作出版基金、国家自然科学基金项目（52004214）、西安石油大学油气井理论与技术创新团队的资助，在此表示感谢。

本书引用了大量前人的研究成果（包括图、照片和视频等），在此一并表示感谢。有些引用没有直接在正文中给出，而是列在教材最后的参考文献中。

由于编者的水平所限，其中不免有不当和错误之处，诚请使用本教材的师生和广大读者批评指正。

编者
2022 年 4 月

Contents/ 目录

1. **Overview of Drilling/ 绪论** ·· 1
   - 1.1 Introduction/ 引言 ·· 1
   - 1.2 History of Drilling/ 钻井的历史 ·· 4
   - 1.3 Drilling Rig Systems/ 钻机系统 ·· 7
   - 1.4 The Well Construction Process/ 建井过程 ································· 30
   - 1.5 Offshore Drilling/ 海洋钻井 ·· 36
   - 1.6 Exercise/ 习题 ·· 38

2. **Geological Conditions for Drilling Engineering/ 钻井工程地质条件** ········ 40
   - 2.1 Geomechanics in Drilling/ 钻井地质力学 ·································· 40
   - 2.2 Formation Pressures/ 地层压力 ··· 67
   - 2.3 Exercise/ 习题 ·· 100

3. **Drilling Tools/ 钻井工具** ·· 105
   - 3.1 Drilling Bits/ 钻头 ··· 105
   - 3.2 The Drillstring/ 钻柱 ··· 134
   - 3.3 Drillstring Design/ 钻柱设计 ··· 158
   - 3.4 Exercise/ 习题 ·· 162

4. **Drilling Fluids/ 钻井液** ·· 163
   - 4.1 Functions of Drilling Fluids/ 钻井液功能 ································· 164
   - 4.2 Drilling-Fluid Categories/ 钻井液类型 ····································· 167
   - 4.3 Rheological Models of Drilling Fluids/ 钻井液流变模型 ········ 178
   - 4.4 Exercise/ 习题 ·· 190

5. **Drilling Hydraulics/ 钻井水力学** ··· 192
   - 5.1 Introduction to Drilling Hydraulics/ 钻井水力学简介 ·············· 192
   - 5.2 Hydrostatic Pressure Calculations/ 静水压力计算 ··················· 193
   - 5.3 Steady Flow of Drilling Fluids/ 钻井液稳态流动 ····················· 203
   - 5.4 Exercise/ 习题 ·· 217

6 Directional Drilling/ 定向钻井 ········· 219

6.1 Fundamentals of Trajectory Design/ 井眼轨道设计基础 ········· 219
6.2 Deviation Control/ 井眼轨迹控制 ········· 244
6.3 Exercise/ 习题 ········· 255

7 Well Control/ 井控 ········· 258

7.1 Introduction/ 引言 ········· 258
7.2 Well Control Principles/ 井控原则 ········· 259
7.3 Warning Indicators of Kicks/ 井涌预警 ········· 267
7.4 Secondary Control/ 二级井控 ········· 276
7.5 Well Killing Procedures/ 压井程序 ········· 292
7.6 Exercise/ 习题 ········· 300

8 Casing & Cementing/ 固井 ········· 302

8.1 Casing Program/ 井身结构 ········· 302
8.2 Casing/ 套管柱 ········· 311
8.3 Cementing/ 注水泥 ········· 327
8.4 Exercise/ 习题 ········· 360

9 Drilling Risks/ 钻井事故与复杂井况 ········· 361

9.1 Pipe Sticking/ 卡钻 ········· 361
9.2 Loss of Circulation/ 井漏 ········· 386
9.3 Hole Deviation/ 井斜 ········· 395
9.4 Borehole Instability/ 井壁失稳 ········· 397
9.5 Exercise/ 习题 ········· 398

References/ 参考文献 ········· 400

Overview of Drilling

绪 论

1.1 Introduction

As we known, petroleum is not located in a place of easy access. Concealed beneath millions of tons of rocks, trapped deep below ground, many thousands of meters under Earth's surface, this energy

Audio 1.1

often is located in areas of very difficult access or even offshore, under very deep waters. Nevertheless, due to its importance in our lives, no matter the location or the depth, in some way we will have to find a means to establish a path between the energy source and the surface. Somehow, in a notable work of engineering, a well will be drilled connecting the energy source to the surface. This well will have the necessary strength to withstand the enormous pressures produced by the adjacent rocks. It will prevent any damage to the surrounding environment. It will be outfitted with modern equipment to allow production in a safe and nondetrimental way. The planning and execution of this remarkable work of engineering is the job of drilling engineers.

However, before drilling an exploration well, an oil company will have to obtain a production licence. This licence will allow the company to drill exploration wells in the area of interest. The licence may be acquired by an oil company directly from the government, during the licence rounds are announced, or at any other time by farming-into an existing licence. Before the exploration wells are drilled the licencee may shoot extra seismic lines, in a closer grid pattern than it had done previously. This will

1.1 引言

众所周知，石油储藏在地下数千米、重达数百万吨的岩石下方，通常位于人类难以到达的偏远地区，甚至是海洋的深水水域之下，开采起来比较困难。然而，石油对于我们的生活却十分重要，无论在什么地方、什么深度，都需要想方设法地打开它与地面之间的连接通道，而打开这一连接通道就是通过一项浩大的工程，也就是钻井来实现。所钻的井必须拥有足够的强度来承受井周岩石产生的巨大压力，并能防止对周围环境的污染。钻井需配备现代设备来实现安全无害的生产作业，钻井工程师的工作就是计划并实施这一项重大的工程。

石油公司通常需要先获得生产许可证，才能进行探井钻井。生产许可证是石油公司在所属地区进行探井作业的许可文件。石油公司可以在许可证公布期间直接向政府申请，也可以在其他任何时间通过收购获得已有的许可证。在钻探井之前，作业方可能还需要采用更密集的布线方式进一步进行

provide more detailed information about the prospect and will assist in the definition of an optimum drilling target. Despite improvements in seismic techniques the only way of confirming the presence of hydrocarbons is to drill an exploration well. Drilling is very expensive, and if hydrocarbons are not found there is no return on the investment, although valuable geological information may be obtained. With only limited information available a large risk is involved. Having decided to go ahead and drill an exploration well proposal is prepared. The objectives of this well will be:

(1) To determine the presence of hydrocarbons;

(2) To provide geological data (cores, logs) for evaluation;

(3) To flow test, the well to determine its production potential, and obtain fluid samples.

The life of an oil or gas field can be sub-divided into the following phases: (1) Exploration; (2) Appraisal; (3) Development; (4) Maintenance; (5) Abandonment (Fig 1.1)

The length of the exploration phase will depend on the success or otherwise of the exploration wells. There may be a single exploration well or many exploration wells drilled on a prospect. If an economically attractive discovery is made on the prospect, then the company enters the Appraisal phase of the life of the field. During this phase more seismic lines may be shot and more wells will be drilled to establish the lateral and vertical extent of (to delineate) the reservoir. These appraisal wells will yield further information, on the basis of which future plans will be based. The information provided by the appraisal wells will be combined with all of the previously collected data and engineers will investigate the most cost effective manner in which to develop the field. If the prospect is deemed to be economically attractive a Field Development Plan will be submitted for approval. It must be noted that the oil company is

地震勘探。这样会获得更详细的勘探信息，并且能帮助作业方确定一个最佳的钻井目标。尽管地震技术已经有了较大发展，但钻探井仍然是确定油气存在的唯一方法。钻井成本通常很高，如果发现不了油气，即便可以获得有价值的地质信息，其结果仍然是投资失败。一旦决定进行钻探，就需要准备探井的钻井方案。探井的主要目的有：

（1）确定油气的存在；

（2）为地层评价提供地质资料（岩心，测井数据）；

（3）通过试井确定产能，并获得流体样品。

油气田的生命周期可分为以下几个阶段：（1）勘探阶段；（2）评价阶段；（3）开发阶段；（4）维护阶段；（5）弃井阶段（图1.1）。

在勘探阶段，需要部署一口或多口探井，勘探时间的长短取决于探井的数量和能否成功发现油气。如果发现预期经济上可观的油气，将进入评价阶段。在评价阶段油公司会通过布置更多的地震测线、打更多的评价井来明确（或划定）油藏的横、纵向范围。这些评价井将提供更详尽的信息，作为未来开发计划的制订基础。工程师们将这些评价井的信息和之前收集的所有数据结合起来，研究最具经济效益的油田开发方式。如果预期经济效益可观，石油公司会提交一份油田开发计划申请批准。必须指

only a licensee and that the oilfield is the property of the state. The state must therefore approve any plans for development of the field. If approval for the development is received, then the company will commence drilling development wells and constructing the production facilities according to the Development Plan. Once the field is on-stream the companies' commitment continues in the form of maintenance of both the wells and all of the production facilities.

出的是，石油公司只是一个许可证持有者，而油田是国家的财产。因此，油田的任何开发计划都必须得到国家的批准。如果获得批准，就进入开发阶段，石油公司就会开始打开发井，并根据开发计划建设地面的生产设施。一旦油田开始投产，石油公司将持续承担维护油井和所有生产设施的任务。

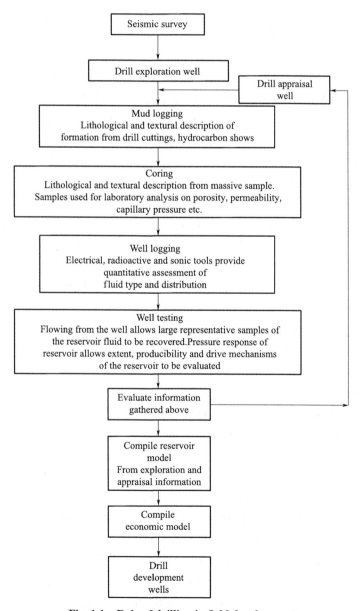

Fig. 1.1 Role of drilling in field development

After many years of production, it may be found that the field is yielding more or possibly less hydrocarbons than initially anticipated at the Development Planning stage and the company may undertake further appraisal and subsequent drilling in the field.

At some point in the life of the field the costs of production will exceed the revenue from the field and the field will be abandoned. All of the wells will be plugged and the surface facilities will have to be removed in a safe and environmentally acceptable fashion.

1.2　History of Drilling

Audio 1.2

Drilling for oil and gas is not new. Even though the modern techniques, equipment, and methods that are going to be described in this book are completely different from the ones used in the prehistoric era of the industry, there is evidence of wells purposely drilled for production of hydrocarbons as early as A.D. 347 in China. Also, there are reports of oil well drilling activities in Japan by A.D. 600.

The first oil well drilled in Europe dates back to 1745 in Pechelbronn, France, where petroleum mining from oil sands had been taking place since 1498. After that, many wells were drilled, mostly using rudimentary hand tools, in Europe, North America, and Asia where an oil well was drilled in 1848 on the Aspheron Peninsula northeast of Baku. Following the Baku well, various shallow oil wells were drilled in Europe during the next decade.

There are many different versions about where the first well of the modern oil industry was drilled. Depending on the historian, locations and dates will vary widely. Also, there are many different depictions about what should be considered a modern well and what would differentiate it from a well. In this chapter, without

油田生产多年以后，实际的产量可能会超出或不如开发计划阶段的最初预期。这时，石油公司需要进一步评价油田，并部署相应的钻井作业。

当油田的开发生产不再具有经济效益时，油田就会进入弃井阶段，这时所有的井将被封堵，地面的设施也会按照安全和环保的要求进行拆除。

1.2　钻井的历史

油气钻井是一项历史悠久的技术。本书将要讲述的现代钻井内容与工业史前时代所使用的技术、设备和方法完全不同。追溯起来，有证据表明，早在公元347年，中国就针对油气生产进行过钻井作业。除此以外，也有报道表明，日本在公元600年也有过钻井活动。

欧洲钻的第一口油井可追溯到1745年，位于法国的佩彻布朗。这个地区自1498年开始就有人从油砂中开采石油。之后，在欧洲、北美和亚洲又打了许多井，大多使用手工工具完成。1848年在亚洲巴库东北部的阿斯菲隆半岛上钻了一口油井。在这之后的十年里，人们又在欧洲钻了各种浅井。

现代石油工业第一口井究竟在哪里打的，有许多不同的观点。不同的历史学家给出的时间和地点有很大的区别。至于什么样的井应该被看作是一口现代井，它和其他井的区别

dueling about whom, where, and when, which would be rather ineffectual, we will mention some important milestones and pioneers that have contributed to the advancement of the modern oil industry in those early days.

In 1858, in Oil Springs (then part of the township of Enniskillen), Ontario, Canada, a rudimentary 49ft well was dug by James Miller Williams with the intention to produce "kerosene" for lamps. Even though this is considered to be one of the pioneer ventures of the North American oil industry, the Williams well did not represent any significant advance as far as drilling technology is concerned.

A true milestone for the drilling industry and probably the world's most widely recognized drilling milestone occurred in 1859. In that year, in Titusville, Pennsylvania, USA, Edwin L. Drake drilled what is, so far as known or documented, the first well purposely planned for oil in the United States. Even though there is evidence of oil and gas wells that had been drilled in the United States for as long as 40 years prior to the Drake well, most of those early wells were actually originally drilled in search of potable water or brine.

In the nineteenth century, cable tool rigs were in widespread use in North America. Cable tool rigs pounded through soil and rock to drill the well by repeatedly dropping a heavy iron bit attached to a cable. Using a 6hp steam engine to power his cable tool drilling equipment, Drake drilled a well 69.5 ft deep that initially produced oil at a rate believed to be from 8 to 10 bbl/d. Immediately after that first success, many other wells followed, causing the oil industry—and particularly its drilling segment—to experience an unprecedented growth.

在哪里，有很多不同的观点。在这一章，我们不讨论谁、在哪里、在何时打了第一口井。我们要阐述一些重要的里程碑和先驱们，他们在早期为现代石油工业的进步做出了很大贡献。

1858年，在加拿大安大略省的油泉村（当时是恩尼斯基伦镇的一部分），詹姆斯·米勒·威廉姆斯打了一口49ft❶深的油井，打算生产用于点灯用的"煤油"，被认为是北美石油工业的先驱之一。但就钻井技术而言，这口井并没有体现出任何有意义的进步。

钻井行业真正的里程碑或者说是世界上最广泛认可的钻井里程碑发生在1859年。那一年，在美国宾夕法尼亚州的泰特斯维尔，埃德温·L.德雷克钻了美国第一口以获取石油为目的的油井，即德雷克油井。尽管有证据表明，在德雷克油井之前，美国已经打了长达40年的石油和天然气井，但实际上这些早期井的最初目的大多数是为了寻找饮用水或盐水。

19世纪，在北美广泛使用的钻井工具是顿钻钻机。它通过缆绳反复提升和下放一个较重的铁质钻头，连续撞击土壤和岩石，形成井眼。德雷克用6hp❷的蒸汽机驱动他的顿钻钻井设备，钻成了一口69.5ft深的井，这口井最初的产油量为每天8～10bbl❸。第一次成功之后，德雷克紧跟着又钻成了许多油井，推动石油工业，特别是钻井环节经历了前所未有的发展。

---

❶ 1ft=0.3048in； ❷ 1hp=746W； ❸ 1bbl=158.98L。

Percussion drilling was widely used by the oil industry until the 1930s. Even though there are archaeological records of the Egyptians using rotary drilling mechanisms as early as 3000 B.C., the process was little used in the early days of the industry because of its complexity compared to percussion drilling.

During the 1890s, Patillo Higgins, a mechanic and self-taught geologist, tried to prove his theory about the existence of oil at a depth of approximately 1,000 ft below a salt dome formation south of Beaumont in eastern Jefferson County, Texas, USA. He gathered funding from various partners and drilled three wells, including one using a cable tool rig, but all three holes did not reach the proposed final depth because unconsolidated sand at a depth of approximately 400 ft caused all three wells to be lost. In 1899 Higgins partnered with Captain Anthony Francis Lucas, an engineer and navy officer who had immigrated to America a few years earlier, to try again to drill a proper test well in the region. Their first try, a well drilled in July 1899, was yet another disappointment, with the wellbeing lost due to the same unstable sands that caused the previous failures. Their second well, however, the fifth try in the region, finally succeeded in reaching the objective.

With operations initiated on 27 October 1900, the Lucas well reached its final depth on 8 January 1901, and two days later, when a new drilling bit was being run into the well, it started flowing with an amazing and unprecedented rate of nearly 100,000 bbl/d, which at that time represented more than the entire oil production of the United States. The Lucas well, besides establishing the presence of large hydrocarbon deposits in the region, also demonstrated the viability

直到20世纪30年代，石油工业一直广泛使用顿钻钻井。尽管有考古记录，早在公元前3000年埃及人就已经使用了旋转钻井机械装置，但由于这种工艺比顿钻钻井更加复杂，因此在工业时代早期很少使用。

19世纪90年代，一位名叫帕蒂洛·希金斯的机械师（也是位自学成才的地质学家），试图证明他的成油理论，认为在美国得克萨斯州杰斐逊县东部博蒙特以南的盐丘下方，1000ft深的地层中存在石油。为此，他从合伙人那里筹集资金，钻了三口井，其中一口井使用了顿钻钻井。但这三口井都没有钻达目的深度，因为在大约400ft深度处，松散的砂岩导致井眼报废。1899年，希金斯与几年前移民美国的工程师、海军军官安东尼·弗朗西斯·卢卡斯船长合作，再次尝试在这一地区进行钻井。1899年7月他们合作的第一口井仍然失败了，原因和之前一样，是不稳定的砂岩所致。然而，他们的第二口井，也是该区域的第五口井——卢卡斯井，最终成功地钻达了目标层。

卢卡斯井于1900年10月27日开钻，1901年1月8日钻达最终深度。两天后，在准备换钻头继续钻进时，井内流体开始以惊人的速度自喷，产量达到近10×10^4bbl/d，这在当时超过了全美的石油总产量。卢卡斯井除了证实这一地区存在着大量的石油以外，也证明了使用旋转钻机

of using rotary drilling rigs to drill oil wells in soft formations where cable tools could not be effectively used except at very shallow depths. There is some disagreement about the final depth of the Lucas well. Captain Lucas stated that the final depth was 1,160 ft, while Al Hamill, the drilling contractor for the well, in an article written 50 years later reported the final depth as 1,020 ft. Nevertheless, more significant than the controversy is the importance of this well for the oil industry. The well was drilled using the most advanced technology known, and after its initial uncontrolled geyser-like production, it was capped, allowing production to restart in a safe and controlled way. The Lucas well is considered by many as the birth of the modern oil industry. The fact that it had its operations initiated by the end of the last year of the nineteenth century and concluded on the very first days of the twentieth century just adds more symbolic meaning to this true milestone of an industry that changed the world.

1.3 Drilling Rig Systems

1.3.1 Drilling Personnel

Modern well drilling is an activity that involves many specialists and usually various companies. The expertise and number of engineers and technicians involved in the planning and execution of a drilling operation will depend on the type of well being drilled, its purpose, the well location, its depth, and the complexity of the operation.

Audio 1.3

A well drilled with the purpose of discovering a new petroleum reservoir is called an exploration (or wildcat) well. Wildcat wells are the very first ones drilled in a certain unexplored area. After a wildcat well has shown the potential of a reservoir to be productive, appraisal wells may be drilled to obtain more information about the reservoir and its extension. Once a newly discovered reservoir is considered economically viable, a development plan is established and development wells are drilled to produce the oil and gas present in the reservoir.

Besides the most common exploration and development wells, special wells may be drilled for a variety of purposes including stratigraphic tests and blowout relief.

Drilling a well requires many different skills and involves many companies. The oil company who manages the drilling and/or production operations is known as the operator. In joint ventures one company acts as operator on behalf of the other partners.

There are many different management strategies for drilling a well but in virtually all cases the oil company will employ a drilling contractor to actually drill the well. The drilling contractor owns and maintains the drilling rig and employs and trains the personnel required to operate the rig. During the course of drilling the well certain specialised skills or equipment may be required (e.g. logging, surveying). These are provided by service companies. These service companies develop and maintain specialist tools and staff and hire them out to the operator, generally on a day-rate basis.

The contracting strategies for drilling a well or wells range from day-rate contracts to turnkey contracts. The most common type of drilling contract is a day-rate contract. In the case of the day-rate contract the operator prepares a detailed well design and program of work for the drilling operation and the drilling contractor simply provides the drilling rig and personnel to drill the well. The contractor is paid a fixed sum of money for every day that he spends drilling the well. All consumable

多口评价井，以获得更多关于油藏本身和油藏范围的信息。一旦认定新发现区域的油藏在经济上具备开采可行性，石油公司就会制定出开发计划，部署开发井进行生产。

除了最常见的探井和开发井以外，还有一些不同目的的特殊井，包括层序测试井和救援井。

钻一口井需要许多不同的技术工作，涉及许多专业公司。管理钻井和（或）生产作业的石油公司称为作业者。在合资企业中，通常由一家公司代表其他合伙人担任作业者。

钻一口井有许多不同的管理策略，但几乎在所有的情况下，石油公司都会雇用钻井承包商进行实际钻井作业。钻井承包商拥有钻机并负责维护钻机，还负责雇用、培训操作钻机所需的人员。在钻井过程中，可能还需要某些专门的技术或装备（如测井、测斜），这些由服务公司提供。服务公司负责开发和维护专业工具，雇用工作人员，并将他们租赁给作业者，通常按日计算费用。

钻井的承包方式分为日费制和总承包制，最常见的是日费制。在这种合同制下，作业者为钻井作业准备详细的钻井设计和工作程序，钻井承包商只需要提供钻机和作业人员实施钻井。钻井承包商每天在钻井作业上的花费是固定的，所有消耗品、运输和支持服务等费用都由作业者提供。

items, transport and support services are provided by the operator.In the case of the turnkey contract the drilling contractor designs the well, contracts the transport and support services and purchases all of the consumables, and charges the oil company a fixed sum of money for whole operation. The role of the operator in the case of a turnkey contract is to specify the drilling targets, the evaluation procedures and to establish the quality controls on the final well. In all cases the drilling contractor is responsible for maintaining the rig and the associated equipment.

The operator will generally have a representative on the rig (sometimes called the "company man") to ensure drilling operations go ahead as planned, make decisions affecting progress of the well, and organise supplies of equipment. He will be in daily contact with his drilling superintendent who will be based in the head office of the operator. There may also be an oil company drilling engineer and/or a geologist on the rig. The drilling contractor will employ a toolpusher to be in overall charge of the rig. He is responsible for all rig floor activities and liaises with the company man to ensure progress is satisfactory. The manual activities associated with drilling the well are conducted by the drilling crew. Since drilling continues 24 hours a day, there are usually 2 drilling crews. Each crew works under the direction of the driller. The crew will generally consist of a derrickman (who also tends the pumps while drilling), 3 roughnecks (working on rig floor), plus a mechanic, an electrician, a crane operator and roustabouts (general labourers). Service company personnel are transported to the rig as and when required.Sometimes they are on the rig for the entire well (e.g. mud engineer) or only for a few days during particular operations (e.g. directional drilling engineer). An overall view of the personnel involved in drilling is shown in Fig. 1.2.

如果采用总承包制，则由钻井承包商进行钻井设计，按合同提供运输和支持服务，购买所有的消耗品，并向石油公司收取全部作业的固定费用。在这种情况下，作业者主要是规定具体的钻井目标、评价作业程序，并进行全井质量控制。无论哪种情况，钻井承包商都要负责维护钻机和相关设备。

石油公司通常会派驻一位代表到现场（有时称为"公司代表"），他的职责是确保钻井作业按计划进行，做出影响生产进度的决定，组织设备供应，和石油公司总部的钻井主管保持日常联系。钻井平台上可能还有石油公司的钻井工程师和地质工程师。钻井承包商将雇用一名钻井队长全面负责钻机运行。钻井队长负责所有的钻台活动，并与公司代表保持联络，确保钻井按进度进行。与钻井有关的手工操作则由钻井人员完成。由于钻井作业是24小时的连续作业，因此通常有2个班的钻井人员。每个班的钻井人员都在司钻的指挥下工作。一个班通常由一名井架工（钻井时还负责钻井泵）、三名钻工（在钻台上工作）、一名机械师、一名电工、一名起重机操作工和一名普通员工组成。服务公司的工作人员只在需要时到达钻井现场，有时他们需要整个钻井过程都在现场（如钻井液工程师），有时只在特定操作期间待几天（如定向钻井工程师）。钻井作业涉及的人员分工如图1.2所示。

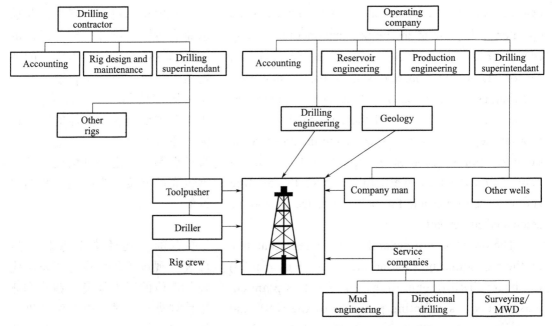

Fig. 1.2 Personnel involved in drilling a well

1.3.2 The Drilling Proposal and Drilling Program

The proposal for drilling the well is prepared by the geologists and reservoir engineers in the operating company and provides the information upon which the well will be designed and the drilling program will be prepared. The proposal contains the following information:

(1) Objective of the well;

(2) Depth, and location (longitude and latitude) of target;

(3) Geological cross section;

(4) Pore pressure profile prediction.

The drilling program is prepared by the drilling engineer and contains the following:

(1) Drilling rig to be used for the well;

(2) Proposed location for the drilling rig;

(3) Hole sizes and depths;

(4) Casing sizes and depths;

(5) Drilling fluid specification;

(6) Directional drilling information;

(7) Well control equipment and procedures;

1.3.2 钻井方案和钻井设计

钻井方案由作业公司的地质工程师和油藏工程师制定，为油井设计和钻井设计提供所需的信息。钻井方案包含以下信息：

（1）钻井目标；

（2）目的层深度和位置（经度、纬度）；

（3）地质剖面；

（4）孔隙压力预测剖面。

钻井设计由钻井工程师制定，主要包括以下内容：

（1）钻机选型；

（2）建议井位；

（3）井眼尺寸和深度；

（4）套管尺寸和深度；

（5）钻井液设计；

（6）定向井参数；

（7）井控设备和程序；

(8) Bits and hydraulics program.

1.3.3 Rotary Drilling Equipment

The first planned oil well was drilled in 1859 by Colonel Drake at Titusville, Pennsylvania USA. This well was less than 100 ft deep and produced about 50 bbl/d. The cable-tool drilling method was used to drill this first well. The term cable-tool drilling is used to describe the technique in which a chisel is suspended from the end of a wire cable and is made to impact repeatedly on the bottom of the hole, chipping away at the formation. When the rock at the bottom of the hole has been disintegrated, water is poured down the hole and a long cylindrical bucket (bailer) is run down the hole to collect the chips of rock. Cable-tool drilling was used up until the 1930s to reach depths of 7,500 ft.

In the 1890s the first rotary drilling rigs were introduced. Rotary drilling is the technique whereby the rock cutting tool is suspended on the end of hollow pipe, so that fluid can be continuously circulated across the face of the drillbit cleaning the drilling material from the face of the bit and carrying it to surface. This is a much more efficient process than the cable-tool technique. The cutting tool used in this type of drilling is not a chisel but a relatively complex tool (drill bit) which drills through the rock under the combined effect of axial load and rotation. The first major success for rotary drilling was at Spindletop, Texas in 1901 where oil was discovered at 1,020ft and produced about 100,000 bbl/d.

There are many individual pieces of equipment on a rotary drilling rig (Fig.1.3). These individual pieces of equipment can however be grouped together into six sub-systems. These systems are: the power system; the hoisting system; the circulating system; the rotary system; the well control system and the well monitoring system. Although the pieces of equipment associated with these systems will vary in design, these systems will be found on all drilling rigs. The equipment discussed below will be found on both land-based and offshore drilling rigs.

（8）钻头和水力参数。

1.3.3 旋转钻井设备

1859年德雷克上校在美国宾夕法尼亚州泰特斯维尔钻成了第一口油井——德雷克井。这口井井深不到100ft，日产油约50bbl，是第一口采用顿钻钻井方法打成的油井。顿钻钻井技术用缆绳的一端悬挂顿钻钻头，通过反复提拉和下放钻头，不断地撞击井底岩石，破碎岩石以后，将水倒入井筒中，下放一个圆柱形的水桶（打捞筒）用来收集岩石碎屑。顿钻钻井一直使用到20世纪30年代，可钻井深达到7500ft。

石油行业在19世纪90年代才开始使用第一台旋转钻机。旋转钻井技术是一种连续破岩技术，它将破岩钻头连接在空心钻柱的末端，并使流体通过钻头连续循环，清洗钻头并携带岩屑。整个钻井过程比顿钻钻井效率更高。破岩工具不再是凿子，而是相对复杂的钻头，能在轴向力和旋转力的共同作用下钻穿岩石。1901年，旋转钻井技术第一次在得克萨斯州的斯宾多托普山成功应用，在1020ft的深度处发现了石油，日产油约10×10^4bbl。

旋转钻机由很多单独的设备组成（图1.3）。这些设备可组合成六个子系统，分别是动力系统、提升系统、循环系统、旋转系统、井控系统和监测系统。尽管不同的钻机在设计上会有所不同，但所有的钻机都包含这六个子系统。下面讨论的是陆地钻机和海上钻井平台都有的设备。

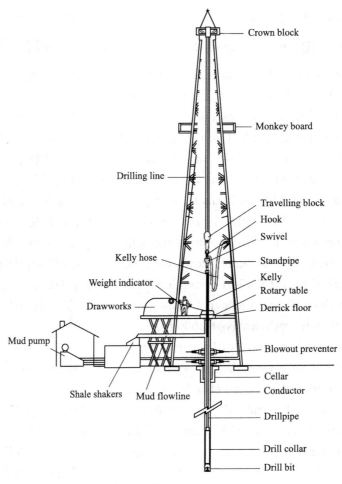

Fig. 1.3 Drilling rig components

1.3.3.1 Power System

Most drilling rigs are required to operate in remote locations where a power supply is not available. They must therefore have a method of generating the electrical power which is used to operate the systems mentioned above. The electrical power generators are driven by diesel powered internal combustion engines (prime movers). Electricity is then supplied to electric motors connected to the drawworks, rotary table and mud pumps. The rig may have, depending on its size and capacity, up to 4 prime movers, delivering more than 3000 horsepower. Horsepower (hp) is an old, but still widely used, unit of power in the drilling industry.

Older rigs used steam power and mechanical transmission systems but modern drilling rigs use electric

1.3.3.1 动力系统

大多数钻机都要在没有电力供应的偏远地区作业。因此，钻机必须配备发电机，为6个子系统的操作提供动力。发电机由柴油内燃机（原动机）驱动，将电力供应给电动机，电动机再与绞车、转盘和钻井泵相连，驱动钻机运行。根据规格和承载能力不同，钻机可配备4台发电机，提供超过3000hp的动力。hp是一种古老的动力单位，但目前钻井行业仍广泛应用。

老式钻机使用蒸汽动力和机械传动系统，而现代钻机则使

transmission. The drawworks and the mud pumps are the major users of power on the rig, although they are not generally working at the same time.

1.3.3.2 Hoisting System

The hoisting system is a large pulley system which is used to lower and raise equipment into and out of the well. In particular, the hoisting system is used to raise and lower the drillstring and casing into and out of the well. The components parts of the hoisting system are shown in Fig. 1.4. The drawworks consists of a large revolving drum, around which a wire rope (drilling line) is spooled. The drum of the drawworks is connected to an electric motor and gearing system. The driller controls the drawworks with a clutch and gearing system when lifting equipment out of the well and a brake (friction and electric) when running equipment into the well. The drilling line is threaded (reeved) over a set of sheaves in the top of the derrick, known as the crown block and down to another set of sheaves known as the travelling block. Having reeved the drilling line around the crown block and travelling block, one end of the drilling line is secured to an anchor point somewhere below the rig floor. Since this line does not move it is called the deadline. The other end of the drilling line is wound onto the drawworks and is called the fastline. The drilling line is usually reeved around the blocks several times. A large hook with a snap-shut locking device is suspended from the travelling block. This hook is used to suspend the drillstring. A set of clamps, known as the elevators, used when running, or pulling, the drillstring or casing into or out of the hole, are also connected to the travelling block.

The tensile strength of the drilling line and the number of times it is reeved through the blocks will depend on the load which must be supported by the hoisting system. It can be seen from Fig. 1.4 that the tensile load (lbs.) on the drilling line, and therefore on the fast line, F_f and dead line F_d in a frictionless system can be determined from the total load supported by the drilling lines, W and the number of lines, N reeved

1.3.3.2 提升系统

提升系统是一组大型滑轮系统，用来将设备从井中提出或下放，特别是对钻柱和套管的提升与下放。提升系统的主要部件如图1.4所示。绞车是提升系统的驱动装置，由一个大的旋转滚筒组成，钢丝绳（大绳）缠绕在这个滚筒上。滚筒与电动机和传动系统连接。司钻通过离合器和传动系统控制绞车将设备提出井口，通过刹车（依靠摩擦和电力）将设备下入井中。钢丝绳绕过（穿过）井架顶部一组定滑轮（称为天车），再向下穿过一组动滑轮组（称为游动滑车）。大绳的一端穿过天车和游动滑车后，会固定在钻台下方的死绳固定器上。由于大绳的这一端不能移动，因此被称为死绳。而缠绕在绞车上的另一端，称为快绳。大绳通常要在天车和游动滑车之间缠绕多次。游动滑车下方悬挂着一个带有止动锁定装置的大钩，用来悬挂钻柱。游动滑车上还连接了一套称为吊卡的夹具，用来提升和下放井中的钻柱或套管。

提升系统所要承受的载荷决定了大绳的抗拉强度和缠绕次数。从图1.4可以看出，大绳的拉伸载荷，即快绳上的F_f和死绳的F_d，由大绳支撑的总载荷W和绕天车、游动滑车的有效绳数N决定，其关系如下：

around the crown and travelling block:

$$F_f = F_d = W/N \tag{1.1}$$

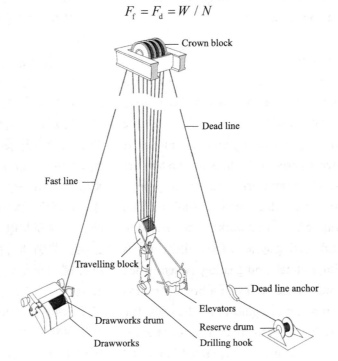

Fig. 1.4　Hoisting system

There is however inefficiency in any pulley system. The level of inefficiency is a function of the number of lines. An example of the efficiency factors for a particular system is shown in Table 1.1. These efficiency factors are quoted in API RP 9B - Recommended Practice on Application, Care and Use of Wire Rope for Oilfield Services. The tensile load on the drilling line and therefore on the fast line will then be:

然而，任何滑轮系统都存在效率问题。效率 E 是有效绳数的函数。API RP 9B 推荐的效率见表1.1。服务公司使用和维护大绳时通常用这个标准。大绳上的拉伸载荷，即快绳所受的拉力为：

$$F_f = W/(EN) \tag{1.2}$$

where E is the efficiency of the from Table 1.1. The load on the deadline will not be a function of the inefficiency because it is static.

式中 E 是效率（表1.1）。死绳上的载荷与效率无关，因为它是静态的。

Table 1.1　Efficiency factors for wire rope reeving, for multiple sheave blocks (API RP 9B)

Number of Lines (N)	6	8	10	12	14
Efficiency (E)	0.874	0.842	0.811	0.782	0.755

The power output by the drawworks, HP_d will be proportional to the drawworks load, which is equal to the

绞车输出的功率 HP_d 与绞车载荷成正比，它等于快绳上的拉力

load on the fast line F_f, times the velocity of the fast line v_f (ft/min.)

$$HP_d = \frac{F_f v_f}{3300} \qquad (1.3)$$

Eight lines are shown in Fig. 1.5 but 6, 8, 10, or 12 lines can be reeved through the system, depending on the magnitude of the load to be supported and the tensile rating of the drilling line used. The tensile capacity of some common drilling line sizes are given in Table 1.2. If the load to be supported by the hoisting system is to be increased then either the number of lines reeved, or a drilling line with a greater tensile strength can be used. The number of lines will however be limited by the capacity of the crown and travelling block sheaves being used.

F_f乘以快绳的速度 v_f（ft/min）。

图1.5中有8条有效绳穿过滑轮组，但穿过滑轮组的有效绳数还可以是6、10或12，这取决于需要承载的载荷大小和所使用钢丝绳的拉伸极限。表1.2给出了一些常见尺寸钢丝绳的断裂强度。如果需要增加提升系统的承载载荷，可以增加有效绳数，或使用更大抗拉强度的钢丝绳。有效绳数会受到所用天车和游车上滑轮组数量的限制。

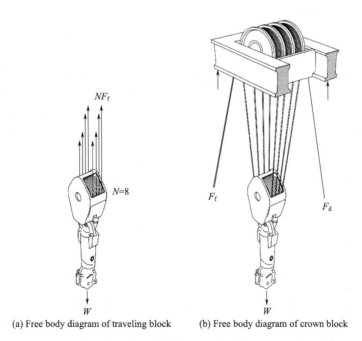

(a) Free body diagram of traveling block   (b) Free body diagram of crown block

Fig. 1.5 Drilling line tension

The drilling line does not wear uniformly over its entire length whilst drilling. The most severe wear occurs when picking up the drillstring, at the point at which the rope passes over the top of the crown block sheaves. The line is maintained in good condition by regularly conducting a slip or a slip and cut operation. In the case

钻井时，钢丝绳会受到不均匀的磨损。最严重的磨损在起钻时，钢丝绳穿过天车滑轮顶部的位置。通过定期滑大绳或倒大绳作业，可以使钢丝绳保持良好状态。在滑大绳操作时，将游车下

of the slipping operation the travelling block is lowered to the drillfloor, the dead line anchor is unclamped and some of the reserve line is threaded through the sheaves on the travelling block and crown block onto the drawworks drum. This can only be performed two or three times before the drawworks drum is full and a slip and cut operation must be performed. In this case the travelling block is lowered to the drillfloor, the dead line anchor is unclamped and the line on the drawworks is unwound and discarded before the reserve line is threaded through the system onto the drawworks drum.

The decision to slip or slip and cut the drilling line is based on an assessment of the work done by the line. The amount of work done by the drilling line when tripping, drilling and running casing is assessed and compared to the allowable work done, as shown in Table 1.2. The work done is expressed in Ton-miles and is calculated as follows:

放至钻台面，松开死绳固定器，将一段备用绳穿过游车和天车的滑轮缠绕到绞车的滚筒上。在绞车滚筒卷满之前只能进行两到三次滑大绳操作，而后就必须进行倒大绳作业。倒大绳作业时，先将游车下放至钻台面，松开死绳固定器，然后松开并丢弃绞车上的钢丝绳，最后将备用绳穿过游车和天车的滑轮缠绕到绞车滚筒上。

通过钢丝绳的工作评估来决定是选择滑大绳作业，还是倒大绳作业。估算钢丝绳在起下钻、钻进和下套管时所做的工作量，并与允许的工作量进行比较，见表1.2。工作量用 t·mile❶ 表示，并计算如下：

Table 1.2　Allowable work and nominal breaking strength of drilling line

Nominal Diameter	Ton-miles between cuts t·mile	Improved Plowed Steel lb❷	Extra Improved Plowed Steel lb
1in	8	89,800	103,400
1⅛in	12	113,000	130,000
1¼in	16	138,800	159,800
1⅜in	20	167,000	192,000
1½in	24	197,800	228,000
Nominal Breaking strength of 6×19 I.W.R.C(Independant Wire Rope Core) Blockline			

（1）Round Trip Operations: The greatest amount of work is done by the drilling line when running and pulling the drillstring from the well. The amount of work done per round trip (running the string in hole and pulling it out again) can be calculated from the following:

（1）起下钻作业：钢丝绳在起下钻时完成的工作量最大。每次起下钻（将钻柱下入井眼并再次起出）完成的工作量可以由以下公式计算得到：

$$T_r = \frac{D(L_s + D)W_m}{10560000} + \frac{D(M + 0.5C)}{2640000} \qquad (1.4)$$

---

❶ 1mile=1.609km；❷ 1lb=453.59g。

（2）Drilling Ahead: The amount of work done whilst drilling ahead is expressed in terms of the work performed in making trips. Analysis of the cycle of operations performed during drilling shows that the work done during drilling operations can be expressed as follows:

$$T_d = 3(T_2 - T_1) \quad (1.5)$$

If reaming operations and pulling back the kelly to add a single or double are ignored then the work becomes:

$$T_d = 2(T_2 - T_1) \quad (1.6)$$

（3）Running Casing: The amount of work done whilst running casing is similar to that for round tripping pipe but since the casing is only run in hole it is one half of the work. The amount of work done can be expressed as:

$$T_c = \frac{D(L_c + D)W_c + 4DM}{21120000} \quad (1.7)$$

（4）Short Trips: The amount of work done in pulling the drillstring back to the previous casing shoe and running back to bottom, for example to ream the hole can be expressed as in terms of the round trips calculated above:

$$T_{ST} = 2(T_4 - T_3) \quad (1.8)$$

where, T_r=Ton-miles for Round Trips; T_{ST}=Ton-miles for Short Trips; T_d=Ton-miles whilst drilling; T_c=Ton-miles for Casing Operations; D=Depth of hole (ft); L_s=Length of drillpipe stand (ft); L_c=Length of casing joint (ft); W_m=wt/ft of drillpipe in mud (lb/ft); W_c=wt/ft of casing in mud (lb/ft); M=wt. of blocks and elevators (lb); C=wt. of collars-wt. of drillpipe (for same length in mud); T_1=Ton miles for 1 round trip at start depth (D_1); T_2=Ton miles for 1 round trip at final depth(D_2); T_3=Ton miles for 1 round trip at depth D_3; T_4=Ton miles for 1 round trip at depth D_4.

（2）钻进作业：钢丝绳在钻进时完成的工作量是以起下钻时的工作量来表示的。对钻井作业周期的分析表明，钻井作业过程中所做的工作可表示如下：

(1.5)

如果忽略扩眼作业和为了接单根或双根钻杆而上提方钻杆，工作量变为：

(1.6)

（3）下套管：下套管作业类似于起下钻，但由于套管只下入井中，所以其工作量是起下钻的一半。下套管完成的工作量可以表示为：

(1.7)

（4）短起下钻：将钻柱提升至上一层套管的套管鞋以上，并在之后下放回井底，例如划眼，此时完成的工作量可以用起下钻的工作量来计算：

(1.8)

式中，T_r 为起下钻的工作量，t·mile；T_{ST} 为短程起下钻的工作量，t·mile；T_d 为钻进作业的工作量，t·mile；T_c 为下套管作业的工作量，t·mile；D 为井深，ft；L_s 为立管的长度，ft；L_c 为套管接头的长度，ft；W_m 为钻杆的单位长度浮重，lb/ft；W_c 为套管的单位长度浮重，lb/ft；M 为游车和吊卡的重量，lb；C 为钻铤浮重－钻杆浮重（等长度）；T_1 为起下钻起始井深的工作量，t·mile；T_2 为起下钻终止井深

The selection of a suitable rig generally involves matching the derrick strength and the capacity of the hoisting gear. Consideration must also be given to mobility and climatic conditions. The standard derrick measures 140ft high, 30ft square base, and is capable of supporting 1,000,000 lbs weight.

The maximum load which the derrick must be able to support can be calculated from the loads shown in Fig. 1.5.

1.3.3.3　Circulating System

Video 1.1

The circulating system is used to circulate drilling fluid down through the drillstring and up the annulus, carrying the drilled cuttings from the face of the bit to surface. The two main functions of the drilling fluid are:

（1）To clean the hole of cuttings made by the bit;

（2）To exert a hydrostatic pressure sufficient to prevent formation fluids entering the borehole.

Drilling fluid (mud) is usually a mixture of water, clay, weighting material (barite) and chemicals. The mud is mixed and conditioned in the mud pits and then circulated downhole by large pumps (slush pumps). The mud is pumped through the standpipe, kelly hose, swivel, kelly and down the drillstring. At the bottom of the hole the mud passes through the bit and then up the annulus, carrying cuttings up to surface. On surface the mud is directed from the annulus, through the flowline (or mud return line) and before it re-enters the mud pits the drilled cuttings are removed from the drilling mud by the solids removal equipment. Once the drilled cuttings have been removed from the mud it is re-circulated down the hole. The mud is therefore in a continuous circulating system. The properties of the mud are checked continuously to ensure that the desired properties of the mud are maintained. If the properties of the mud change

的工作量，t·mile；T_3 为起下钻深度 D_3 的工作量，t·mile；T_4 为起下钻深度 D_4 的工作量，t·mile。

选择合适的钻机时通常需要将井架的强度和提升系统的承载能力相匹配，同时还必须考虑可移动性和气候条件。标准井架高140ft，基座面积 30ft^2，并能支撑 100×10^4lb 的重量。

井架能够支撑的最大载荷可以从图 1.5 所示的载荷中计算出来。

1.3.3.3　循环系统

循环系统用来将钻井液通过钻柱向下循环，并经环空向上返出，同时将井眼中的钻屑从钻头处携带到地面。钻井液的两大作用是：

（1）清洁井眼，携带岩屑；

（2）利用钻井液柱静液压力平衡地层压力，防止地层流体侵入井眼。

钻井液（俗称泥浆）通常是水、黏土、加重材料（重晶石）和化学剂的混合物。钻井液在钻井液池中混合和调节，然后用大型泵（钻井泵）推动其在井下循环。钻井液通过立管、水龙带、水龙头和方钻杆，并从钻柱内向下流动。到达井底后，经钻头喷出，进入环空向上流动，同时携带岩屑返回地面。在地面，钻井液通过回流管线从环空流出，在重新进入钻井液池之前，需要用固控设备去除其中的岩屑。清除岩屑后，就可以重新循环。所以，钻井液会连续循环。为了确保钻井液能够保持预期的性能，需要在钻井过程中不断对钻井液性能进

then chemicals will be added to the mud to bring the properties back to those that are required to fulfil the functions of the fluid. These chemicals will be added whilst circulating through the mud pits or mud with the required properties will be mixed in separate mud pits and slowly mixed in with the circulating mud.

When the mud pumps are switched off, the mud will stop flowing through the system and the level of the mud inside the drillstring will equal the level in the annulus. The level in the annulus will be equal to the height of the mud return flowline. If the mud continues to flow from the annulus when the mud pumps are switched off then an influx from the formation is occurring and the well should be closed in with the blowout preventer stack. If the level of fluid in the well falls below the flowline when the mud pumps are shut down losses are occurring (the mud is flowing into the formations downhole). Losses will be discussed at length in a subsequent chapter.

The mud pits are usually a series of large steel tanks, all interconnected and fitted with agitators to maintain the solids, used to maintain the density of the drilling fluid, in suspension. Some pits are used for circulating (e.g. suction pit) and others for mixing and storing fresh mud. Most modern rigs have equipment for storing and mixing bulk additives (e.g. barite) as well as chemicals (both granular and liquid). The mixing pumps are generally high volume, low pressure centrifugal pumps.

At least 2 slush pumps are installed on the rig. At shallow depths they are usually connected in parallel to deliver high flow rates. As the well goes deeper the pumps may act in series to provide high pressure and lower flowrates.

Positive displacement type pumps are used (reciprocating pistons) to deliver the high volumes and high pressures required to circulate mud through the drillstring and up the annulus. There are two types of positive displacement pumps in common use: (1) Duplex (2 cylinders) - double acting;

行检查。如果钻井液性能发生变化，就需要向钻井液中添加化学剂，来确保钻井液性能良好。化学剂在钻井液循环经过钻井液池时加入，或者单独在一个钻井液池中配制好后，再缓慢地和正在循环使用的钻井液混合。

钻井泵关闭，钻井液就会停止循环，此时钻柱内和环空中的钻井液液面高度相等。如果环空中的液面高度等于钻井液返回管线的高度，或者钻井泵关闭时仍有钻井液继续从环空流出，这就意味着有地层流体侵入井筒，这时应利用防喷器组合进行关井。如果钻井泵关闭时，井中的流体高度低于回流管线，则表示发生了漏失（钻井液流入井下地层）。漏失将在后续章节中详细讨论。

钻井液池通常由一套大型钢罐组成，钢罐相互连通，并装有搅拌器，用来保持钻井液固相悬浮。一部分钻井液池用于循环（例如吸入池），而其他钻井液池则用于混合和储存配好的钻井液。大多数现代钻机都配备了储存及混合块状添加剂（例如重晶石）和化学剂（颗粒状、液体状）的设备。混合泵一般用高容积的低压离心泵。

一台钻机至少配备2台钻井泵。钻较浅井段时，常将钻井泵并联起来保证大排量。随着井深的增加，需要将泵串联起来，提供高泵压和低排量。

通过钻柱和环空循环钻井液时需要活塞泵（往复式活塞）提供大容积和高压。常用的活塞泵有两种：（1）双缸—双作用；（2）三缸—单作用。

(2) Triplex (3 cylinders) - single acting.

Triplex pumps are generally used in offshore rigs and duplex pumps on land rigs. Duplex pumps (Fig. 1.6) have two cylinders and are double-acting (i.e. pump on the up- stroke and the down-stroke). Triplex pumps (Fig. 1.7) have three cylinders and are single-acting (i.e. pump on the up-stroke only). Triplex pumps have the advantages of being lighter, give smoother discharge and have lower maintenance costs.

海上钻机一般采用三缸泵，陆地钻机一般使用双缸泵。双缸泵（图1.6）有两个液缸，实现双作用（即上下冲程完成两次吸排）。三缸泵（图1.7）有三个液缸，实现单作用（即只有上冲程泵出）。三缸泵的优点是重量轻，排放顺畅，维护成本低。

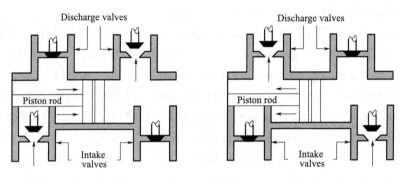

Fig. 1.6　Duplex pump

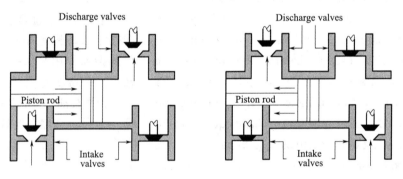

Fig. 1.7　Triplex pump

The discharge line from the mud pumps is connected to the standpipe-a steel pipe mounted vertically on one leg of the derrick. A flexible rubber hose (kelly hose) connects the top of the standpipe to the swivel via the gooseneck.

Once the mud has been circulated round the system it will contain suspended drilled cuttings, perhaps some gas and other contaminants. These must be removed before the mud is recycled. The mud passes over a shale shaker, which is basically a vibrating screen. This will remove the larger particles, while allowing the

钻井泵的出口管线连接到立管（垂直安装在井架一条腿上的钢管）上，再用一条柔性橡胶软管（水龙带）通过鹅颈管连接到水龙头。

一旦钻井液开始循环，就会含有悬浮的钻屑，可能还会有一些气体和其他污染物。这些都必须在钻井液再循环之前去除。从井中返回的钻井液要先通过振动筛去除较大的颗粒物，然后携带

residue to pass into settling tanks. The finer material can be removed using other solids removal equipment. If the mud contains gas from the formation it will be passed through a degasser which separates the gas from the liquid mud. Having passed through all the mud processing equipment the mud is returned to the mud tanks for recycling.

1.3.3.4 Rotary System

The rotary system is used to rotate the drillstring, and therefore the drillbit, on the bottom of the borehole. The rotary system includes all the equipment used to achieve bit rotation.

The swivel is positioned at the top of the drillstring. It has 3 functions:

(1) Supports the weight of the drill string;

(2) Permits the string to rotate;

(3) Allows mud to be pumped while the string is rotating.

The hook of the travelling block is latched into the bail of the swivel and the kelly hose is attached to the gooseneck of the swivel.

The kelly is the first section of pipe below the swivel. It is normally about 40ft long, and has an outer hexagonal cross-section. It must have this hexagonal (or sometimes square) shape to transmit rotation from the rotary table to the drillstring. The kelly has a right hand thread connection on its lower end, and a left hand thread connection on its upper end. A short, inexpensive piece of pipe called a kelly saver sub is used between the kelly and the first joint of drillpipe. The kelly saver sub prevents excessive wear of the threads of the connection on the kelly, due to continuous make-up and breakout of the kelly whilst drilling. Kelly cocks are valves installed at either end of the kelly to isolate high pressures and prevent backflow from the well if an influx occurs at the bottom of the well. The rotary table is located on the drill floor and can be turned in both clockwise and anti-clockwise directions. It is controlled from the drillers console. This rotating table has a square recess and four post holes. A large cylindrical sleeve, called a master

残留物进入沉淀池。较细的颗粒物可以用其他固控设备去除。如果钻井液中含有来自地层的气体，则需要通过除气器将气体与液体钻井液分离。经过所有处理设备后，钻井液返回钻井液罐再次循环。

1.3.3.4 旋转系统

旋转系统用于带动钻柱旋转，并由此带动井底的钻头旋转。旋转系统包括用于带动钻头旋转的所有设备。

旋转系统中的水龙头位于钻柱顶部，有3个主要功能：

（1）支撑钻柱的重量；

（2）允许钻柱旋转；

（3）允许钻井液在钻柱旋转时循环。

游车上的大钩闩锁在水龙头的提环中，水龙带连接到水龙头的鹅颈管上。

方钻杆是水龙头下方的第一段钻柱，通常长约40ft，其横截面内为圆形，外为六边形（有时为四边形），这样才能将旋转动力从转盘传递给钻柱。方钻杆的下端接头为右旋外螺纹接头，上端接头为左旋内螺纹接头。通常用一廉价短管连接方钻杆和第一根钻杆，称为方钻杆保护接头。方钻杆保护接头能够防止钻井时连续接、卸方钻杆造成接头螺纹的过度磨损。方钻杆阀是安装在方钻杆上下两端的阀门，用来隔离高压，并防止井底发生井侵时液体回流。转盘位于钻台上，可顺时针和逆时针旋转，通过司钻操作台进行控制。转盘有一个方形凹槽和四个柱孔。通常用大圆柱形套筒来保护转盘，称为主补心。

bushing, is used to protect the rotary table.

The torque from the rotary table is transmitted to the kelly through a device known as the kelly bushing. The kelly bushing has 4 pins, which fit into the post holes of the rotary table. When power is supplied to the rotary table torque is transmitted from the rotating table to the kelly via the kelly bushing. The power requirements of the rotary table can be determined from:

$$P_{rt} = \frac{\omega T}{2\pi} \tag{1.9}$$

where, P_{rt}=Power (hp); ω=Rotary Speed (r/min); T=Torque (ft·lbf).

Slips are used to suspend pipe in the rotary table when making or breaking a connection. Slips are made up of three tapered, hinged segments, which are wrapped around the top of the drillpipe so that it can be suspended from the rotary table when the top connection of the drillpipe is being screwed or unscrewed. The inside of the slips has a serrated surface, which grips the pipe.

To unscrew (or "break") a connection, two large wrenches (or tongs) are used. A stand (3 lengths of drillpipe) of pipe is raised up into the derrick until the lowermost drillpipe appears above the rotary table. The roughnecks drop the slips into the gap between the drillpipe and master bushing in the rotary table to wedge and support the rest of the drillstring. The breakout tongs are latched onto the pipe above the connection and the makeup tongs below the connection. With the make-up tong held in position, the driller operates the breakout tong and breaks out the connection.

To make a connection the make-up tong is put above, and the breakout tong below the connection. This time the breakout tong is fixed, and the driller pulls on the make-up tong until the connection is tight. Although the tongs are used to break or tighten up a connection to the required torque, other means of screwing the connection together, prior to torqueing up, are available:

转盘通过方补心将扭矩传递给方钻杆。方补心有4个销子，它们分别插入转盘的4个柱孔中。当转盘转动时，扭矩通过方补心传递到方钻杆。转盘所需的功率为：

式中，P_{rt}为功率，hp；ω为转速，r/min；T为扭矩，ft·lbf❶。

卡瓦是用来接、卸钻杆时悬挂钻柱的设备，由三片锥形铰链销轴联结而成，包裹在钻杆顶部，当钻杆顶部的接头在进行紧扣或卸扣时，可以使钻杆悬挂在转盘上。卡瓦的内部有一个锯齿状的表面，这样可以抓住钻柱。

两个吊钳（或称大钳）用于卸开（或断开）接头。起钻时，钻杆以一个立柱（3根钻杆）为单位提出井口并码放到井架内，直到最下端的钻杆起出转盘。钻工将卡瓦放入转盘中钻杆与主补心之间的间隙内，楔入并支撑井中其余部分的钻柱。卸扣钳锁定接头上方的钻杆，上扣钳位于接头的下方。在上扣钳就位后，司钻操作卸扣钳卸开接头。

接单根时，上扣钳在上，卸扣钳在下。此时，司钻固定好卸扣钳，并启动上扣钳将接头完全上紧。虽然吊钳用于提供上、卸接头所需的扭矩，在扭转之前也可用其他方法对接头进行上扣：

---

❶ 1lbf=4.45N。

(1) For making up the kelly, the lower tool joint is fixed by a tong while the kelly is rotated by a kelly spinner. The kelly spinner is a machine which is operated by compressed air.

(2) A drillpipe spinner (power tongs) may be used to make up or backoff a connection (powered by compressed air).

(3) For making up some subs or special tools (e.g. MWD subs) a chain tong is often used.

1. Procedure for Adding Drillpipe When Drilling Ahead

When drilling ahead the top of the kelly will eventually reach the rotary table (this is known as kelly down). At this point a new joint of pipe must be added to the string in order to drill deeper. The sequence of events when adding a joint of pipe is as follows (Fig. 1.8):

(1) Stop the rotary table, pick up the kelly until the connection at the bottom of the kelly saver sub is above the rotary table, and stop pumping.

(2) Set the drillpipe slips in the rotary table to support the weight of the drillstring, break the connection between the kelly saver sub and first joint of pipe, and unscrew the kelly.

（1）方钻杆上扣时，用大钳固定下端的工具接头，同时用方钻杆旋扣器旋转方钻杆。方钻杆旋扣器是一种由压缩空气驱动的设备。

（2）钻杆旋扣器（动力钳）可用于接头的上扣或倒扣（由压缩空气驱动）。

（3）一些接头或特殊工具（如 MWD 接头）的上扣通常使用链钳。

1. 钻进时接单根的程序

钻进时，方钻杆的顶部最终会到达转盘（称为方入）。此时为了钻得更深，必须在钻柱顶部增加一根新钻杆。接单根的操作顺序如下（图1.8）：

（1）停止转盘，上提方钻杆直到方钻杆保护接头底部接头位于转盘之上，停泵。

（2）将钻杆卡瓦放置在转盘中，用以支撑钻柱的重量，卸开方钻杆保护接头和第一根钻杆之间的接头，并卸开方钻杆。

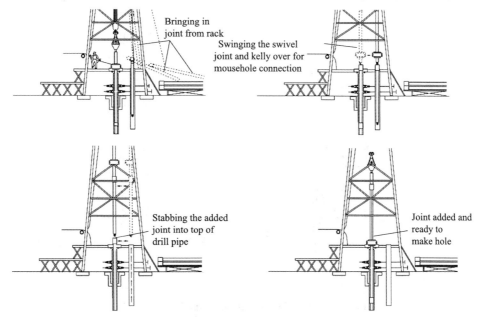

Fig. 1.8 Proceedure for adding drill pipe to the drillstring

(3) Swing the kelly over to the next joint of drillpipe which is stored in the mousehole (an opening through the floor near the rotary table).

(4) Stab the kelly into the new joint, screw it together and use tongs to tighten the connection.

(5) Pick up the kelly and new joint out of the mousehole and swing the assembly back to the rotary table.

(6) Stab the new joint into the connection above the rotary table and make-up the connection.

(7) Pick up the kelly, pull the slips and run in hole until the kelly bushing engages the rotary table.

(8) Start pumping, run the bit to bottom and rotate and drill ahead.

This procedure must be repeated every 30ft as drilling proceeds.

2. Procedure for Pulling the drillstring from the Hole

When the time comes to pull out of the hole the following procedure is used (Fig. 1.9):

(1) Stop the rotary, pick up the kelly until the connection at the bottom of the kelly saver sub is above the rotary table, and stop pumping.

（3）将方钻杆悬摆至即将接入钻杆的上方，该钻杆存放在小鼠洞中（钻台上转盘附近的一个孔洞）。

（4）将方钻杆插入新钻杆顶部的接头，并用大钳上扣。

（5）从鼠洞中提出方钻杆和新钻杆，并将它们悬摆至转盘。

（6）将新钻杆接入转盘上方的接头中，接单根。

（7）上提方钻杆，取出卡瓦，然后下入井中，直到方钻杆补心与转盘衔接。

（8）开泵，将钻头下到井底，旋转并继续钻进。

在钻井过程中，接单根作业必须每30ft重复一次。

2. 起钻的程序

起钻时使用以下程序（图1.9）：

（1）停止转盘，上提方钻杆直到方钻杆保护接头底部的接头位于转盘之上，停泵。

Fig. 1.9 Procedure for pulling pipe from the hole

(2) Set the drillpipe slips, break out the kelly and set the kelly back in the rat-hole (another hole in the rig floor which stores the kelly and swivel when not in use).

(3) Remove the swivel from the hook (i.e. kelly, kelly bushing, swivel and kelly hose all stored in rat hole).

(4) Latch the elevators onto the top connection of the drillpipe, pick up the drillpipe and remove the slips. Pull the top of the drillpipe until the top of the drillpipe is at the top of the derrick and the second connection below the top of the drillpipe is exposed at the rotary table. A stand (3 joints of pipe) is now exposed above the rotary table.

(5) Roughnecks use tongs to break out the connection at the rotary table and carefully swings the bottom of the stand over to one side. Stands must be stacked in an orderly fashion. .

(6) The Derrickman, on the monkey board, grabs the top of the stand, and sets it back in fingerboard.

When running pipe into the hole it is basically the same procedure in reverse.

3. Iron Roughneck

On some rigs a mechanical device known as an iron roughneck may be used to make-up and break-out connections. This machine runs on rails attached to the rig floor, and is easily set aside when not in use. Its mobility allows it to carry out mouse hole connections when the tracks are correctly positioned. The device consists of a spinning wrench and torque wrench, which are both hydraulically operated. Advantages offered by this device include controlled torque, minimal damage to threads (thereby increasing the service life of the drillpipe) and reducing crew fatigue.

4. Top Drive Systems

Most offshore drilling rigs now have top drive systems installed in the derrick. A top drive system consists of a power swivel, driven by a 1,000 hp dc electric motor. This power swivel is connected to the

（2）放钻杆卡瓦，方钻杆卸扣并放回大鼠洞（钻台上另一个洞，用于存放暂时不使用的方钻杆和水龙头）。

（3）从大钩上卸下水龙头（方钻杆、方补心、水龙头和水龙带都放在大鼠洞中）。

（4）将吊卡固定在钻杆顶部接头上，上提钻杆取出卡瓦。将钻杆顶部上提至井架顶部，并使钻杆顶部下方的第二个接头露出转盘。此时，一个立柱（3根钻杆）露出转盘面。

（5）钻工用大钳把转盘面上的接头卸开，并小心地把立柱底部摆放到一边。立柱必须有序排放。

（6）操作台上的井架工，抓住立柱的顶端，并把它放回指梁中。

下钻时的程序基本与起钻相同，只是顺序正好相反。

3. 铁钻工

有的钻井过程用一个称为铁钻工的机械装置来进行上扣、卸扣。它通过钻台上的专用轨道运行，不用时易于搁置，通过设定轨道可以用它从大鼠洞中取出连接管柱。铁钻工由旋扣钳和主钳背钳组成，均采用液压操作，它的优点是可控制扭矩，对螺纹伤害小（从而增加了钻杆的使用寿命），减少作业者的疲劳。

4. 顶驱系统

现在大多数海洋钻机都安装了顶驱系统。顶驱系统的关键部件之一是水龙头钻井马达总成，它由1000hp的直流电动机驱动，

travelling block and both components run along a vertical guide track which extends from below the crown block to within 3 metres of the rig floor. The electric motor delivers over 25,000 ft • lb torque and can operate at 300 rpm. The power swivel is remotely controlled from the driller's console, and can be set back if necessary to allow conventional operations to be carried out.

A pipe handling unit, which consists of a 500 ton elevator system and a torque wrench, is suspended below the power swivel. These are used to break out connections. A hydraulically actuated valve below the power swivel is used as a kelly cock.

A top drive system replaces the functions of the rotary table and allows the drillstring to be rotated from the top, using the power swivel instead of a kelly and rotary table. The power swivel replaces the conventional rotary system, although a conventional rotary table would generally, also be available as a back up.

The advantages of this system are:

(1) It enables complete 90 stands of pipe to be added to the string rather than the conventional 30 singles. This saves rig time since 2 out of every 3 connections are eliminated. It also makes coring operations more efficient.

(2) When tripping out of the hole the power swivel can be easily stabbed into the string to allow circulation and string rotation when pulling out of hole, if necessary (e.g. to prevent stuck pipe).

(3) When tripping into the hole the power swivel can be connected to allow any bridges to be drilled out without having to pick up the Kelly.

The procedures for adding a stand, when using a top drive system is as follows:

(1) Suspend the drillstring from slips, as in the conventional system, and stop circulation.

(2) Break out the connection at the bottom of the power sub.

(3) Unlatch the elevators and raise the block to the

与游车相连，并一起沿垂直导轨运行。垂直导轨从天车下方一直延伸到离钻台面3m的位置。电动机能够提供超过25000ft•lb的扭矩，并可以300r/min的转速作业。水龙头钻井马达总成通过司钻操作台进行远程控制，如果必要也可设置回常规操作模式。

钻杆处理装置悬挂在水龙头钻井马达总成下方，由能负重500t的吊卡和扭矩钳组成，可用来连接和断开钻柱。水龙头钻井马达总成下方的液压驱动阀相当于方钻杆阀。

顶驱系统取代了转盘的功能，它从顶部旋转钻柱，用水龙头钻井马达总成取代了方钻杆和转盘。尽管水龙头钻井马达总成取代了传统的旋转系统，但传统的转盘通常也可作为备用。

顶驱系统的优点是：

（1）用传统方式接30根单根的时间，用顶驱可以接90根，这样省了2/3的接单根时间。同时顶驱也可以使取心作业更高效。

（2）起钻时，动力水龙头可以很容易地插入钻杆，这样可以在必要时（如预防卡钻）开启循环和钻柱旋转。

（3）下钻时，可以连接水龙头钻井马达装置，在不接方钻杆的情况下钻过砂桥。

使用顶驱系统接立柱的作业程序如下：

（1）与常规系统一样，先用卡瓦悬挂钻柱，并停止循环。

（2）卸开马达接头底部的接头。

（3）松开吊卡，将游车提升

top of the derrick.

(4) Catch the next stand in the elevators, and stab the power sub into the top of the stand.

(5) Make up the top and bottom connections of the stand.

(6) Pick up the string, pull slips, start pumps and drill ahead.

1.3.3.5 Well Control System

The function of the well control system is to prevent the uncontrolled flow of formation fluids from the wellbore. When the drill bit enters a permeable formation the pressure in the pore space of the formation may be greater than the hydrostatic pressure exerted by the mud column. If this is so, formation fluids will enter the wellbore and start displacing mud from the hole. Any influx of formation fluids (oil, gas or water) in the borehole is known as a kick.

The well control system is designed to:

(1) Detect a kick;

(2) Close-in the well at surface;

(3) Remove the formation fluid which has flowed into the well;

(4) Make the well safe.

Failure to do this results in the uncontrolled flow of fluids - known as a blow-out - which may cause loss of lives and equipment, damage to the environment and the loss of oil or gas reserves. Primary well control is achieved by ensuring that the hydrostatic mud pressure is sufficient to overcome formation pressure. Hydrostatic pressure is calculated from:

$$p = 0.052 MW \cdot TVD \quad (1.10)$$

where, p=hydrostatic pressure (psi); MW=mud weight (lbm/gal); TVD=vertical height of mud column (ft).

Primary control will only be maintained by ensuring that the mud weight is kept at the prescribed value, and keeping the hole filled with mud. Secondary well control

至井架顶部。

（4）用吊卡卡住待接入立柱，并把马达接头插入立柱顶端。

（5）连接立柱的上下接头。

（6）上提钻柱，取出卡瓦，开泵钻进。

1.3.3.5 井控系统

井控系统的作用是防止地层流体不受控制地侵入井筒。当钻头进入渗透性地层时，地层孔隙压力可能会大于钻井液液柱施加的静液压力。这样的话，地层流体就会侵入井筒并开始顶替钻井液。井眼中任何地层流体（石油、天然气或水）的侵入都称为溢流。

井控系统的目的有：

（1）发现溢流；

（2）地面关井；

（3）清除流入井内的地层流体；

（4）确保井的安全。

上述操作失败将导致地层流体不受控制地流出，即所谓的井喷。这可能造成生命和设备的损失、环境的破坏及油气储层的损失。一级井控是确保钻井液的静液压力足以平衡地层压力。静液压力由下式计算：

式中，p 为静水压力，psi❶；MW 为钻井液密度，lbm/gal❷；TVD 为钻井液液柱的垂直高度，ft。

只有确保井筒内充满钻井液且钻井液密度保持在规定值之上，才能维持一级井控。二级井控是

---

❶ 1psi=6894.8Pa；❷ 1lbm/gal=0.1198g/cm³。本书 lbm 和 lb 均指质量（磅），lbf 指磅力。

is achieved by using valves to prevent the flow of fluid from the well until the well can be made safe.

1. Detecting a Kick

There are many signs that a driller will become aware of when a kick has taken place. The first sign that a kick has taken place could be a sudden increase in the level of mud in the pits. Another sign may be mud flowing out of the well even when the pumps are shut down (i.e. without circulating). Mechanical devices such as pit level indicators or mud flowmeters, which trigger off alarms to alert the rig crew that an influx has taken place, are placed on all rigs. Regular pit drills are carried out to ensure that the driller and the rig crew can react quickly in the event of a kick

2. Closing in the Well

Blow out preventors (BOPs) must be installed to cope with any kicks that may occur. BOPs are high-pressure valves, which seal off the top of the well. On land rigs or fixed platforms the BOP stack is located directly beneath the rig floor. On floating rigs, the BOP stack is installed on the seabed. In either case, the valves are hydraulically operated from the rig floor.

Video 1.3

There are two basic types of BOP.

(1) Annular preventor: designed to seal off the annulus between the drillstring and the side of hole (may also seal off open hole if kick occurs while the pipe is out of the hole). These are made of synthetic rubber which, when expanded, will seal off the cavity.

(2) Ram type preventor: designed to seal off the annulus by ramming large rubber-faced blocks of steel together. Different types are available:

① blind rams: seal off in open hole;

② pipe rams: seal off around drillpipe;

③ shear rams: sever drillpipe (used as last resort).

使用阀门来防止流体流出井筒，确保油气井能够安全生产。

1. 发现溢流

司钻可以通过许多征兆发现溢流。其中一个征兆就是钻井液池中的液面高度突然增加，还有就是停泵（即停止循环）后钻井液仍从井中流出。所有钻机都安装了钻井液池液面指示仪或钻井液流量计这样的机械装置，当溢流发生时它们会触发警报，提醒作业人员。此外，还要定期进行井控演练，确保司钻和作业人员在发生溢流时能迅速反应。

2. 关井

为了应对可能发生的溢流，必须安装防喷器组合。从本质上讲，防喷器组合是一种高压阀门，它能密封油井顶部。在陆地钻机或固定式平台上，防喷器组合安装在钻井平台的正下方。使用浮式钻井平台时，防喷器组合一般安装在海床上。两种情况下的防喷器组合都是从钻台上通过液压操作控制。

防喷器组合有两种基本类型：

（1）环形防喷器：用于密封钻柱和井壁之间的环空（如果发生溢流时井内无钻柱，也可密封整个井眼）。其密封件由合成橡胶制成，膨胀时可封闭井眼。

（2）闸板防喷器：通过推动钢制的闸板密封胶芯封闭环空。其主要类型有：

① 全封闸板防喷器：封闭全井眼；

② 半封闸板防喷器：封闭钻柱外的环形空间；

③ 剪切闸板防喷器：剪断钻杆，封闭全井眼（作为最后手段）。

Normally the BOP stack will contain both annular and ram type preventors (Fig. 1.10).

To stop the flow of fluids from the drillpipe, the kelly cock valves can be closed, or an internal BOP (basically a non-return check valve preventing upward flow) can be fitted into the drillstring.

防喷器组合通常同时包含环形防喷器和闸板防喷器（图1.10）。

为了阻止从钻杆内流出流体，可以关闭方钻杆阀，或者在钻柱中安装内部防喷器（本质上是一个防止流体向上流动的止回阀）。

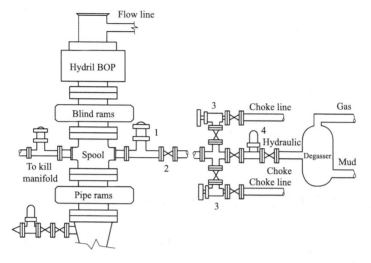

Fig. 1.10 BOP stackup
1—Hydraulic valve; 2—Plug valve; 3—Adjustable choke; 4—Pressure guage

3. Circulating out a Kick

To remove the formation fluids now trapped in the annulus a high pressure circulating system is used. A choke manifold with an adjustable choke is used to control flow rates during the circulation. Basically heavier mud must be pumped down the drillpipe to control the formation pressure, and the fluids in the annulus circulated to surface. As the kick starts moving up the hole the choke opening is restricted to hold enough back pressure on the formation to prevent any further influx. The fluids are circulated out via the choke line, through the choke manifold out to a gas/mud separator and a flare stack. Once the heavier mud has reached surface the well should be dead.

1.3.3.6 Well Monitoring System

Safety requires constant monitoring of the drilling process. If drilling problems are detected early remedial action can be taken quickly, thereby avoiding major problems. The driller must be aware of how drilling

3. 循环压井

为了去除已经进入环空的地层流体，需要使用高压循环系统。循环时用带有可调式节流阀的节流管汇来控制流量。基本上是通过钻杆泵入加重钻井液来控制地层压力，并将环空中的流体顶替到地面。当溢流流向井口时，有控制地打开节流阀，对井底产生足够的回压，防止地层流体进一步侵入。环空流体从节流管线流出，通过节流管汇流到液气分离器和放喷管线。一旦加重钻井液返至地面，压井结束。

1.3.3.6 监测系统

为了确保安全生产，需要连续监测钻井过程。如果能及早发现钻井事故，迅速采取补救措施，就可以避免重大事故的发生。司

parameters are changing (e.g. WOB, RPM, pump rate, pump pressure, gas content of mud etc.). For this reason there are various gauges installed on the driller's console where he can read them easily.

Another useful aid in monitoring the well is mudlogging. The mudlogger carefully inspects rock cuttings taken from the shale shaker at regular intervals. By calculating lag times the cuttings descriptions can be matched with the depth and hence a log of the formations being drilled can be drawn up. This log is useful to the geologist in correlating this well with others in the vicinity. Mud loggers also monitor the gas present in the mud by using gas chromatography.

钻必须时刻注意钻井参数的变化（如钻压、转速、泵速、泵压、钻井液含气量等）。因此，司钻操作台上安装有各种监测仪表，便于参数读取。

还有一种油气井监测方法是钻井液录井。钻井液录井人员定期检验从振动筛处采集的岩屑。通过计算迟到时间，可将岩屑类型与井深相匹配，从而绘制出正钻地层的录井曲线。地质学家可以用录井曲线将本井与邻井联系起来。此外，录井人员可通过气相色谱监测钻井液中的气体。

1.4 The Well Construction Process

1.4 建井过程

Audio 1.4

The operations involved in drilling a well can be best illustrated by considering the sequence of events involved in drilling the well shown in Fig.1.11. The dimensions (depths and diameters) used in this example are typical of those found in the North Sea but could be different in other parts of the world. For simplicity the process of drilling a land well will be considered below. The following description is only an overview of the process of drilling a well (the construction process). The design of the well, selection of equipment and operations involved in each step will be dealt with in greater depth in subsequent chapters of this manual.

图1.11描述了钻井相关作业的井眼、套管尺寸。图中实例井使用的井眼尺寸（井深和井眼直径）在北海地区比较典型，但在其他地区可能会有所不同。为了简单起见，下面介绍陆地钻井的过程，仅仅是钻一口井的过程概述（建井过程）。有关一口井的设计、设备的选型及每一步所涉及的操作都将在本书的后续章节中进行更深入的讨论。

1.4.1 Installing the 30in Conductor

1.4.1 下30in导管

The first stage in the operation is to drive a large diameter pipe to a depth of approximately 100ft below ground level using a truck mounted pile-driver. This pipe (usually called casing or, in the case of the first pipe installed, the conductor) is installed to prevent the unconsolidated surface formations from collapsing whilst

钻井作业的第一个阶段是用车载打桩机将大直径的导管下入到地面以下大约100ft的深度。安装这种管道（通常称为套管，或者在初次安装时被称为导管）的目的是在向下的钻井过程中防止

drilling deeper. Once this conductor, which typically has an outside diameter (O. D.) of 30in is in place the full sized drilling rig is brought onto the site and set up over the conductor, and preparations are made for the next stage of the operation.

松散的地层发生坍塌。一旦导管（外径通常为30in）安装就位，整套的大型钻机就会被运至现场，安装在导管之上，为下一阶段的作业做好准备。

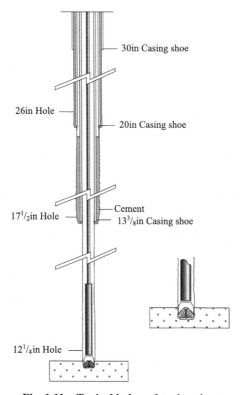

Fig. 1.11 Typical hole and casing sizes

1.4.2 Drilling and Casing the 26in Hole

1.4.2 钻 26in 井眼并下套管

The first hole section is drilled with a drillbit, which has a smaller diameter than the inner diameter (I.D.) of the conductor. Since the I.D. of the conductor is approximately 28in, a 26in diameter bit is generally used for this hole section. This 26in hole will be drilled down through the unconsolidated formations, near surface, to approximately 2,000ft.

If possible, the entire well, from surface to the reservoir would be drilled in one-hole section. However, this is generally not possible because of geological and formation pressure problems which are encountered whilst drilling. The well is therefore drilled in sections, with casing being used to isolate the problem formations

第一个井段用直径小于导管内径的钻头进行钻进。由于导管的内径大约是28in，因此该井段通常使用直径为26in的钻头。这段26in的井眼将向下穿过松散的地层，距地面大约2000ft。

如果可能的话，从地面到储层整口井用一个井段来完成。但是，这通常是不可能的，因为在钻井过程中会遇到地质和地层压力的一些问题。所以，钻井作业通常会分段进行，一旦钻穿复杂

31

once they have been penetrated. This means however that the wellbore diameter gets smaller and smaller as the well goes deeper and deeper. The drilling engineer must assess the risk of encountering these problems, on the basis of the geological and formation pressure information provided by the geologists and reservoir engineers, and drilling experience in the area. The well will then be designed such that the dimensions of the borehole that penetrates the reservoir, and the casing that is set across the reservoir, will allow the well to be produced in the most efficient manner possible. In the case of an exploration well the final borehole diameter must be large enough to allow the reservoir to be fully evaluated.

Whilst drilling the 26in hole, drilling fluid is circulated down the drillpipe, across the face of the drillbit, and up the annulus between the drillpipe and the borehole, carrying the drilled cuttings from the face of the bit to surface. At surface the cuttings are removed from the mud before it is circulated back down the drillpipe, to collect more cuttings.

When the drillbit reaches approximately 2,000ft the drillstring is pulled out of the hole and another string of pipe (surface casing) is run into the hole. This casing, which is generally 20in O.D., is delivered to the rig in 40ft lengths (joints) with threaded connections at either end of each joint. The casing is lowered into the hole, joint by joint, until it reaches the bottom of the hole. Cement slurry is then pumped into the annular space between the casing and the borehole. This cement sheath acts as a seal between the casing and the borehole, preventing cavings from falling down through the annular space between the casing and hole, into the subsequent hole and/or fluids flowing from the next hole section up into this annular space.

1.4.3　Drilling and Casing the 17½ in Hole

Once the cement has set hard, a large spool called a wellhead housing is attached to the top of the 20in

地层，就用套管将其封隔。这就意味着，随着井深的增加，井眼直径会越来越小。钻井工程师必须根据地质学家和油藏工程师所提供的地质及地层压力信息，再考虑到地区的经验，评估遇到复杂情况的风险大小，然后进行井身结构设计。设计时，储层井眼和套管尺寸应能够确保油气井以最有效的方式进行生产。打探井时，最终的井眼要有足够的尺寸，以便后续能够对储层进行充分的评估。

钻26in井眼时，钻井液从钻杆内向下循环经过钻头，并沿着钻杆和井壁之间的环空上返，将井眼中的岩屑从钻头处携带到地面。在地面，滤除岩屑后的钻井液会继续从钻杆内向下循环，从而可以从井中携带出更多的岩屑。

钻到2000ft左右时，起出钻柱，下入一段套管（表层套管）。表层套管的外径一般为20in，单根套管的长度为40ft，两端有带螺纹的接头。下套管时，将套管逐根连接下入井底。然后将水泥浆泵入套管与井壁之间的环空。套管和井壁之间的水泥环起密封的作用，防止下一井段钻井过程中，从套管和井壁之间的环空向下坍塌掉块，同时也会防止下一井段作业时，钻井液向上流入这个环形空间。

1.4.3　钻17½in井眼并下套管

一旦水泥固结，就要在20in套管的顶部连接一个叫套

casing. This wellhead housing is used to support the weight of subsequent casing strings and the annular valves known as the Blowout prevention (BOP) stack which must be placed on top of the casing before the next hole section is drilled. Since it is possible that formations containing fluids under high pressure will be encountered whilst drilling the next (17½ in) hole section a set of valves, known as a Blowout prevention (BOP) stack, is generally fitted to the wellhead before the 17½in hole section is started. If high pressure fluids are encountered they will displace the drilling mud and, if the BOP stack were not in place, would flow in an uncontrolled manner to surface. This uncontrolled flow of hydrocarbons is termed a Blowout and hence the title Blowout Preventers (BOPs). The BOP valves are designed to close around the drillpipe, sealing off the annular space between the drillpipe and the casing. These BOPs have a large I.D. so that all of the necessary drilling tools can be run in hole.

When the BOPs have been installed and pressure tested, a 17½in hole is drilled down to 6,000 ft. Once this depth has been reached the troublesome formations in the 17½in hole are isolated behind another string of casing (13⅝in intermediate casing). This casing is run into the hole in the same way as the 20in casing and is supported by the 20in wellhead housing whilst it is cemented in place.

When the cement has set hard the BOP stack is removed and a wellhead spool is mounted on top of the wellhead housing. The wellhead spool performs the same function as a wellhead housing except that the wellhead spool has a spool connection on its upper and lower end whereas the wellhead housing has a threaded or welded connection on its lower end and a spool connection on its upper end. This wellhead spool supports the weight of the next string of casing and the BOP stack which is required for the next hole section.

管头的大型四通，用来支撑后续套管串和防喷器组合的重量。防喷器组合由多个阀门组成，在钻下一井段之前，必须在套管顶部安装防喷器组合。这是由于在钻下一井段（17½in）时有可能会遇到含有高压流体的地层，因此在17½in井段开始之前，通常要将一防喷器组合安装在井口上。如果钻遇高压流体时防喷器组合没有安装到位，就会出现高压流体驱替钻井液，并不受控制地流到地面的现象。这种不受控制的油气流动称为井喷，因此这组井控阀门称为防喷器组合。防喷器组合用于封闭钻杆和套管之间的环空。它的内径一般较大，以便将钻井所需的所有工具下入井中。

当防喷器组合已经安装就位并完成压力测试后，就可以钻进17½in的井眼，可以钻到6000ft深度处。一旦钻达这个深度，17½in井段的复杂地层就需要用一段套管串（13⅝in的中间套管）进行封隔。中间套管的下入方式与20in套管相同，并由20in套管头提供支撑，同时用水泥胶结。

水泥固化以后，将防喷器组合移开，并将井口四通安装在套管头的顶部。井口四通与套管头具有相同的功能，但井口四通的上端和下端各有一个四通接头，而套管头的下端是一个螺纹或焊接接头，上端是一个四通接头。由于在下一段钻进时仍然需要防喷器组，井口四通就要有能力支撑下一组套管串和防喷器组合的重量。

1.4.4 Drilling and Casing the 12¼in Hole

When the BOP has been re-installed and pressure tested a 12¼in hole is drilled through the oil bearing reservoir. Whilst drilling through this formation oil will be visible on the cuttings being brought to surface by the drilling fluid. If gas is present in the formation it will also be brought to surface by the drilling fluid and detected by gas detectors placed above the mud flow line connected to the top of the BOP stack. If oil or gas is detected the formation will be evaluated more fully.

The drillstring is pulled out and tools which can measure for instance: the electrical resistance of the fluids in the rock (indicating the presence of water or hydrocarbons); the bulk density of the rock (indicating the porosity of the rocks); or the natural radioactive emissions from the rock (indicating the presence of non-porous shales or porous sands) are run in hole. These tools are run on conductive cable called electric wireline, so that the measurements can be transmitted and plotted (against depth) almost immediately at surface. These plots are called petrophysical logs and the tools are therefore called wireline logging tools.

In some cases, it may be desireable to retrieve a large cylindrical sample of the rock known as a core. In order to do this the conventional bit must be pulled from the borehole when the conventional drillbit is about to enter the oil-bearing sand. A donut shaped bit is then attached a special large diameter pipe known as a core barrel is run in hole on the drillpipe. This coring assembly allows the core to be cut from the rock and retrieved. Porosity and permeability measurements can be conducted on this core sample in the laboratory.

In some cases, tools will be run in the hole which will allow the hydrocarbons in the sand to flow to surface in a controlled manner. These tools allow the fluid to flow in much the same way as it would when the well is on production. Since the produced fluid is allowed to flow through the drillstring or, as it is sometimes called, the drilling string, this test is termed a drill-stem test or DST.

If all the indications from these tests are good then the oil company will decide to complete the well. If the tests are negative or show only slight indications of oil, the well will be abandoned.

1.4.5 Completing the Well

If the well is to be used for long term production, equipment which will allow the controlled flow of the hydrocarbons must be installed in the well. In most cases the first step in this operation is to run and cement production casing (9⅝ in O.D.) across the oil producing zone. A string of pipe, known as tubing (4½in O.D.), through which the hydrocarbons will flow is then run inside this casing string. The production tubing, unlike the production casing, can be pulled from the well if it develops a leak or corrodes. The annulus between the production casing and the production tubing is sealed off by a device known as a packer. This device is run on the bottom of the tubing and is set in place by hydraulic pressure or mechanical manipulation of the tubing string. When the packer is positioned just above the pay zone its rubber seals are expanded to seal off the annulus between the tubing and the 9⅝ in casing. The BOP's are then removed and a set of valves (Christmas Tree) is installed on the top of the wellhead. The Xmas tress is used to control the flow of oil once it reaches the surface. To initiate production, the production casing is "perforated" by explosive charges run down the tubing on wireline and positioned adjacent to the pay zone. Holes are then shot through the casing and cement into the formation. The hydrocarbons flow into the wellbore and up the tubing to the surface.

具可以使流体按照油井投产时相同的方式流动。由于此时所产流体通过钻柱流到地面，所以这种测试被称为钻杆试验（DST）或中途测试。

如果这些测试的所有迹象都表现良好，那么石油公司将决定对这口井进行完井。如果测试结果并不乐观或只显示少量的油气，则该井将被废弃。

1.4.5 完井

如果一口井要进行长期生产，就必须在井中安装可以控制油气流动的设备。大多数情况下，完井第一步是下入生产套管（外径9⅝in）并注水泥封固，生产套管贯穿整个产层。然后，在生产套管中下入一段称为油管（外径4½in）的管串，它是油气流到地面的通道。生产油管不同于生产套管，如果发生泄漏或腐蚀，它可以从井中取出更换。生产套管和生产油管之间的环空由封隔器密封。封隔器下入油管底部，并通过液压或油管柱的机械操纵进行坐封。当封隔器到产层上方时，橡胶密封部件膨胀打开，密封油管和9⅝in套管之间的环空。然后卸掉防喷器组合，在井口顶部安装一组阀门（采油树）用来控制流量。为了生产，还需要对生产套管进行射孔，射孔是用电缆将聚能射孔弹下入油管中正对产层的位置；然后，引燃射孔弹依次射穿套管、水泥环并进入地层一段距离。这样，油气流体就能流入井筒，并通过油管向上输送到地面。

1.5 Offshore Drilling

Audio 1.5

About 25% of the world's oil and gas is currently being produced from offshore fields (e.g. North Sea, Gulf of Mexico, China Bohai Sea). Although the same principles of rotary drilling used onshore are also used offshore there are certain modifications to procedures and equipment which are necessary to cope with a more hostile environment.

In the North Sea, exploration wells are drilled from a jack-up (Fig. 1.12) or a semisubmersible (Fig. 1.13) drilling rig. A jack-up has retractable legs which can be lowered down to the seabed. The legs support the drilling rig and keep the rig in position. Such rigs are generally designed for water depths of up to 350 ft water depth. A semi-submersible rig is not bottom supported but is designed to float (such rigs are commonly called "floaters"). Semi-submersibles can operate in water depths of up to 3,500 ft. In very deep waters (up to 7,500 ft) drillships (Fig. 1.14) are used to drill the well. Since the position of floating drilling rigs is constantly changing relative to the seabed special equipment must be used to connect the rig to the seabed and to allow drilling to proceed.

If the exploration wells are successful, the field may be developed by installing large fixed platforms from which deviated wells are drilled (Fig. 1.15). There may be up to 40 such wells drilled from one platform to cover an entire oilfield. For the very large fields in the North Sea (e.g. Forties, Brent) several platforms may be required. These deviated wells may have horizontal displacements of 10,000 ft and reach an inclination of 70 degrees or more. For smaller fields a fixed platform may not be economically feasible and alternative methods must be used (e.g. floating production system on the Balmoral field). Once the development wells have been

1.5 海洋钻井

目前，世界上大约25%的油气产自海上油田（例如欧洲北海、墨西哥湾、中国渤海等）。虽然海洋钻井用的是和陆地钻井原理一样的旋转钻井，但为了应付更加恶劣的环境，钻井程序和设备需要有一定的改进。

在北海，用自升式钻井平台（图1.12）或半潜式钻井平台（图1.13）进行探井钻井。自升式钻井平台有可伸缩的桩腿，可以下放到海底。桩腿支撑钻机，并保持钻机平稳就位。通常，这种钻机的设计水深可达350ft。半潜式钻机不是底部支撑的，而是漂浮式设计（这种钻机通常被称为"浮式钻井平台"）。半潜式钻井平台的工作水深可以达到3500ft。在超深水域（水深高达7500ft），就要使用钻井船（图1.14）。由于浮式钻机的位置相对于海底时常发生变化，因此必须使用特殊的设备将钻机连接到海底，以使钻井作业正常进行。

如果探井钻探成功，就可以安装大型固定式平台，并从这些平台上钻若干口斜井来开发整个油田（图1.15）。为了覆盖整个油田，从一个平台上钻出的斜井数可能多达40口。对于北海地区的超大型油田（例如福蒂斯油田和布伦特油田）可能需要几个平台。这些斜井的水平位移可达10000ft，井斜角可达70°以上。对于较小的油田，采用固定式平台可能不够经济，所以必须使用替代方法

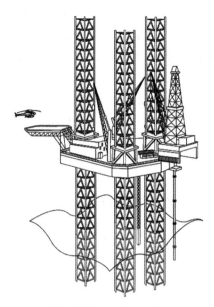

Fig. 1.12　Jack-up rig

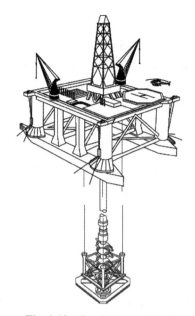

Fig. 1.13　Semisubmersible rig

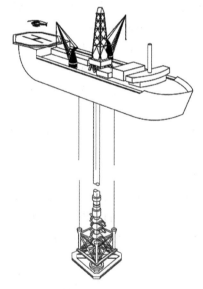

Fig. 1.14　Drillship

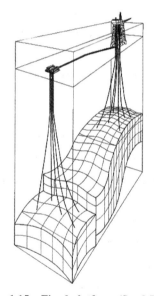

Fig. 1.15　Fixed platform (Steel Jacket)

drilled the rig still has a lot of work to do. Some wells may require maintenance (workovers) or sidetracks to intersect another part of the reservoir(re-drill). Some wells may be converted from producers to gas injectors or water injectors.

A well drilled from an offshore rig is much more

（例如巴尔莫勒油田采用浮式生产系统）。开发井钻完以后，钻机仍然有许多工作要做，例如，油井维护（修井），或为了贯穿油藏的另一部分而进行侧钻（再钻）；有的井还可以从生产井转换为注气或注水井。

海洋钻井比相同深度的陆

expensive than a land well drilled to the same depth. The increased cost can be attributed to several factors, e.g. specially designed rigs, subsea equipment, loss of time due to bad weather, expensive transport costs (e.g. helicopters, supply boats). A typical North Sea well drilled from a fixed platform may cost around $10 million. Since the daily cost of hiring an offshore rig is very high, operating companies are very anxious to reduce the drilling time and thus cut the cost of the well.

1.6 Exercise

(1) What are the major differences between percussion drilling and the modern rotary-drilling process?

(2) List the classification of wells according to their objective, trajectory, and environment.

(3) List the main types of rotary-drilling rigs for onshore and offshore environments.

(4) What are the main systems present in a drilling rig?

(5) A drillstring with a buoyant weight of 200,000lb must be pulled from the well. A total of 8 lines are strung between the crown block and the travelling block. Assuming that a four sheave, roller bearing system is being used. Calculate: ① the tension in the fast line; ② the tension in the deadline; ③ the vertical load on the rig when pulling the string.

(6) Consider a triplex pump having 6in. liners and 11in. stroke operating at 120 spm and a discharge pressure of 3000 psi. Calculate: ① the volumetric output at 100% efficiency; ② the Horsepower output of the pump when operating under the conditions above.

(7) An intermediate casing string is to be cemented in place at a depth of 10,000ft. The well contains 10.5 lbm/gal mud when the casing string is placed on bottom. The cementing operation is designed so that the 10.5lbm/gal mud will be displaced from the annulus by ① 300ft of

8.5 lbm/gal mud flush, ② 1,700 ft of 12.7 lbm/gal filler cement, and ③ 1,000 ft of 16.7 lbm/gal high-strength cement. The high-strength cement will be displaced from the casing with 9lbm/gal brine. Calculate the pump pressure required to completely displace the cement from the casing.

(8) List the main parameters that should be controlled and measured by the well-monitoring system while drilling a well.

lbm/gal 的填充水泥和 1000ft 16.7 lbm/gal 的高强度水泥从环空中将 10.5lbm/gal 钻井液顶替出来,用 9 lbm/gal 盐水顶替套管内高强度水泥。计算从套管内完全顶替水泥所需的泵压。

(8) 列出钻井时,监测系统应控制和测量的主要参数。

Geological Conditions for Drilling Engineering 2
钻井工程地质条件

2.1 Geomechanics in Drilling

2.1.1 Borehole Stability Analysis for Vertical Wells

2.1.1.1 Description of the Problem

Audio 2.1

Stability of boreholes became an important issue in the early 1980s when long, highly inclined wells were evolving, to be able to drain large reservoirs from single offshore platforms. Bradley is considered the person who introduced analytic borehole stability analysis to the oil industry. From that time, geomechanics has evolved as a petroleum discipline, and today a geomechanical analysis is often conducted for more complex wells, in order to reduce risk and cost. In this section, we will focus on the understanding of the physics of borehole stability. One objective is to promote physical understanding. This is an introductory text to borehole mechanics seen from a drilling perspective. The major drilling challenge that relates to borehole mechanics is the stability of the wellbore. Several well problems often arise during drilling:

(1) A circulation loss occurs when the volume of returned mud is less than the volume of mud pumped. Circulation losses are unplanned events that usually must be resolved before drilling can continue. Circulation losses also may lead to loss of well control, resulting in a blowout, or lead to difficulty in cleaning the borehole,

2.1 钻井地质力学

2.1.1 直井的井壁稳定性分析

2.1.1.1 问题描述

在20世纪80年代早期，井壁稳定性就已经成为一个重要的问题。当时要用海洋平台开发大型油藏，需要钻多口大斜度、长井段的井。通常认为是布拉德利将井壁稳定性分析引入石油工业中。从那时起，地质力学发展成为一门石油学科。而今天，为了降低风险和成本，人们通常要对复杂井进行地质力学分析。本节将重点介绍井壁稳定性的物理学原理，主要目标就是要从钻井的角度来介绍井筒力学，从而提高对物理学原理的理解。与井筒力学有关的主要钻井问题就是井壁的稳定性。钻井过程中经常会出现以下几个问题：

（1）当返出的钻井液体积小于泵入的钻井液体积时，就意味着发生了井漏。井漏是意外事故，通常必须在继续钻进之前得以解决。否则，可能导致井筒失控，引发井喷；或者导致井眼清洗困

which may eventually lead to a stuck drillstring. One remedy is to reduce the mud weight.

(2) Mechanical borehole collapse often occurs at low borehole pressures; such as happens with too low mud weight or during circulation losses or if the well is swabbed in while tripping pipe. The remedy is often to increase wellbore pressure, usually by increasing the mud weight.

(3) Particularly in shales, chemical effects may induce hole enlargement or collapse. When water-based drilling fluids are used, the shale may react with the mud filtrate (fluid that penetrates the wellbore wall), deteriorating the borehole. Oil-based muds are often better on hole collapse, but more difficult if circulation losses arise.

There are many publications presenting various empirical correlations for borehole stability, mainly addressing fracturing. However, in the last decades an analytic approach has emerged, wherein the problems are analyzed using the principles of classical mechanics. The advantage is that the various problems can be seen from a common reference frame. This section will provide an introduction to the mechanics' approach.

Fig. 2.1 illustrates some common drilling problems. The mud weight or the bottomhole pressures are often a compromise between well control and borehole stability.

难，最终使钻柱遇卡。补救办法之一就是降低钻井液密度。

（2）井壁坍塌通常发生在井眼压力较低的时候，例如钻井液密度过低，或井漏期间，又或者是在起钻过程中井眼受到抽汲作用的影响。其补救措施一般是通过增加钻井液密度来提高井筒压力。

（3）化学效应可能会引起井眼扩大或坍塌，尤其是在页岩地层中。使用水基钻井液时，页岩可能与钻井液滤液（渗透到井壁的流体）发生反应，导致井壁失稳。油基钻井液更适合解决此类井壁坍塌，但如果出现井漏，处理起来更加困难。

许多文献针对井壁稳定性提出了各种经验关系式，它们主要是解决地层破裂问题。然而，在过去的几十年里，出现了一种利用经典力学原理分析问题的方法。这种方法的优点是可以从一个共同的参考体系中看出各种问题。本节将介绍这种力学分析方法。

图2.1描述了一些常见的钻井问题。通常，钻井液密度或井底压力选择要综合考虑井控和井壁稳定两方面因素来进行。

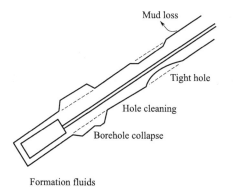

Fig. 2.1 Typical borehole problems

It is observed that 10% ~ 20% of the time spent on a well is due to unplanned events. These events often have a root in borehole stability. Knowing that the worldwide drilling budgets are many billion dollars, we understand therefore that borehole instability is a very costly problem. Table 2.1 shows the unplanned time spent on an exploration well in the North Sea. The unplanned events in Table 2.1 are mostly related to borehole stability. Some wells have lower downtime, but if severe problems arise, they are often very time-consuming to solve.

据统计，一口井处理意外事件的时间占 10% ~ 20%。这些意外事件常常源于井壁稳定性问题。全世界的钻井预算高达数十亿美元，由此可知井壁失稳的花费巨大。表 2.1 给出了北海地区一口探井处理意外事件花费的时间，其中的意外事件主要与井壁稳定性有关。一些井的停钻时间虽短，但如果出现严重问题，则需要耗费大量的时间来解决。

Table 2.1 Example of unplanned events

Unplanned Events	Mud Losses	Tight hole, reaming	Squeeze Cementing	Fishing	Total time Loss	Percent of Well Time
Time Used to Cure	2.5d	0.3d	2.5d	0.3d	5.6d	5.6/30=19%

2.1.1.2 Units and Equations

First we will define our reference frames. Pressures are defined in terms of the hydrostatic head at a given depth, or

$$p = \rho g Z$$

where, p is the wellbore pressure, ρ is the mud density, g is the gravity constant, and Z is the true vertical depth. The drilling industry uses the mud density as a reference. For simple comparison to the mud weight, we use equivalent density instead of pressure. Another advantage of the equivalent density is that it takes out the depth element. It is defined as

$$\rho_e = \frac{p}{gZ}$$

In metric units, the gradient equation becomes

$$\gamma_e = \frac{p}{0.098 \times Z}$$

where

$$\gamma_e = \frac{\rho_e}{\rho_{\text{water}}}$$

For most drilling applications, the gradient is

2.1.1.2 单位和方程

首先，定义参考体系。压力是根据给定深度的静液压力来定义的，表示为：

(2.1)

式中，p 是井筒压力，ρ 是钻井液密度，g 是重力常量，Z 是实际的垂直深度。钻井行业用钻井液密度作为参考。为了简化与钻井液密度的对比，使用等效密度来代替压力。等效密度的另一个优点是忽略了深度。其定义为：

(2.2)

公制单位下压力梯度方程为：

(2.3)

其中

对于大多数钻井应用来说，

preferred because it is depth-independent and can be directly compared to the static mud weight. However, during transient fluid processes such as cement displacement and circulating out a kick, it is advised to use pressures.

2.1.1.3　In-Situ Stresses

All rocks are subjected to stresses at any depth. It is a convention in the petroleum industry to define these stresses as follows:

(1) A vertical principal stress, usually the overburden stress σ_v. This results from the cumulative weight of the sediments above a given point. Usually this is obtained from bulk density logs or from the density of cuttings.

(2) Two principal horizontal stresses, the maximum horizontal stress σ_H and the minimum horizontal stress σ_h. The magnitudes of these are obtained from leak off data or other measurements. Note that we use the sign convention of compressive stress as positive, a convenience since we usually are dealing with compressive stresses. We assume that these principal stresses are always vertical and horizontal. Furthermore, we assume the following stress states:

(3) In a relaxed depositional-basin environment, the two horizontal stresses are smaller than the overburden stress, a so-called normal fault stress state: $\sigma_h < \sigma_H < \sigma_v$. The two horizontal stresses are often similar and equal to 70–90% of the overburden stress.

(4) Tectonic stresses may arise due to faulting or plate tectonics. Two different states may exist:

① Strike-slip fault stress state: $\sigma_h < \sigma_v < \sigma_H$;

② Reverse fault stress state: $\sigma_v < \sigma_h < \sigma_H$;

Most oil fields are located in sedimentary basins and are in normal fault stress states. This will form the basis for the following development. However, before looking at the actual borehole mechanics, we need to consider other properties of porous media. Regarding stresses, Terzaghi defined the effective stress principle: The total stress is the sum of the pore pressure and the stress in the rock matrix, or

$$\sigma_{\text{total}} = \sigma' + p_{\text{pore}}$$

(2.4)

where, we have indicated the effective stress as σ'. This is illustrated in Fig. 2.2. Imagine that the total load on the wellbore wall is the mud pressure inside the borehole. This load is taken up by the stresses in the rock matrix plus the pore pressure. When we study failure of rock, we always compute the effective stresses, which apply to the rock itself.

虑多孔介质的其他性质。关于应力，太沙基定义了有效应力：即总应力是孔隙压力和基岩应力的总和，表示为：

式中，σ' 为有效应力，如图 2.2 中所示。设想井壁所受的总荷载是井筒内液柱所产生的压力，这一载荷由基岩应力和孔隙压力共同承担。当研究岩石破坏时，计算的是有效应力，它适用于岩石本身。

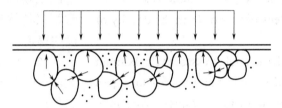

Fig. 2.2 Illustration of effective stresses

The effective stresses are the stresses acting on the rock matrix when we exclude the pore pressure. Because we are concerned with failure of the rock matrix, we have to use effective stresses.

2.1.1.4 Pore Pressures

Sedimentary rocks are usually porous. The pores usually are filled with water, which is the most abundant fluid. In oil and gas reservoirs, the water is locally replaced with hydrocarbons.

The pore pressure is an important parameter for several reasons. First, it is important for the production of hydrocarbons and for determining whether the reservoir can be produced naturally or if artificial lift is required. The pore pressure is also important for borehole stability because of the effective stress principle (Eq. 2.4).

In sedimentary rocks such as sandstones, the pore pressure can be measured directly with logging tools. However, shales are nearly impermeable, so there exist no direct methods to measure the pore pressure here. Instead, the pore pressure is inferred from drilling data

有效应力是排除了孔隙压力后作用于岩石基质上的应力。因为关心的是岩石基质的破坏，所以要使用有效应力。

2.1.1.4 孔隙压力

沉积岩通常是多孔介质。孔隙中充满了水，水是地球上最充足的流体。在油气藏中，烃类只是部分取代了水。

孔隙压力是一个重要的参数。首先，它对于油气的生产很重要，因为它决定了油藏生产能否自喷，还是需要人工举升。另外，通过有效应力原理[式(2.4)]可知，孔隙压力对于井壁稳定性也很重要。

在砂岩等沉积岩中，可以利用测井工具直接测量孔隙压力。然而，对于几乎不渗透的页岩，没有测量孔隙压力的直接方法。需要从钻井数据和各种测井数据

and from various logs.

It should be noted that there is a large uncertainty in the pore pressure prediction of these indirect methods. In the reservoir, direct measurements are considered accurate. Because the pore pressure profile has a direct bearing on the selection of the casing depths, the uncertainty should be understood. Remember also that if a homogeneous tight shale has a high pore pressure, it cannot flow and therefore cannot lead to a well-control incident.

In general, a cap rock is required to create overpressure. There are several different mechanisms that create abnormal pore pressures. Some mechanisms are

(1) Buoyancy, where the lightest fluid moves to the top and the heaviest to the bottom;

(2) Rock compaction of a closed volume;
(3) Consolidation effects;
(4) Chemical effects.

The buoyancy effect is considered a dominating mechanism, and it is always present in a reservoir. In the following examples, we will explore the pressures throughout an oil and gas reservoir as shown in Fig. 2.3.

应该注意的是，这些间接预测孔隙压力的方法存在很大的不确定性，还是直接在储层中测量更加准确。由于孔隙压力剖面与套管下深有直接关系，所以应该了解孔隙压力的不确定性。对于均匀致密的页岩，虽然具有较高的孔隙压力，但由于它的不渗透性，也不会导致井控事故。

一般来说，盖层岩石是产生超压状态的必备条件。产生异常孔隙压力的不同机理有以下几种：

（1）浮力作用，其中最轻的流体运移到顶部，而最重的流体转移到底部；

（2）封闭体积的岩石压实；
（3）固结作用；
（4）化学效应。

浮力作用是一种主要的机理，它总是存在于储层中。在下面的例子中，探讨整个油气藏的压力，如图 2.3 所示。

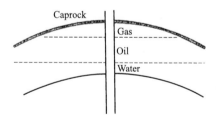

Fig. 2.3 Trapped hydrocarbons

2.1.1.5 Fracturing Pressures

In this section, we will present the borehole mechanics model used in the oil industry. The first part assumes simple conditions, as found in relaxed sedimentary basins, such as equal horizontal in-situ stresses and also a vertical borehole. Later we will present more-complex scenarios.

There are two different mechanics approaches used

2.1.1.5 破裂压力

下面介绍用于石油工业的井筒力学模型。首先假设简单的条件，即在水平地应力相等的疏松沉积盆地中有一个垂直井眼。稍后将介绍更复杂的情况。

在石油工业中有两种不同的

in the oil industry:

(1) Classical mechanics approach. We assume an infinite plate with a hole in the middle. This hole represents the wellbore (Fig. 2.4.). For fracturing and collapse analysis during drilling, this is the method used.

(2) Fracture mechanics approach, assuming that a fracture already exists. This is used in stimulation operations where massive fracturing and reservoir stimulation take place, and relates to boreholes that are already fractured. This will not be pursued in this chapter.

The plate in Fig. 2.4 is subjected to external loading defined by the in-situ stresses. The borehole is the hole in the middle. At the borehole wall, for the special case $\sigma_H = \sigma_h$, three different stresses exist as illustrated in Fig. 2.5:

(1) The radial stress is given by the mud pressure:

$$\sigma_r = p_w \quad (2.5a)$$

(2) The tangential stress, or hoop stress:

$$\sigma_\theta = 2\sigma_h - p_w \quad (2.5b)$$

(3) The axial stress, or vertical stress:

$$\sigma_z = \text{Constant} \quad (2.5c)$$

力学方法：

（1）经典力学方法。假设一个无限大的板块中间有一个洞，这个洞代表井眼（图2.4）。钻井过程中的地层破裂和坍塌分析使用该方法。

（2）断裂力学方法，假设断裂已经存在。这一方法用于大规模压裂和储层增产作业，以及已经破裂的井眼。本章不讨论这一点。

图2.4中的板块受到地应力决定的外部载荷作用。中间是井眼。对于特殊情况 $\sigma_H=\sigma_h$，在井壁上存在三种不同的应力，如图2.5所示：

（1）由液柱压力产生的径向应力：

（2）切向应力或周向应力：

（3）轴向应力或垂直应力：

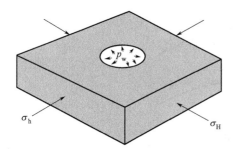

Fig. 2.4 Classical mechanics approach: an infinite plate with a hole in the middle

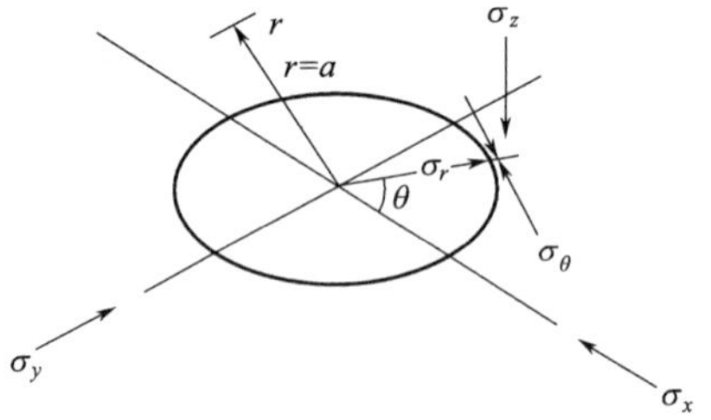

Fig. 2.5 Stresses acting on the borehole wall

The tangential stress depends on the horizontal stress. The factor 2 (Eq.2.5b) is called a stress concentration factor and is due to the circular geometry of the borehole. If the borehole has an oval shape or some other noncircular shape, higher stress concentration factors often arise. If $\sigma_H \neq \sigma_h$, then σ_θ would vary with θ.

切向应力取决于水平应力，式（2.5b）中系数2被称为应力集中系数，它由井眼的圆形几何形状所致。如果井眼具有椭圆或其他非圆形形状，则往往会产生更高的应力集中系数。如果 $\sigma_H \neq \sigma_h$，

Also, observe that the borehole pressure directly affects the tangential stress.

For this special case, we observe from Eqs. 2.5a and 2.5b that the sum of the radial and tangential borehole stresses is constant. A consequence of this is that at high borehole pressures the tangential stress is low, whereas at low borehole pressure the tangential stress is high.

Fig. 2.6 visualizes the effects of varying the borehole pressure. If the mud weight applies the same load as the stresses before the hole was drilled, there is no disturbance, as illustrated in Fig. 2.6(a). If the borehole pressure is lower than the in-situ stresses, the borehole will shrink, or actually fail in collapse [Fig. 2.6(b)], because of the high hoop stress that is created with low borehole pressures. Finally, Fig. 2.6(c) illustrates that with a high borehole pressure, the hole will expand until it fails or fractures.

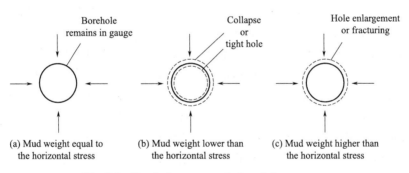

Fig. 2.6 Borehole response to borehole pressure

Mud losses may occur through the resulting fracture. Borehole fracturing is a tensile failure, because the tangential stress goes into tension. Rocks generally have low tensile strength. Often tiny cracks and fissures exist in the rock body, or are created during the drilling operation. It is therefore a common assumption to neglect rock tensile strength. Fracturing pressure is defined as the pressure at which the effective hoop stress is zero. For a vertical well with equal horizontal in-situ stresses,

$$\sigma_\theta - p_o = 2\sigma_h - p_w - p_o = 0 \quad (2.6a)$$

or

$$p_{wf} = 2\sigma_h - p_o \quad (2.6b)$$

For unequal horizontal in-situ stresses, the fracturing pressure becomes

$$\sigma_\theta - p_o = 3\sigma_h - \sigma_H - p_w - p_o \quad (2.7a)$$

or

$$p_{wf} = 3\sigma_h - \sigma_H - p_o \quad (2.7b)$$

where, σ_θ = tangential stress; σ_h = minimum horizontal stress; σ_H = maximum horizontal stress; p_o = pore pressure; p_w = borehole pressure; p_{wf} = fracturing pressure.

Often we perform a leak off test (LOT) or a formation integrity test after each casing string is cemented in place. The purpose is to ensure that the formation is sufficiently strong, and that the cement has sufficient integrity such that the next section can be drilled. This leak off test is our main parameter to estimate the magnitude of the in-situ stresses.

Fig. 2.7 shows the general trends for the fracture gradient in a depositional basin. The fracturing pressure increases with depth, and it decreases with borehole inclination. The reason for the latter is that the in-situ stresses are different. For a vertical well, there are two nearly equal horizontal stresses acting on the wellbore. For a horizontal well, the overburden stress and the horizontal stress are acting on the borehole, creating an anisotropic stress state.

1. The LOT

This pressure test is very important for the drilling of wells. After each casing is installed and cemented in place, a hydraulic test is performed. This is called the LOT. When the cement is hardened around the casing, the casing shoe is drilled 4–6 m into the new formation below the shoe. Then the well is shut in and the borehole is pressurized, usually using the cement pump. The pressure that builds up in the annulus is shown in Fig. 2.8. The initial linear slope is due to the compressibility of the drilling fluids in the borehole. When the pressure buildup deviates from the straight line, we assume that a fracture initiates in the borehole wall. This is commonly defined as the LOT point. Beyond this point, the pump is stopped and the pressure drop is observed. It is common to assume that the point where the pressure curve changes slope indicates the minimum horizontal stress; however, this interpretation is debated.

裂缝。该点通常被定义为漏失点。一旦超过这个点，立即停泵，观察压降。通常假设压力曲线斜率发生变化的点表示最小水平地应力；然而，这种解释存在争议。

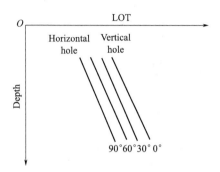

Fig. 2.7 Fracture gradients for relaxed depositional basin

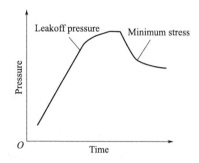

Fig. 2.8 Pressure plot during leakoff test

2. The Optimal Mud Weight

In a typical well, we have a pore pressure prognosis, an overburden stress prognosis, and several LOT data. To develop this information into a predictive tool, we must estimate the horizontal stresses. From Eq. 2.6, we obtain

$$\sigma_h = \frac{1}{2}(p_{wf} + p_o)$$

With the assumptions given, the horizontal stress is actually the midpoint between the fracture pressure and the pore pressure. For this reason, it is often called the median-line principle. Fig. 2.9 shows an example. The oil industry has commonly used a mud weight barely exceeding the pore pressure, as shown in the left stepped curve in Fig. 2.9. When borehole stability analysis became invoked, a high mud weight like the right stepped curve was often recommended to reduce the tangential stress and, hence, the collapse potential of the well. This often led to fluid-loss problems instead. The middle stepped curve gave better results because it is based on the idea of minimum disturbance of the stresses acting on the borehole. This is explained in the following.

2. 最佳钻井液密度

对于一口典型的井，有孔隙压力预测、上覆岩层压力预测和一些漏失试验数据。为了将这些信息开发成为一种预测工具，必须估计水平应力。由式（2.6）可得：

(2.8)

在给定的假设条件下，水平应力实际是破裂压力和孔隙压力的中点。因此，这通常被称为中线原理。图 2.9 给出了一个示例。如图中左侧阶梯曲线所示，石油行业通常使用的钻井液密度只略微超过了孔隙压力。当进行井壁稳定性分析时，通常建议使用高密度钻井液（如右侧的阶梯曲线）来降低切向应力，从而降低井眼坍塌的可能性。然而，这通常会导致漏失问题。选择中间的阶梯曲线时效果更好，因为它的基本思想是使作用于井眼上的应力波动最小。下面将对此进行解释。

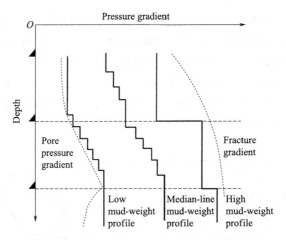

Fig. 2.9 Alternative mud-weight schedules

Before the well was drilled, a horizontal stress state σ_h existed in the rock. During drilling, the rock that was in the hole is replaced with drilling mud. If the drilling mud creates the same stress, there is no disturbance in stresses. However, a lower mud weight sets up a compressive tangential stress, and a higher mud weight sets up a lower tangential stress. A mud weight equal to the horizontal stress level is actually the optimal mud weight for the well. We therefore often start a new well program using the median-line principle for mud weight. This mud weight must be modified for several reasons:

(1) The mud density in a well section is constant, but the stresses and pore pressure change with depth. The mud weight is therefore a compromise over a depth interval.

(2) Wells often have potential fluid-loss zones. The mud weight selection must consider this.

(3) In directional wells, differential sticking (higher wellbore pressure presses the drillstring against lower pore pressure) can result in a stuck drillstring. This risk can be reduced by reducing the mud weight.

(4) In exploration wells, sometimes the mud weight is kept close to the pore pressure (tagging the pore pressure). This is one method to establish a more correct

在钻井之前，岩石中存在水平应力 σ_h。在钻井过程中，井眼中的岩石被钻井液所替代。如果钻井液产生同样的应力，就不会产生应力波动。然而，钻井液密度越低，切向应力越大；钻井液密度越高，切向应力越小。实际上与水平应力相等的钻井液密度是该井的最佳钻井液密度。因此，常用中线原理来确定钻井液密度。但由于以下原因必须调整钻井液密度：

（1）同一井段钻井液密度是恒定的，但水平应力和孔隙压力会随深度变化。因此，当超过一定深度的井段时钻井液密度就需要调整。

（2）选择钻井液密度时必须考虑潜在的漏失层。

（3）在定向井中，压差卡钻（井筒压力与孔隙压力间的正压差，使钻柱紧贴到井壁上）会导致钻柱遇卡。通过减小钻井液密度可以降低这一风险。

（4）在探井中，有时使钻井液密度接近于孔隙压力（标记孔隙压力）。这是一种比预测更为准

pore pressure curve than the prognosis.

(5) Mud cost may be of concern.

Fig. 2.9 shows three mud weight selection principles: low mud weight, median-line mud weight, and high mud weight. Recent experience favors the median-line method.

The median-line principle is a simple tool to establish an optimal mud-weight schedule, taking into account the concerns discussed above. Aadnoy reports a reduction in tight holes and back reaming after invoking this principle.

The mud weight should not be changed continuously. From a practical perspective, the mud engineer may increase mud weight every 4–6 hours. Propose 3 to 4 mud weight increases in the interval. At the top of the new well section, it is common to start with a mud weight below the median line.

2.1.1.6 Borehole Collapse

Borehole collapse typically takes place at lower borehole pressures. The high stress contrast between the high hoop stress and the low borehole pressures gives rise to a high shear stress. Therefore, collapse is defined as a shear failure. Sometimes tight holes occur, which may require frequent wiper trips or reaming. This can, in certain wells, lead to stuck drillstring or difficulties in landing the casing string. There are many reasons for a tight hole; for example, dogleg severity (high wellbore curvature) can contribute, or simply inward creep of the borehole wall, also aided by shale swelling.

Audio 2.2

Most boreholes will enlarge over time. This is often a time-dependent collapse phenomenon. Problems caused by hole enlargement include difficulties in removing rock fragments and drilled cuttings from the borehole, or a reduced quality of the logging operation or cement placement behind casing strings. It is important to understand that a tight hole and borehole collapse are similar events; in one case, the hole may yield, while in the latter case, an abrupt failure may occur. If rock cavings are seen in the mud returns, the correcting action is usually to increase the mud weight, thereby reducing the hoop stress.

Fig. 2.10 shows a typical collapse failure. The shear failure planes are curved because of the circular geometry of the hole. As shown in Fig. 2.10(a), the shear planes connect, resulting in rock fragments falling into the borehole.

崩落的岩石，那么通常要增加钻井液密度，从而降低周向应力。

典型的坍塌破坏如图 2.10 所示。由于井眼是圆形的几何形状，所以剪切破坏面是弯曲的。剪切面相互连接，导致岩屑落入井眼中，如图 2.10（a）所示。

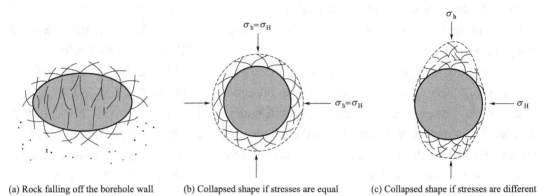

(a) Rock falling off the borehole wall (b) Collapsed shape if stresses are equal (c) Collapsed shape if stresses are different

Fig. 2.10　Collapse of borehole wall

If the external borehole stresses are equal, the collapsed hole will retain a circular shape, as seen in Fig. 2.10(b). However, if the stresses are different, an elongated borehole will result. For a vertical well, the longer hole axis will point in the direction of the minimum in-situ stress[Fig.2.10(c)]. This is often used as a method to assess the direction of the minimum horizontal stress from caliper logs, and it is called breakout analysis. This method usually is not applied for deviated wells because it is believed that the drillstring rotation may provide an upward bias for an elongated borehole.

1. Shear Failure

Before analyzing borehole collapse, we must define the failure mechanism, which is a shear failure. Strength data are obtained from cores as shown in Fig. 2.11. Core samples are subjected to a constant confining pressure and loaded axially until they fail. This process is repeated for various confining pressures. The failure behavior depends on the loading state—that is, the confining pressure level. An example of such laboratory tests is shown in Table 2.2.

如果井眼的外部应力相等，则坍塌的井眼将保持圆形，如图 2.10（b）所示。如果应力不相等，则会导致井眼被拉长（椭变）。对于直井，长轴指向最小地应力的方向 [图 2.10（c）]。这种方法常用于从井径测井中评定最小水平应力的方向，称为破坏分析。但这种方法不适用于斜井，因为钻柱旋转可能会使拉长的井眼产生向上的倾斜。

1. 剪切破坏

在分析井眼坍塌之前，必须明确其破坏机理是一种剪切破坏。从岩心获得的强度数据如图 2.11 所示。岩样在恒定的围压下受到轴向载荷，直到失效。这个过程在不同的围压下反复进行。失效的表现方式取决于载荷的状态，即围压水平。实验室试验的一个示例见表 2.2。

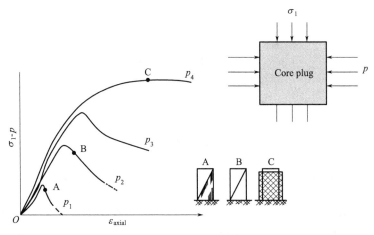

Fig. 2.11 Shear strength from core samples

Table. 2.2 Triaxial strength data for leuders limestone

Test No.	1	2	3	4	5	6
Confining Pressure, σ_3 (bar)	0	41	69	139	207	310
Yield Strength, σ_1 (bar)	690	792	938	1069	1248	1448

There are many details that must be considered when testing core plugs. This discussion will not be pursued here, but the reader is referred to Cook and Edwards. For a given rock plug test, the data from the test are the maximum compressive stress; the minimum stress, which is the confining pressure; and the pore pressure inside the plug. Here we will show how these pressures are used for modeling.

The failure data from Table 2.2 are plotted in Fig. 2.12. Six core plugs are tested to failure for various confining pressures. We show the data in a Mohr-Coulomb plot. Along the horizontal axis, the failure and the confining pressure for each test are marked, and a circle is made between these points. A line is drawn on top of all circles. This is the failure line, which we will use in our collapse analysis.

If the stress state for an application falls below the failure line, the specimen is intact. However, crossing the failure line, the specimen will fail. The laboratory-obtained data from Table 2.2 can be modeled with many other failure models. We will restrict the discussion to the most common model.

在测试岩心时，有许多细节必须考虑。这里不作讨论，读者可以参考库克和爱德华兹的有关文献。对于给定的岩心试验，从试验得到的数据有最大压应力、最小应力即围压，以及岩心内部的孔隙压力。下面展示如何用这些压力进行建模。

将表2.2中的数据绘制如图2.12所示的图形。6个岩心在不同的围压下进行测试直至破坏。下面用莫尔—库仑图展示这些数据。沿水平轴标出每一次试验的破坏点和围压，并在这些点之间画一个圆（莫尔圆）。在所有的圆上面画一条线。这就是将在坍塌分析中用到的破坏包络线。

如果应力状态在破坏包络线以下，则样本完好无损。而当应力超过破坏包络线时，样本就会失效。从表2.2中获得的实验室数据也可以用其他破坏模型建模。这里只讨论最常见的一种模型。

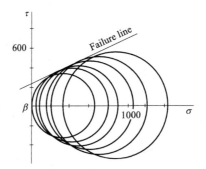

Fig. 2.12 Mohr-Coulomb failure model for data of Table 2.3

2. The Mohr-Coulomb Failure Model

In a 2D stress state, the stresses can be described by means of Mohr's circle. This is done by constructing a circle with a diameter equal to the difference between the maximum and the minimum stress at failure.

The Mohr-Coulomb failure model is this failure line, which mathematically can be expressed as

$$\tau = \tau_0 + \sigma' \tan \phi$$

where, τ is shear stress; τ_0 is cohesive rock strength; σ' is rock effective stress; ϕ is angle of internal friction for rocks.

The failure line is established from laboratory-obtained data as shown above. To apply this failure model to a well, we must derive expressions for the stresses acting on the wellbore. Fig. 2.13 shows the stresses at failure.

2. 莫尔—库仑破坏模型

在二维应力状态下，应力可以通过构造一个莫尔圆来描述，该圆的直径等于破坏时最大和最小应力之差。

莫尔—库仑破坏模型就是这条破坏包络线，可用数学公式表示为：

(2.9)

式中，τ 为剪切应力；τ_0 为黏性岩石强度；σ' 为有效应力；ϕ 为岩石内摩擦角。

破坏包络线是由上述实验室数据建立起来的。为了将这一破坏模型应用到实际中，必须推导出作用在井筒上的应力表达式。破坏时的应力如图2.13所示。

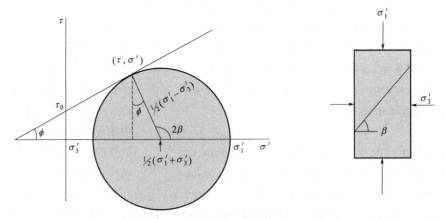

Fig. 2.13 Stresses at failure for the Mohr-Coulomb failure model

In the Fig. 2.13, we use effective stresses. Inspection of the figure reveals that the coordinates (τ, σ') at failure are defined by the following equations:

$$\tau = \frac{1}{2}(\sigma_1' - \sigma_3')\cos\phi \tag{2.10a}$$

$$\sigma' = \frac{1}{2}(\sigma_1' + \sigma_3') - \frac{1}{2}(\sigma_1' - \sigma_3')\sin\phi \tag{2.10b}$$

The models above are expressed in terms of principal stresses. The maximum principal stress is the tangential stress and the minimum principal stress is the wellbore pressure. Inserting Eq. 2.10 into the Mohr-Coulomb failure model, Eq. 2.9, and the borehole stresses from Eq. 2.5 (equal horizontal stresses), results in the following equation for the critical collapse pressure:

$$p_{wc} = \sigma_h(1 - \sin\phi) - \tau_0\cos\phi + p_o\sin\phi \tag{2.11}$$

The Mohr-Coulomb model contains two material properties. The angle ϕ is defined as the angle of internal friction. Sandstone, for example, will exhibit friction along a shear plane because the grains will restrict motion. This is true whether the sand grains are cemented or not. The cohesive strength τ_0, on the other hand, reflects the degree of cementation of the material.

Although simple, Eq. 2.11 shows the interrelationships that cause mechanical wellbore collapse. A high formation stress and a high pore pressure induce collapse. High cohesive strength or the cementation of the rock actually resists collapse. Loose sands have no cohesive strength and therefore have a high collapse pressure. Therefore, the borehole might be subjected to sand production. A high angle of internal friction also opposes collapse.

The fracture angle on the plug specimen shown in Fig. 2.13 can be determined from the following expression:

$$\beta = 45° + \frac{\phi}{2} \tag{2.12}$$

We will show the mechanisms that lead to

mechanical hole collapse by referring to the three borehole stresses given in Eq. 2.5. Because we are considering porous media, we will define them in terms of effective stresses:

Radial effective stress:

$$\sigma'_r = p_w - p_o \tag{2.13a}$$

Tangential effective stress:

$$\sigma'_\theta = 2\sigma_h - p_w - p_o \tag{2.13b}$$

Vertical effective stress:

$$\sigma'_v = \sigma_v - p_o \tag{2.13c}$$

三种井眼应力来说明导致井眼坍塌的机理。鉴于是多孔介质,用有效应力来定义它们:

径向有效应力:

切向有效应力:

垂直有效应力:

2.1.2 Borehole Stability Analysis for Inclined Wells

We have so far discussed vertical wells. Many wells today are deviated. This complicates the picture, which is now three-dimensional, and one has to properly account for the effects of wellbore deviation. In the following sections, the general methodology will be presented. We will first define the general equations for stresses around a borehole.

2.1.2.1 The Kirsch Equations

In the previous derivation, we have studied the stresses at the borehole wall. Now we will investigate the stress state in the rock formation. The following equations define this:

2.1.2 斜井井壁稳定性分析

前面讨论的都是直井,然而现在很多井都是斜井。斜井的三维应力图更加复杂,并且必须适当考虑井斜的影响。下面,将介绍一般的方法。首先定义井眼周围应力的一般方程。

2.1.2.1 基尔希方程

在前面的推导中,研究了井壁的应力。以下公式给出了岩层中应力状态的定义:

$$\sigma_r = \frac{1}{2}(\sigma_x + \sigma_y)\left(1 - \frac{a^2}{r^2}\right) + \frac{1}{2}(\sigma_x - \sigma_y)\left(1 + 3\frac{a^4}{r^4} - 4\frac{a^2}{r^2}\right)\cos 2\theta + \tau_{xy}\left(1 + 3\frac{a^4}{r^4} - 4\frac{a^2}{r^2}\right)\sin 2\theta + \frac{a^2}{r^2}p_w \tag{2.14a}$$

$$\sigma_\theta = \frac{1}{2}(\sigma_x + \sigma_y)\left(1 + \frac{a^2}{r^2}\right) - \frac{1}{2}(\sigma_x - \sigma_y)\left(1 + 3\frac{a^4}{r^4}\right)\cos 2\theta - \tau_{xy}\left(1 + 3\frac{a^4}{r^4}\right)\sin 2\theta - \frac{a^2}{r^2}p_w \tag{2.14b}$$

$$\sigma_z = \sigma_{zz} - 2\upsilon(\sigma_x - \sigma_y)\frac{a^2}{r^2}\cos 2\theta - 4\upsilon\tau_{xy}\frac{a^2}{r^2}\sin 2\theta \tag{2.14c}$$

$$\tau_{r\theta} = \left[\frac{1}{2}(\sigma_x - \sigma_y)\sin 2\theta + \tau_{xy}\cos 2\theta\right]\left(1 - 3\frac{a^4}{r^4} + 2\frac{a^2}{r^2}\right) \tag{2.14d}$$

$$\tau_{rz} = (\tau_{xz}\cos\theta + \tau_{yz}\sin\theta)\left(1 - \frac{a^2}{r^2}\right) \tag{2.14e}$$

$$\tau_{\theta z} = (-\tau_{xz}\cos\theta + \tau_{yz}\sin\theta)\left(1 + \frac{a^2}{r^2}\right) \tag{2.14f}$$

At the borehole wall (r=a), the above equations reduce to

Radial stress:

$$\sigma_r = p_w \tag{2.15a}$$

Tangential stress:

$$\sigma_\theta = \sigma_x + \sigma_y - p_w - 2(\sigma_x - \sigma_y)\cos 2\theta - 4\tau_{xy}\sin 2\theta \tag{2.15b}$$

Axial stress, plane strain:

$$\sigma_z = \sigma_z - 2\upsilon(\sigma_x - \sigma_y)\cos 2\theta - 4\upsilon\tau_{xy}\sin 2\theta \tag{2.15c}$$

Axial stress, plane stress:

$$\sigma_z = \sigma_{zz} \tag{2.15d}$$

Shear stress:

$$\tau_{\theta z} = 2(\tau_{yz}\cos\theta - \tau_{xz}\sin\theta) \tag{2.15e}$$

$$\tau_{rz} = \tau_{r\theta} = 0 \tag{2.15f}$$

where, σ_r = radial stress; σ_x = normal stress, x coordinate direction; σ_y = normal stress, y coordinate direction; σ_z = normal stress, z coordinate direction; σ_{zz} = normal stress, z coordinate direction; a = borehole radius; θ = position on borehole wall from x axis; τ_{xy} = shear stress, x plane in the y direction; τ_{rz} = shear stress, r plane in the z direction; $\tau_{r\theta}$ = shear stress, r plane in the θ direction; τ_{yz} = shear stress, y plane in the z direction; $\tau_{\theta z}$ = shear stress, θ plane in the z direction; υ = Poisson's ratio for the rock matrix.

Eq. 2.15 is used for most borehole stability analysis because the formation fails at the borehole wall. That is, the stresses are usually highest on the borehole wall, so therefore it will fail here first. The plane stress solution (vertical stresses remain constant) for the axial stress usually is used because of simplicity. The difference between the plane stress and plane strain solutions (wellbore displaces only in the horizontal plane) is usually negligible.

式中，σ_r 为径向应力；σ_x 为 x 坐标方向应力分量；σ_y 为 y 坐标方向应力分量；σ_z 为 z 坐标方向应力分量；σ_{zz} 为 z 坐标方向应力分量；a 为井径；θ 为井壁从 x 轴偏离的角度；τ_{xy} 为 y 方向 x 平面的剪应力；τ_{rz} 为 z 方向 r 平面的剪应力；$\tau_{r\theta}$ 为 θ 方向 r 平面的剪应力；τ_{yz} 为 z 方向 y 平面的剪应力；$\tau_{\theta z}$ 为 z 方向 θ 平面的剪应力；υ 为岩石基质的泊松比。

由于地层坍塌发生在井壁，所以大多数井壁稳定性分析都使用式（2.15）。也就是说，通常井壁上的应力是最高的，因此这里会首先失效。为了简化计算，通常使用轴向应力的平面应力解（垂直应力保持常数）。平面应力和平面应变解（井筒位移仅在水平面上）之间的差通常忽略不计。

2.1.2.2 Deviated Boreholes and Stresses in Three Dimensions

Deviated boreholes are in general subjected to a more complex stress state than vertical wells, even in a sedimentary basin. The reason is that if horizontal stresses are equal for a vertical well, the stresses normal to the hole will change when the well becomes inclined.

In applications of the Kirsch equations given above, often one assumes a horizontal and vertical in-situ stress field. The borehole, however, may take any orientation. Therefore, one must define equations to transform the in situ stresses to the orientation of the borehole.

It is common in the oil industry to assume three principal in-situ stresses: the vertical or overburden stress σ_v, and the maximum and minimum horizontal stresses, σ_H and σ_h. Fig. 2.14 shows the most important stresses. The input stresses are the in-situ stresses σ_v, σ_H, and σ_h.

Because the borehole may take any orientation, these stress must be transformed to a new coordinate system, x, y, z, where we observe stresses as σ_x, σ_y, σ_z. The directions of the new stress components are given by the borehole inclination from vertical, φ, the geographical azimuth, α, and the position on the borehole wall from the x axis, θ. One of the properties of this transformation is that the y axis is always parallel to the plane formed by σ_H and σ_h.

The following equations define all transformed stress components:

$$\sigma_x = (\sigma_H \cos^2 \alpha + \sigma_h \sin^2 \alpha)\cos^2 \varphi + \sigma_v \sin^2 \varphi \quad (2.16a)$$

$$\sigma_y = (\sigma_H \sin^2 \alpha + \sigma_h \cos^2 \alpha) \quad (2.16b)$$

$$\sigma_z = (\sigma_H \cos^2 \alpha + \sigma_h \sin^2 \alpha)\sin^2 \varphi + \sigma_v \cos^2 \varphi \quad (2.16c)$$

$$\tau_{yz} = \frac{1}{2}(\sigma_h - \sigma_H)\sin 2\alpha \sin \varphi \quad (2.16d)$$

$$\tau_{xz} = \frac{1}{2}(\sigma_H \cos^2\alpha + \sigma_h \sin^2\alpha - \sigma_v)\sin 2\varphi \qquad (2.16e)$$

$$\tau_{xy} = \frac{1}{2}(\sigma_h - \sigma_H)\sin 2\alpha \cos\varphi \qquad (2.16f)$$

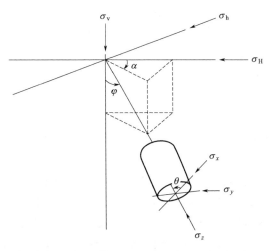

Fig. 2.14 Orientation of borehole relative to the in-situ stresses

2.1.3 General Methodology for Analysis of Wellbore Stability

In this section, the general methodology for borehole stability analysis will be presented for both fracturing and collapse. This is valid for all stress states (normal, strike-slip, and reverse) and for all borehole orientations. The calculation procedure is as follows:

Audio 2.3

(1) Calculate the stresses in the direction of the borehole.

(2) Insert these data into the borehole stress equations.

(3) Determine the point on the borehole wall where failure will occur.

(4) Implement a failure model.

(5) Compute borehole pressure at failure.

In the following sections, the general methodology for failure analysis will be presented both for fracturing and collapse.

Assuming that we know the in-situ stress state, we

2.1.3 井壁稳定性分析的一般方法

下面介绍对井眼破裂和坍塌都有效的井壁稳定性分析的一般方法。它适用于所有应力状态（正断层、走滑断层和逆断层）和所有的井眼方向。计算过程如下：

（1）计算井眼方向上的应力。

（2）将这些数据代入井眼应力方程中。

（3）确定井壁上可能发生失效的点。

（4）建立失效模型。

（5）计算失效时的井眼压力。

在下面的章节中，将介绍破裂和坍塌失效分析的一般方法。

假设已知地应力状态，必须

must transform these to the orientation of the borehole by using Eq. 2.16. The transformed stresses may then be inserted into the stress equations (Eq. 2.15).

Now we have expressions for the stresses at the borehole wall, or the stress state in the adjacent formation. Usually the borehole pressure is unknown at this stage. The object is often to determine the critical pressure that leads to failure of the borehole.

2.1.3.1 Principal Borehole Stresses

To solve for the critical pressure, the stresses discussed above are inserted into the failure criteria for the borehole. Remember, however, to use effective stresses by subtracting the pore pressure. This applies only for normal stresses, not for shear stresses.

The borehole wall is subjected to normal stresses and shear stresses. It is common to find the maximum normal stress by defining a direction (plane) where the shear stresses vanish. The resulting stresses are called principal stresses. At the borehole wall, these are:

$$\sigma_1 = p_w \tag{2.17a}$$

$$\sigma_2 = \frac{1}{2}(\sigma_\theta + \sigma_z) + \frac{1}{2}\sqrt{(\sigma_\theta - \sigma_z)^2 + 4\tau_{\theta z}^2} \tag{2.17b}$$

$$\sigma_3 = \frac{1}{2}(\sigma_\theta + \sigma_z) - \frac{1}{2}\sqrt{(\sigma_\theta - \sigma_z)^2 + 4\tau_{\theta z}^2} \tag{2.17c}$$

After calculating the principal stresses above, the subscripts are often interchanged such that σ_1 always refers to the maximum compressive principal stress, σ_2 to the intermediate, and σ_3 to the least principal stress. Typical principal stresses are

Fracturing:

$$\sigma_1 = p_w$$
$$\sigma_3 = \frac{1}{2}(\sigma_\theta + \sigma_z) - \frac{1}{2}\sqrt{(\sigma_\theta - \sigma_z)^2 + 4\tau_{\theta z}^2} \tag{2.18a}$$

Collapse:

$$\sigma_1 = \frac{1}{2}(\sigma_\theta + \sigma_z) + \frac{1}{2}\sqrt{(\sigma_\theta - \sigma_z)^2 + 4\tau_{\theta z}^2} \tag{2.18b}$$
$$\sigma_3 = p_w$$

利用式（2.16）将其转换到井眼的方向，然后将转换后的应力代入应力方程[式（2.15）]。

现在，就有了井壁上的应力表达式或相邻地层的应力状态。在这个阶段，井眼压力通常是未知的。我们的目的是要确定导致井眼失效的临界压力。

2.1.3.1 井眼主应力

为了求解临界压力，将上述应力代入井眼破坏准则中。但是要记住，使用有效应力时要扣除孔隙压力。而且这只适用于正应力，不适用于剪应力。

井壁承受正应力和剪应力。一般通过确定剪应力消失的方向（平面）来确定最大正应力，所产生的应力称为主应力，在井壁上，这些主应力分别是：

计算完上面的主应力后，下标常常互换为：σ_1 指最大主应力，σ_2 指中间应力，σ_3 指最小主应力。典型的主应力有：

破裂：

坍塌：

2.1.3.2 Borehole Fracturing

The borehole will fracture when the minimum effective principal stress reaches the tensile rock strength σ_t. This is expressed as

$$\sigma_3' = \sigma_3 - p_o = \sigma_t \qquad (2.19)$$

Inserting Eq. 2.18a into Eq. 2.19, the critical tangential stress is given by

$$\sigma_\theta = \frac{\tau_{\theta z}^2}{\sigma_z - \sigma_t - p_o} + p_o + \sigma_t \qquad (2.20)$$

Inserting the equation for the tangential stress, Eq. 2.15b, the critical borehole pressure is given by

$$p_w = \sigma_x + \sigma_y - 2(\sigma_x - \sigma_y)\cos 2\theta - 4\tau_{xy}\sin 2\theta - \frac{\tau_{\theta z}^2}{\sigma_z - \sigma_t - p_o} - p_o - \sigma_t \qquad (2.21)$$

There is another unknown for the general case. The fracture may not arise in the direction of the *x* or *y* axis because of shear effects. To resolve this issue, Eq. 2.21 is differentiated to define the extreme conditions ($dp_w/d\theta=0$).

The normal stresses are in general much larger than the shear stresses. Neglecting second-order terms, Eq. 2.22 defines the position on the borehole wall for the fracture:

$$\tan 2\theta = \frac{2\tau_{xy}}{\sigma_x - \sigma_y} \qquad (2.22)$$

The final fracture equation is obtained by inserting the angle from Eq. 2.22 into Eq. 2.21.

The general fracturing equation is now defined. It is valid for all cases, arbitrary directions, and anisotropic stresses, but must in general be solved by numerical methods.

If symmetric conditions exist, all shear stress components may vanish. In these cases, the fracture may take place at one of the following conditions: $\sigma_H = \sigma_h$; $\gamma = 0°$; $\alpha = 0°, 90°$. It is also common to assume that the rock has zero tensile strength because it may contain cracks or fissures. Inserting these conditions, the fracturing equation becomes

$$p_w = 3\sigma_x - \sigma_y - p_o - \sigma_t, \qquad (\sigma_x < \sigma_y, \quad \theta = 90°) \qquad (2.23a)$$

$$p_w = 3\sigma_y - \sigma_x - p_o - \sigma_t, \qquad (\sigma_y < \sigma_x, \quad \theta = 0°) \qquad (2.23b)$$

通常假设岩石的抗拉强度为零，因为它可能含有裂缝或裂纹。代入这些条件，破裂方程变为：

These equations simply say that a fracture will initiate normal to the least stress and will propagate in the direction of the largest normal stress.

Also, observe that assuming a maximum and a minimum stress normal to the borehole wall and vanishing shear stresses, the general fracturing equation becomes

$$p_w = 3\sigma_{min} - \sigma_{max} - p_o - \sigma_t \qquad (2.24)$$

这些方程说明裂缝将从最小应力的法线方向开始，并沿着最大正应力的方向延伸。

此外，假设最大和最小应力垂直于井壁，且剪应力消失，一般破裂方程就变为：

2.1.3.3 Borehole Collapse

While fracturing occurs at high borehole pressures, collapse is a phenomenon associated with low borehole pressures. This can be seen from Eq. 2.18b. At low borehole pressures, the tangential stress becomes large. Since there now is a considerable stress contrast between the radial and the tangential stress, a considerable shear stress arises. If a critical stress level is exceeded, the borehole will collapse in shear.

The maximum principal stress (from Eq. 2.18b) is dominated by the tangential stress and is given by

$$\sigma_1 = \frac{1}{2}(\sigma_\theta + \sigma_z) + \frac{1}{2}\sqrt{(\sigma_\theta - \sigma_z)^2 + 4\tau_{\theta z}^2}$$

and the minimum principal stress (from Eq. 2.21b) is given by

$$\sigma_3 = p_w$$

Differentiating the maximum principal stress equation, Eq. 2.25, we can determine the position on the borehole wall at which the collapse will occur. This is complicated because there are many implicit functions in Eq. 2.25. For the case above with vanishing shear stresses, it turns out that the collapse position on

2.1.3.3 井眼坍塌

井眼破裂是在较高的井眼压力下发生的，而井眼坍塌则与井眼压力低有关。这可以从（式2.18b）中看出来。在低的井眼压力下，切向应力变大。由于径向应力和切向应力之间存在相当大的应力差，由此产生相当大的剪应力。如果超过临界应力水平，井眼将在剪切作用下发生坍塌。

此时，切向应力为最大主应力 [由式（2.18b）可知]，则有：

(2.25)

最小主应力为 [由式（2.21b）可知]：

(2.26)

对最大主应力方程[式（2.25）]进行微分，可以确定发生井壁坍塌的位置。这很复杂，因为在式（2.25）中有许多隐式函数。对于上述剪切应力消失的情况，可以得出井壁坍塌的位置距离裂缝产

the borehole wall is 90° from the position of fracture initiation. Invoking this angle into Eq. 2.25, the collapse stress is obtained.

If symmetric conditions exist, all shear stress components may vanish. In these cases, the collapse failure may take place at one of the following conditions: $\sigma_H = \sigma_h$, $\varphi = 0°$, $\alpha = 0°$ or 90°. Inserting these conditions into Eq. 2.23, the borehole pressure causing highest tangential stress is

$$\sigma_1 = 3\sigma_y - \sigma_x - p_w \quad (\sigma_x < \sigma_y, \ \theta = 0°) \quad (2.27a)$$

$$\sigma_1 = 3\sigma_x - \sigma_y - p_w \quad (\sigma_y < \sigma_x, \ \theta = 90°) \quad (2.27b)$$

or, in general,

$$\sigma_1 = 3\sigma_{\max} - \sigma_{\min} - p_w \quad (2.28)$$

These equations simply say that borehole collapse will initiate in the direction of the minimum horizontal in-situ stress. Eq. 2.28 is strictly valid if the borehole direction is aligned with the in-situ stress direction.

Having obtained expressions for the maximum and the minimum principal stress, a failure model must be defined.

2.1.4 Empirical Correlations

Many correlations have been used in the oil industry to enable the transfer of knowledge from one well to another. Some of these are just simple correlations, whereas others are based on models or physical principles. Although many correlations are still useful, others are replaced by more fundamental engineering methods that have evolved in recent years. Here, we will briefly discuss some of the classical correlation methods still in use.

The first factor to discuss is the drillability, which is actually a normalized rate of penetration.

2.1.4.1 Drillability Correlations

In its simplest form, the rate of penetration R (m/h, ft/h) is modeled as a function of supplied energy:

生的位置为90°。将该角度代入式（2.25），即可得到坍塌压力。

如果存在对称条件，所有剪应力分量可能消失。在这些情况下，坍塌可能发生在下列条件之一：$\sigma_H = \sigma_h$，$\varphi = 0°$，$\alpha = 0°$ 或 90°。将这些条件代入式（2.23）中，产生最大切向应力的井眼压力为：

或一般有：

这些方程说明，井眼坍塌将从最小水平应力的方向上开始。如果井眼的方向与地应力的方向一致，则式（2.28）严格有效。

在得到最大和最小主应力的表达式后，必须确定破坏模型。

2.1.4 经验公式

在石油行业中，人们常利用经验公式将一口井的经验知识用到另一口井上。其中一些只是简单的相关性，另一些则基于模型或物理学原理。尽管许多经验公式仍然有用，但其他的已被近年来发展起来的更基本的工程方法所取代。这里，将简要地讨论一些仍在使用的经典相关性方法。

首先要讨论的是可钻性，它实际上是一个标准化的机械钻速。

2.1.4.1 可钻性

最简单的模型中，机械钻速 R 是能量的函数：

$$R = k \cdot WOB \cdot N \tag{2.29}$$

where, *WOB* is the bit force, *N* is the rotary speed, and the factor *k* represents the drillability. In a soft rock, *R* is high and *k* is high. Conversely, in a hard rock *R* and *k* are small for the same bit force and rotary speed. The drillability is actually an instant measure of the rock properties at the face of the drillbit. It is presently the only information obtained at the drillbit face during drilling.

By computing the drillability, we can create a log that tells us something about the rock; some interesting information can be obtained if we understand the relationships. Fig. 2.15 shows an example from an underground blowout in the North Sea. The well had been flowing for a year before it was killed with a relief well. The drillabilities for the two wells were compared, and they were identical until the wells were about 6 m apart.

式中，*WOB* 为钻压，*N* 为转速，*k* 为可钻性系数。对于软地层，可钻性系数大，机械钻速高；相反，对于硬地层，相同的钻压和转速，可钻性系数和机械钻速都较小。可钻性实际上是对钻头表面岩石性质的一种即时测量，是钻井时从钻头工作面获得的唯一信息。

通过计算可钻性，可以生成一个记录，它能给出一些关于岩石的信息；如果理解了其中的关系，就可以得到一些有用的信息。图 2.15 所示的是北海一次地下井喷的实例。在利用救援井实施压井之前，这口井已经喷了一年。对比两口井的可钻性发现，直到井间相距约 6m 之前，两口井的可钻性都是相同的。

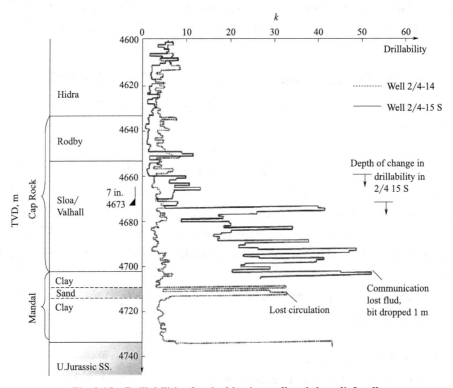

Fig. 2.15 Drillabilities for the blowing well and the relief well

Drillability also has been used to estimate pore pressures for many years. There is usually a considerable uncertainty in pore-pressure estimation; pore pressure in shales cannot be measured, but in permeable rocks like sandstones, real pressure measurements can be used to calibrate pore-pressure curves. A common method is to compute sound transit time from seismic records. During drilling, the drillability is also used as an indicator of pore pressure. Well known is the so-called d-exponent developed by Jorden and Shirley ; during drilling, the following parameter is computed:

$$d = \frac{\lg\left(\dfrac{R}{60N}\right)}{\lg\left(\dfrac{12W}{10^6 B}\right)} \qquad (2.30)$$

The bit diameter is B. A number of corrections are proposed to this equation, and it is still in use in the oil industry.

The d-exponent is actually a drillability equation in logarithmic form. If a weak rock is encountered, both drilling rate and drillability increase. Conversely, a hard rock slows down the drilling rate and causes the drillability to decrease.

During drilling, a trend line is established for the d-exponent. If this starts to deviate from the trend, it can be associated with some of the following factors:

(1) Bit wear. Rock bits wears gradually, and the drilling efficiency may reduce gradually; an abrupt drop in drillability may be caused by drillbit or roller cone failure.

(2) If drillability increases relative to the trend line, it may be a pore-pressure indicator. Reservoir rock below the caprock often has higher drillability because it is weaker and more porous.

Hareland and Hoberock developed a method to estimate the rock strength from drillability data; the result is used as input data for wellbore stability analysis. The

多年来，人们也用可钻性来估算孔隙压力。孔隙压力的估算通常有相当大的不确定性；页岩中的孔隙压力是无法测量的，但在砂岩等可渗透岩石中，实际的压力测量值可以用来校准孔隙压力曲线。一种常用的方法是从地震记录中计算声波传播的时间。在钻井过程中，也用可钻性作为孔隙压力的一个指标。乔登和雪莉提出了d指数法；在钻井过程中，计算如下参数：

式中，B为钻头直径。该方程进行了多次修正，并仍运用于石油行业中。

d指数方程实际上是一个对数形式的可钻性方程。如果遇到软的岩石，钻速和可钻性都会增加。相反，坚硬的岩石使钻速降低，并导致可钻性下降。

在钻井过程中，需要建立一条d指数的趋势线。如果d指数开始偏离趋势线，可能与以下因素有关：

（1）钻头磨损。随着钻头逐渐磨损，钻进效率逐渐降低；可钻性的突然下降可能是钻头或牙轮失效引起的。

（2）如果相对于趋势线，可钻性增加了，这可能是孔隙压力的标志。盖层以下的储集岩往往具有较高的可钻性，因为储层岩石较软且多孔。

哈兰和霍贝罗克开发了一种根据可钻性数据估算岩石强度的方法；其结果可用于井壁稳定性

concept of drillability has been used to a small extent in drilling. For the foreseeable future, there is considerable potential for further applications of drillability; drilling optimization is one such area.

2.1.4.2 Fracture Pressure Correlations

In order to predict fracture pressures and pore pressures, a number of correlations have been derived over the years. Although some of these are valid for the places in which they were derived, a number of correlations based on various physical assumptions have been seen to have a more general applicability; many of these have been developed in the Gulf of Mexico region.

Hubbert and Willis developed a very useful correlation; they assumed that the horizontal stress in a relaxed basin should be one-third to one-half of the overburden stress. Although we today know that this value is too low, their correlation is still effective. Matthew and Kelly modified this model by introducing a "matrix stress coefficient," which implied that the stress ratios were not constant with depth. Pennebaker related the overburden gradient to geologic age and established "effective stress ratio" relationships. Pennebaker also correctly found that the fracture gradient is related to the overburden stress gradient.

Eaton introduced the Poisson effect by defining the horizontal stress as a result of the overburden stress. The correlation coefficient for this case was actually Poisson's ratio. Finally, Christman extended this work to an offshore environment.

At first glance, the five methods listed above look different; however, Pilkington compared these methods and found that they were very similar. By introducing the same correlation coefficient, all five models can be defined by the following equation:

$$p_{wf} = K(\sigma_v - p_o) + p_o$$

Pilkington also used field data and showed that Eq. 2.31 basically gave the same result as each of the

2.1.4.2 破裂压力

为了预测破裂压力和孔隙压力，多年来人们研究出了许多经验公式。虽然一些经验公式在其推导的地方是有效的，但人们认为基于各种物理假设的经验公式更具有普遍的适用性，其中许多是在墨西哥湾地区开发的。

休伯特和韦利斯提出了一个非常有用的经验公式：他们假设疏松盆地的水平应力是上覆岩层压力的三分之一到二分之一。尽管现在已经知道这个值太低了，但它们之间的相关性仍然有效。马修斯和凯利引入"骨架应力系数"对该模型进行了修正，并表明应力比随深度变化。尼贝克将上覆岩层压力梯度与地质年代关联起来，建立了"有效应力比"关系。尼贝克还发现破裂压力梯度与上覆岩层压力梯度相关。

伊顿通过将水平应力定义为上覆岩层压力作用的结果，引入了泊松效应。实际上这种情况下的相关系数就是泊松比。最后，克里斯曼将这一研究扩展到了海洋环境。

初看起来，上面列出的五种方法各不相同；然而，皮尔金顿将这些方法进行了对比，并发现它们非常相似。通过引入相同的相关系数，这五个模型均可由以下公式进行定义：

(2.31)

皮尔金顿代入油田数据后发现，式（2.31）得到的结果与五个模型基

five models. Most of these models were developed in the Texas Louisiana area, where they still may serve; however, in the early 1980s, wellbore inclination increased, and because the empirical correlations could not handle this, continuum mechanics was introduced. In addition to handling the directions of inclined wells it also opened up for various stress states. Any relaxed or tectonic setting can now be handled by using classical mechanics.

2.1.4.3 Pore-Pressure Correlations

Pore pressure is a key factor in petroleum production, and it also has a significant effect on well construction and wellbore stability. Typically, 70% of the rock we drill is shale or clay; as discussed, these rocks are usually impermeable, and it is therefore not possible to measure pore pressure that are measured in the reservoir.

We need a pore-pressure curve to select mud weights and casing points. The pore-pressure curves inferred from many sources are used as absolute; of course, underbalanced drilling in a tight shale will not lead to a well kick. So if we could guarantee that there were no permeable stringers, we could drill the well with a mud weight below the pore pressure. Unfortunately, such a guarantee is unlikely.

From this discussion, it is clear that pore pressure from logs and other sources is not accurate unless it is calibrated (e.g., with a pressure measurement). This is usually not the case, so pore-pressure curves in general have significant uncertainty.

2.2 Formation Pressures

2.2.1 Introduction

The magnitude of the pressure in the pores of a formation, known as the formation pore pressure (or simply formation pressure), is an important

Audio 2.4

本相同。这些模型大部分是在得克萨斯—路易斯安那地区开发的，它们现在仍可使用；然而，在20世纪80年代早期，由于经验公式不能处理井斜角增大带来的问题，人们引入了连续介质力学。它不仅能够处理斜井的方向，还能应对各种应力状态。现在任何松散或构造环境都可以用经典力学来处理。

2.1.4.3 孔隙压力

孔隙压力是石油生产中的一个关键因素，它对建井和井壁稳定性也有重要影响。通常钻遇的岩石有70%是页岩或黏土；如前所述，这些岩石通常是不渗透的，因此不能测量其孔隙压力，压力测量是在储层中进行的。

需要根据孔隙压力曲线来选择钻井液密度和套管下入点。由多源信息推导出的孔隙压力曲线被用作绝对准则；当然，在致密页岩中进行欠平衡钻井是不会导致井涌的。因此，如果能够保证不存在渗透性的地层，就可以用低于孔隙压力的钻井液密度来钻井。然而，这是不可能的。

从这一讨论中可以清楚地看出，来自测井和其他来源的孔隙压力，除非经过校准（如通过压力测量），否则是不准确的。因此孔隙压力曲线一般具有显著的不确定性。

2.2 地层压力

2.2.1 引言

地层孔隙中的压力大小，称为地层孔隙压力（简称地层压力），是油井设计和作业等许多方面都要考虑的一个重要因素。它将影

consideration in many aspects of well planning and operations. It will influence the casing design and mud weight selection and will increase the chances of stuck pipe and well control problems. It is particularly important to be able to predict and detect high pressure zones, where there is the risk of a blow-out.

In addition to predicting the pore pressure in a formation it is also very important to be able to predict the pressure at which the rocks will fracture. These fractures can result in losses of large volumes of drilling fluids and, in the case of an influx from a shallow formation, fluids flowing along the fractures all the way to surface, potentially causing a blowout.

When the pore pressure and fracture pressure for all of the formations to be penetrated have been predicted the well will be designed, and the operation conducted, such that the pressures in the borehole neither exceed the fracture pressure, nor fall below the pore pressure in the formations being drilled.

2.2.2 Formation Pore Pressures

During a period of erosion and sedimentation, grains of sediment are continuously building up on top of each other, generally in a water filled environment. As the thickness of the layer of sediment increases, the grains of the sediment are packed closer together, and some of the water is expelled from the pore spaces. However, if the pore throats through the sediment are interconnecting all the way to surface the pressure of the fluid at any depth in the sediment will be same as that which would be found in a simple column of fluid. The pressure in the fluid in the pores of the sediment will only be dependent on the density of the fluid in the pore space and the depth of the pressure measurement. it will be independent of the pore size or pore throat geometry. The pressure of the fluid in the pore space can be measured and plotted against depth as shown in a figure This type of diagram is known as a P-Z diagram.

The pressure in the formations to be drilled is often expressed in terms of a pressure gradient. This gradient

响套管设计和钻井液密度的选择，并可能增加卡钻和井控问题的发生概率。由于高压层存在井喷的风险，因此预测和监测高压地层尤为重要。

除了预测地层中的孔隙压力外，预测岩石的破裂压力也非常重要。这些裂缝可能会导致大量钻井液漏失，如果浅层地层有流体侵入，则流体就会沿着裂缝一路流向地面，从而可能导致井喷。

当所有待钻地层的孔隙压力和破裂压力预测完成后，将对油井进行设计和钻井作业，钻井期间，井筒内的压力要小于破裂压力且大于所钻地层的孔隙压力。

2.2.2 地层孔隙压力

经过一定时间的侵蚀和沉积，沉积物颗粒通常在一个充满水的环境中不断地相互叠加。随着沉积层厚度的增加，沉积物颗粒被挤压得更加紧密，有一些水会从孔隙空间中排出。如果沉积物的孔隙喉道一直相互连通到地表，那么任何深度处的流体压力都与单一流体的静液柱压力相等。沉积物孔隙中流体的压力取决于孔隙空间中流体的密度和测量深度，而与孔隙的大小或孔喉的几何形状无关。可以测量并绘制孔隙空间中的流体压力与深度的关系图，这种图称为 P-Z 图。

地层压力通常用压力梯度来表示。压力梯度可由一条穿过特

is derived from a line passing through a particular formation pore pressure and a datum point at surface and is known as the pore pressure gradient. The datum which is generally used during drilling operations is the drillfloor elevation but a more general datum level, used almost universally, is Mean Sea Level, MSL. When the pore throats through the sediment are interconnecting, the pressure of the fluid at any depth in the sediment will be same as that which would be found in a simple column of fluid and therefore the pore pressure gradient is a straight line as shown in Fig. 2.16. The gradient of the line is a representation of the density of the fluid. Hence the density of the fluid in the pore space is often expressed in units of psi/ft.

定地层孔隙压力和一个地面基准点的连线推导出来，所以也称为孔隙压力梯度。钻井作业期间，一般以钻台作为基准面，但更为普遍的是使用平均海平面（MSL）作为基准面。当沉积物的孔喉相互连通时，沉积物中任何深度处的流体压力与单一流体的静液柱压力相等，因此孔隙压力梯度是一条直线，如图2.16所示。这条线的梯度代表了流体的密度。因此，孔隙空间中流体的密度常用psi/ft作为单位。

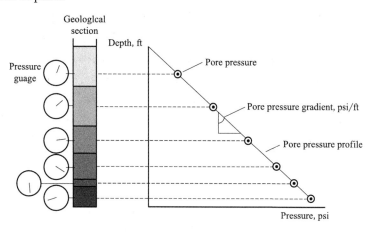

Fig. 2.16 P-Z Diagram representing pore pressures

This is a very convenient unit of representation since the pore pressure for any given formation can easily be deduced from the pore pressure gradient if the vertical depth of the formation is known. Representing the pore pressures in the formations in terms of pore pressure gradients is also convenient when computing the density of the drilling fluid that will be required to drill through the formations in question. If the density of the drilling fluid in the wellbore is also expressed in units of psi/ft then the pressure at all points in the wellbore can be compared with the pore pressures to ensure that the pressure in the wellbore exceeds the pore pressure. The differential between the mud pressure and the pore

这是一个非常便于表述的单位，因为对于任何给定地层，如果已知地层深度，就可以很容易地从孔隙压力梯度推导出孔隙压力。用孔隙压力梯度代表地层孔隙压力，这在计算钻穿地层所需的钻井液密度时很方便。如果井筒内钻井液的密度也以psi/ft为单位，就可以将井筒内各点的压力与孔隙压力进行对比，以确保井筒内的压力大于孔隙压力。在任意给定的深度处，钻井液液柱压力与孔隙压力的差称为该深度的

pressure at any given depth is known as the overbalance pressure at that depth (Fig. 2.17). If the mud pressure is less than the pore pressure, then the differential is known as the underbalance pressure. The fracture pressure gradient of the formations is also expressed in units of psi/ft.

Most of the fluids found in the pore space of sedimentary formations contain a proportion of salt and are known as brines. The dissolved salt content may vary from 0 to over 200,000 mg/kg. Correspondingly, the pore pressure gradient ranges from 0.433 psi/ft (pure water) to about 0.50 psi/ft. In most geographical areas the pore pressure gradient is approximately 0.465 psi/ft (assumes 80,000 mg/kg salt content) and this pressure gradient has been defined as the normal pressure gradient. Any formation pressure above or below the points defined by this gradient are called abnormal pressures (Fig. 2.18). The mechanisms by which these abnormal pressures can be generated will be discussed below. When the pore fluids are normally pressured the formation pore pressure is also said to be hydrostatic.

过平衡压力（图2.17）。如果钻井液液柱压力小于孔隙压力，则称为欠平衡压力。地层的破裂压力梯度也可以用psi/ft单位表示。

在沉积地层的孔隙空间中发现的大多数流体都含有一定比例的盐，它们被称为卤水。溶解盐的含量从0到200000mg/kg不等。相应地，孔隙压力梯度的范围从0.433psi/ft（纯水）到0.50psi/ft左右。大多数地区的孔隙压力梯度约为0.465 psi/ft（假设含盐量为80,000mg/kg），该压力梯度被定义为正常压力梯度。任何高于或低于该梯度的点都被称为异常压力（图2.18）。这些异常压力的产生机理将在后面讨论。当孔隙流体受到正常压力时，地层孔隙压力也称为静水压力。

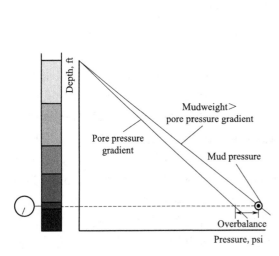

Fig. 2.17 Mud density compared to pore pressure gradient

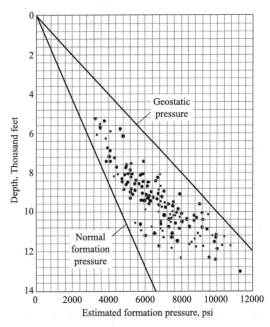

Fig. 2.18 Abnormal formation pressures plotted against depth for 100 US wells

2.2.3 Overburden Pressures

The pressures discussed above relate exclusively to the pressure in the pore space of the formations. It is however also important to be able to quantify the vertical stress at any depth since this pressure will have a significant impact on the pressure at which the borehole will fracture when exposed to high pressures. The vertical pressure at any point in the earth is known as the overburden pressure or geostatic pressure. The overburden gradient is derived from a cross plot of overburden pressure versus depth (Fig. 2.19). The overburden pressure at any point is a function of the mass of rock and fluid above the point of interest. In order to calculate the overburden pressure at any point, the average density of the material (rock and fluids) above the point of interest must be determined. The average density of the rock and fluid in the pore space is known as the bulk density of the rock:

2.2.3 上覆岩层压力

以上讨论的压力只与地层孔隙空间中的压力有关。确定任意深度处的垂直应力也很重要，因为当井眼暴露于高压下时，垂直应力会对井眼的破裂压力产生重大影响。地下任意一点的垂直压力称为上覆岩层压力。上覆岩层压力梯度由上覆岩层压力—深度交会图得到（图2.19）。任意点的上覆岩层压力都是其上方岩石和流体质量的函数。为了计算任意点处的上覆岩层压力，必须确定该点以上物质（岩石和流体）的平均密度。岩石和孔隙空间中流体的平均密度称为岩石的体积密度：

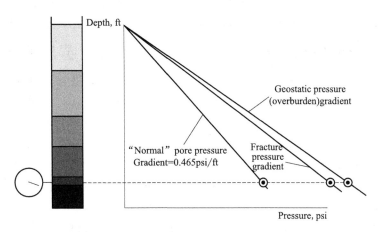

Fig. 2.19 Pore pressure, fracture pressure and overburden pressures for a particular formation

$$\rho_b = \rho_f \times \phi + \rho_m(1-\phi) \tag{2.32}$$

or 或

$$\rho_b = \rho_m - (\rho_m - \rho_f)\phi \tag{2.33}$$

where, ρ_b = bulk density of porous sediment; ρ_m = density of rock matrix; ρ_f = density of fluid in pore space; ϕ = porosity.

式中，ρ_b为多孔沉积物的体积密度；ρ_m为岩石基质的密度；ρ_f为孔隙中流体的密度；ϕ为孔隙度。

Since the matrix material (rock type), porosity, and fluid content vary with depth, the bulk density will also vary with depth. The overburden pressure at any point is therefore the integral of the bulk density from surface down to the point of interest.

The specific gravity of the rock matrix may vary from 2.1 (sandstone) to 2.4 (limestone). Therefore, using an average of 2.3 and converting to units of psi/ft, it can be seen that the overburden pressure gradient exerted by a typical rock, with zero porosity would be:

$$2.3 \times 0.433 = 0.9959 (\text{psi/ft}) \tag{2.34}$$

This figure is normally rounded up to 1 psi/ft and is commonly quoted as the maximum possible overburden pressure gradient, from which the maximum overburden pressure, at any depth, can be calculated. It is unlikely that the pore pressure could exceed the overburden pressure. However, it should be remembered that the overburden pressure may vary with depth, due to compaction and changing lithology, and so the gradient cannot be assumed to be constant.

2.2.4　Abnormal Pressures

Pore pressures which are found to lie above or below the "normal" pore pressure gradient line are called abnormal pore pressures (Fig. 2.20 and Fig. 2.21). These formation pressures may be either Subnormal (i.e. less than 0.465 psi/ft) or Overpressured (i.e. greater than 0.465 psi/ft). The mechanisms which generate these abnormal pore pressures can be quite complex and vary from region to region. However, the most common mechanism for generating overpressures is called Undercompaction and can be best described by the undercompaction model.

The compaction process can be described by a simplified model (Fig. 2.22) consisting of a vessel containing a fluid (representing the pore fluid) and a spring (representing the rock matrix). The overburden stress can be simulated by a piston being forced down on the vessel. The overburden (S) is supported by the stress in the spring (σ) and the fluid pressure (p). Thus:

$$S = \sigma + p \tag{2.35}$$

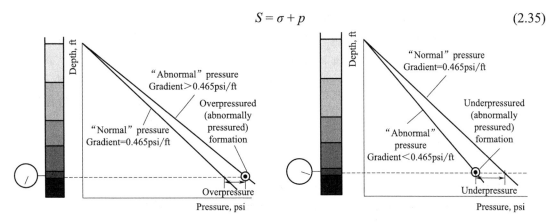

Fig. 2.20 Overpressured formation   Fig.2.21 Vnderpressured (subnormal pressured) formation

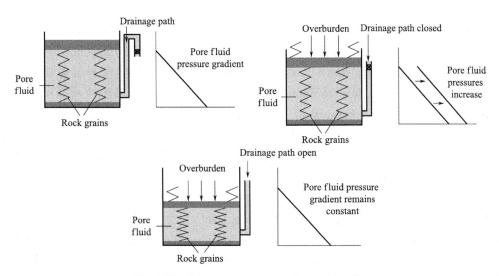

Fig. 2.22 Overpressure generation mechanism

If the overburden is increased (e.g. due to more sediments being laid down) the extra load must be borne by the matrix and the pore fluid. If the fluid is prevented from leaving the pore space (drainage path closed) the fluid pressure must increase above the hydrostatic value. Such a formation can be described as overpressured (i.e. part of the overburden stress is being supported by the fluid in the pore space and not the matrix). Since the water is effectively incompressible the overburden is almost totally supported by the pore fluid and the grain to grain contact stress is not increased. In a formation where the fluids are free to move (drainage path open),

Video 2.1

同承担。因此：

如果上覆岩层压力增加（比如沉积物增多），增加的载荷必须由岩石基质和孔隙流体共同承担。如果此时的地层是一个封闭体系（即排水通道关闭），流体无法从孔隙空间排出，那么流体压力一定会升高至静液压力以上。这种地层可描述为超压地层（即部分上覆岩层压力由孔隙空间中的流体而不是基质来承担）。由于水是不可压缩的，此时增加的上覆岩层压力几乎完全由孔隙流体支撑，而颗粒间的接触应力不会增

the increased load must be taken by the matrix, while the fluid pressure remains constant. Under such circumstances the pore pressure can be described as Normal, and is proportional to depth and fluid density.

In order for abnormal pressures to exist the pressure in the pores of a rock must be sealed in place i.e. the pores are not interconnecting. The seal prevents equalization of the pressures which occur within the geological sequence. The seal is formed by a permeability barrier resulting from physical or chemical action.

A physical seal may be formed by gravity faulting during deposition or the deposition of a fine grained material. The chemical seal may be due to calcium carbonate being deposited, thus restricting permeability. Another example might be chemical diagenesis during compaction of organic material. Both physical and chemical action may occur simultaneously to form a seal (e.g. gypsum-evaporite action).

2.2.4.1 Origin of Subnormal Formation

The major mechanisms by which subnormal (less than hydrostatic) pressures occur may be summarised as follows:

(1) Thermal Expansion. As sediments and pore fluids are buried the temperature rises. If the fluid is allowed to expand the density will decrease, and the pressure will reduce.

(2) Formation Foreshortening. During a compression process there is some bending of strata (Fig. 2.23). The upper beds can bend upwards, while the lower beds can bend downwards. The intermediate beds must expand to fill the void and so create a subnormally pressured zone. This is thought to apply to some subnormal zones in Indonesia and the US. Notice that this may also cause overpressures in the top and bottom beds.

加。在开放体系（排水通道打开）中，流体可以在地层中自由流动，增加的载荷由基质承担，而流体的压力保持不变。在这种情况下，孔隙压力为正常值，并与深度和流体密度成正比。

异常压力一定存在于孔隙之间相互不连通的岩石中。密封层阻止了地质序列中压力的均衡，这是由物理或化学作用产生的渗透性屏障所致。

物理密封层由沉积过程中重力作用下的断层或细颗粒沉积层构成。化学密封层可能是碳酸钙沉积引起渗透性降低所致，也有可能是有机物质压实过程中的化学成因所致。此外，物理和化学作用也可能同时发生形成密封层（例如石膏—蒸发岩作用）。

2.2.4.1 低压地层的成因

异常低压（低于流体静压力）的主要形成机理可以总结如下：

（1）热膨胀。温度随埋深增加而升高。通常情况下，如果液体膨胀，其密度就会降低，压力也会减小。

（2）地层收缩。在压实过程中，岩层会发生一定的弯曲（图2.23）。上岩层可以向上弯曲，下岩层可以向下弯曲，而中间层则必须膨胀以填满空隙，由此形成一个异常低压带。这种情况适用于印尼和美国的一些异常低压地区。注意，这也可能导致顶部和底部地层中的超压。

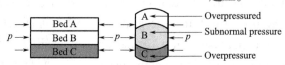

Fig. 2.23 Foreshortening of intermediate beds

(3) Depletion. When hydrocarbons or water are produced from a competent formation in which no subsidence occurs a subnormally pressured zone may result. This will be important when drilling development wells through a reservoir which has already been producing for some time. Some pressure gradients in Texas aquifers have been as low as 0.36 psi/ft.

(4) Precipitation. In arid areas (e.g. Middle East) the water table may be located hundreds of feet below surface, thereby reducing the hydrostatic pressures.

(5) Potentiometric Surface. This mechanism refers to the structural relief of a formation and can result in both subnormal and overpressured zones. The potentiometric surface is defined by the height to which confined water will rise in wells drilled into the same aquifer. The potentiometric surface can therefore be thousands of feet above or below ground level (Fig. 2.24).

(6) Epeirogenic Movements. A change in elevation can cause abnormal pressures in formations open to the surface laterally, but otherwise sealed. If the outcrop is raised this will cause overpressures, if lowered it will cause subnormal pressures (Fig. 2.25).

Pressure changes are seldom caused by changes in elevation alone since associated erosion and deposition are also significant. Loss or gain of water saturated sediments is also important.

（3）压力衰竭。当油气或水从某个地层中采出时，该地层没有发生沉降，这可能会形成异常低压。在已经生产一段时间的油藏中打开发井时，要注意这一点。得克萨斯州的一些含水层压力梯度已经低至 0.36psi/ft。

（4）降水量。在干旱地区（如中东），地下水位可能位于地表以下数百英尺，静水压力降低。

（5）测势面。这种机理针对由地层构造起伏导致的异常低压和超压地层。测势面是指在相同的含水层中，承压水在井中上升的高度。因此，测势面可以高于或低于地面数千英尺（图 2.24）。

（6）造山运动。海拔的变化导致与地面横向连通的地层中会引起压力异常，而对于封闭地层则不会。如果露头升高，则会造成高压；如果露头降低，则会造成低压（图 2.25）。

仅由海拔变化引起的压力变化非常少，因为海拔变化的沉积和侵蚀作用，以及饱和水沉积物的增减对压力的变化也很重要。

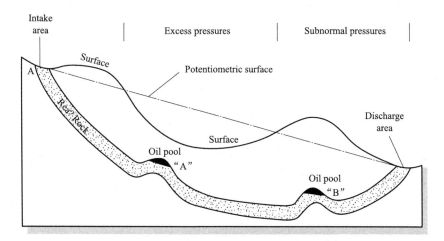

Fig. 2.24 The effect of the potentiometric surface in relationship to the ground surface causing overpressures and subnormal pressures

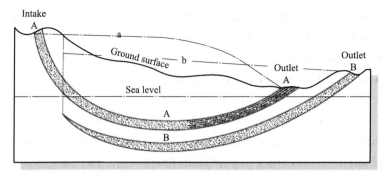

Fig. 2.25 Section through a sedimentary basin showing two potentiometric surfaces relating to the two reservoirs A and B

The level of underpressuring is usually so slight it is not of any practical concern. By far the largest number of abnormal pressures reported have been overpressures, and not subnormal pressures.

2.2.4.2 Origin of Overpressured Formations

These are formations whose pore pressure is greater than that corresponding to the normal gradient of 0.465 psi/ft. As shown in Fig. 2.26 these pressures can be plotted between the hydrostatic gradient and the overburden gradient (1psi/ft). The following examples of overpressures have been reported (Table 2.3).

低压的程度通常微不足道。到目前为止，报道最多的异常压力是超压，低压很少。

2.2.4.2 超压地层的成因

异常高压地层的孔隙压力大于正常压力梯度 (0.465psi/ft)。如图 2.26 所示，这些压力介于静压力梯度和上覆岩层压力梯度（1psi/ft）之间。有报道的异常高压实例见表 2.3。

Table 2.3 Examples of overpressure areas

Gulf Coast	Iran	North Sea	Carpathian Basin
0.8～0.9psi /ft	0.71～0.98psi /ft	0.5～0.9psi /ft	0.8～1.1psi /ft

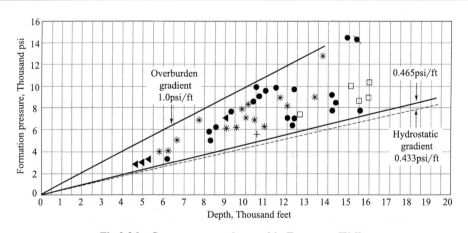

Fig.2.26 Overpressures observed in European Wells

From the above list it can be seen that overpressures occur worldwide. Some results from European fields are given in Fig. 2.26. There are numerous mechanisms which cause such pressures to develop. Some, such as potentiometric surface and formation foreshortening have already been mentioned under subnormal pressures since both effects can occur as a result of these mechanisms. The other major mechanisms are summarised below:

(1) Incomplete Sediment Compaction. Incomplete sediment compaction or undercompaction is the most common mechanism causing overpressures. In the rapid burial of low permeability clays or shales there is little time for fluids to escape. Under normal conditions the initial high porosity (about 50%) is decreased as the water is expelled through permeable sand structures or by slow percolation through the clay/shale itself. If, however the burial is rapid and the sand is enclosed by impermeable barriers (Fig. 2.27), there is no time for this process to take place, and the trapped fluid will help to support the overburden.

从表2.3中可以看出，异常高压发生在世界各地。图2.26给出了欧洲某地区的超压实例。导致异常高压的机理有很多。其中有一些，例如测势面和地层收缩，已经在介绍异常低压时提到了，因为这两种作用都可能导致低压和超压。异常高压的其他主要形成机理概述如下：

（1）沉积物欠压实。沉积物的不完全压实或欠压实是引起异常高压最常见的机理。在快速掩埋的低渗透黏土或页岩中，流体几乎没有时间逸出。在正常情况下，水通过可渗透的砂岩结构或黏土、页岩缓慢渗滤，并随着水的排出，初始的高孔隙度（50%左右）会逐渐降低。但是，如果掩埋的速度过快，砂粒被不渗透层所包围（图2.27），孔隙内的流体没有时间排出，则被圈闭的流体将与基岩一起共同支撑上覆岩层。

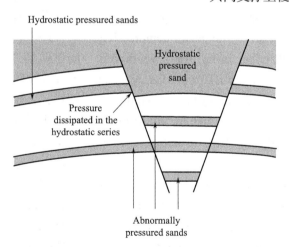

Fig. 2.27 Barriers to flow and generation of overpressured sand

(2) Faulting. Faults may redistribute sediments, and place permeable zones opposite impermeable zones, thus creating barriers to fluid movement. This may prevent water being expelled from a shale, which will cause high porosity and pressure within that shale

（2）断层作用。断层可使沉积物重新分布，使渗透带与不渗透带对接，从而形成流体流动的障碍。这可能会阻止水从页岩中排出，进而导致压实状态下的页

under compaction.

(3) Phase Changes during Compaction. Minerals may change phase under increasing pressure, e.g. gypsum converts to anhydrite plus free water. It has been estimated that a phase change in gypsum will result in the release of water. The volume of water released is approximately 40% of the volume of the gypsum. If the water cannot escape, then overpressures will be generated. Conversely, when anhydrite is hydrated at depth it will yield gypsum and result in a 40% increase in rock volume. The transformation of montmorillonite to illite also releases large amounts of water.

(4) Massive Rock Salt Deposition. Deposition of salt can occur over wide areas. Since salt is impermeable to fluids the underlying formations become overpressured. Abnormal pressures are frequently found in zones directly below a salt layer.

(5) Salt Diaperism. This is the upwards movement of a low density salt dome due to buoyancy which disturbs the normal layering of sediments and produces pressure anomalies. The salt may also act as an impermeable seal to lateral dewatering of clays.

(6) Tectonic Compression. The lateral compression of sediments may result either in uplifting weathered sediments or fracturing (faulting) of stronger sediments. Thus formations normally compacted at depth can be raised to a higher level. If the original pressure is maintained the uplifted formation is now overpressured.

(7) Repressuring from Deeper Levels. This is caused by the migration of fluid from a high to a low presssure zone at shallower depth. This may be due to faulting or from a poor casing、cement job. The unexpectedly high pressure could cause a kick, since no lithology change would be apparent. High pressures can occur in shallow sands if they are charged by gas from lower formations.

(8) Generation of Hydrocarbons. Shales which are deposited with a large content of organic material

岩中存在高孔隙度和高压。

（3）压实过程中的相变。矿物在压力增加的情况下可能会发生相变，例如，石膏会转化为硬石膏和游离水。据估计，相变释放出的游离水大约占石膏体积的40%。如果游离水不能排出，则会在地层中形成超压；当硬石膏在一定深度处水化形成石膏时，其岩石体积将增加40%。此外蒙脱石转化为伊利石时，也会释放出大量的水。

（4）块状岩盐沉积。盐的沉积可以发生在开阔地区。由于盐对流体来说是不渗透的，位于其下方的地层压力会变为超压。因此，在盐层正下方的地层中经常会发现异常压力。

（5）盐丘。低密度的盐丘受浮力作用向上运动，它扰乱了沉积物的正常分层，产生压力异常。盐也可以作为一种不渗透的密封层，阻止黏土的横向脱水。

（6）构造挤压作用。沉积物的侧向挤压可能导致风化沉积物上升或较硬沉积物的破裂（断裂）。这样，在地层深处被压实的地层可以上升至较浅深度。如果保持原来的压力，则隆起的地层就处于超压状态。

（7）来自高压层的补充加压。这是由较浅深度的流体从高压区运移到低压区造成的。它可能是断层作用或不合格的套管、注水泥作业导致的。由于没有明显的岩性变化，意外的高压可能引起井涌。如果有浅层气体的充注，浅层砂体也会产生高压。

（8）烃类物质的产生。在压实作用下，随着有机质的降解，

will produce gas as the organic material degrades under compaction. If it is not allowed to escape the gas will cause overpressures to develop. The organic by-products will also form salts which will be precipitated in the pore space, thus helping to reduce porosity and create a seal.

2.2.5 Drilling Problems Associated with Abnormal Formation Pressures

When drilling through a formation, sufficient hydrostatic mud pressure must be maintained to:

(1) Prevent the borehole collapsing and;
(2) Prevent the influx of formation fluids.

To meet these 2 requirements, the mud pressure is kept slightly higher than formation pressure. This is known as overbalance. If, however, the overbalance is too great this may lead to:

(1) Reduced penetration rates (due to chip hold down effect);
(2) Breakdown of formation (exceeding the fracture gradient) and subsequent lost circulation (flow of mud into formation);
(3) Excessive differential pressure causing stuck pipe.

The formation pressure will also influence the design of casing strings. If there is a zone of high pressure above a low pressure zone the same mud weight cannot be used to drill through both formations otherwise the lower zone may be fractured. The upper zone must be "cased off", allowing the mud weight to be reduced for drilling the lower zone. A common problem is where the surface casing is set too high, so that when an overpressured zone is encountered and an influx is experienced, the influx cannot be circulated out with heavier mud without breaking down the upper zone. Each casing string should be set to the maximum depth allowed by the fracture gradient of the exposed formations. If this is not done an extra string of protective casing may be required. This will not only prove expensive, but will also reduce the wellbore diameter. This may have implications when the well is to be completed since the production tubing size may have to be restricted.

Having considered some of these problems it should be clear that any abnormally pressured zone must be identified and the drilling programme designed to accommodate it.

2.2.6 Transition Zone

Audio 2.5

It is clear from the descriptions of the ways in which overpressures are generated above that the pore pressure profile in a region where overpressures exist will look something like the P-Z diagram shown in Fig. 2.28. It can be seen that the pore pressures in the shallower formations are "normal". That is that they correspond to a hydrostatic fluid gradient. There is then an increase in pressure with depth until the "overpressured" formation is entered. The zone between the normally pressured zone and the overpressured zone is known as the transition zone.

到限制,进而影响油井的完井。

考虑到这些问题,就需要弄清楚是否存在异常压力带,并在钻井设计时加以考虑。

2.2.6 压力过渡带

从以上对超压产生方式的描述可以清楚地看出,超压区域的孔隙压力剖面类似于图2.28。可以看出,较浅地层的孔隙压力是正常的,它们与流体静压力梯度相当。而后,地层压力随深度而增加,直到进入超压地层。存在于正常压力带和超压带之间的地层称为压力过渡带。

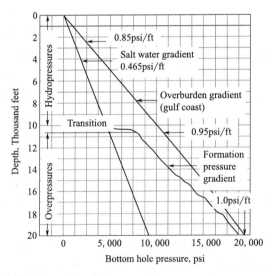

Fig. 2.28 Transition from normal pressures to overpressures

The pressures in both the transition and overpressured zone is quite clearly above the hydrostatic pressure gradient line. The transition zone is therefore the seal or caprock on the overpressured formation. It is important to note that the transition zone shown in Fig. 2.28 is representative of a thick shale sequence. This shale will

过渡带和超压带的压力都明显高于静压力。过渡带是超压地层上的密封层或盖层。值得注意的是,图2.28所示的过渡带代表的是一个厚页岩层序。这种页岩的孔隙度较低,因此孔隙中的流

have some low level of porosity and the fluids in the pore space can therefore be overpressured. However, the permeability of the shale is so low that the fluid in the shale and in the overpressured zone below the shale cannot flow through the shale and is therefore effectively trapped. Hence the caprock of a reservoir is not necessarily a totally impermeable formation but is generally simply a very low permeability formation.

If the seal is a thick shale, the increase in pressure will be gradual and there are techniques for detecting the increasing pore pressure. However, if the seal is a hard, crystalline rock (with no permeability at all) the transition will be abrupt and it will not be possible to detect the increase in pore pressure across the seal.

When drilling in a region which is known to have overpressured zones the drilling crew will therefore be monitoring various drilling parameters, the mud, and the drilled cuttings in an attempt to detect this increase in pressure in the transition zone. It is the transition zone which provides the opportunity for the drilling crew to realise that they are entering an overpressured zone. The key to understanding this operation is to understand that although the pressure in the transition zone may be quite high, the fluid in the pore space cannot flow into the wellbore. When however the drillbit enters the high permeability, overpressured zone below the transition zone the fluids will flow into the wellbore. In some areas operating companies have adopted the policy of deliberately reducing the overbalance so as to detect the transition zone more easily-even if this means taking a kick.It should be noted that the overpressures in a transition zone cannot result in an influx of fluid into the well since the seal has, by definition, an extremely low permeability. The overpressures must therefore be detected in some other way.

2.2.7 Prediction and Detection of Abnormal Pressures

The techniques which are used to predict (before drilling), detect (whilst drilling) and confirm (after

体会受到超压作用。页岩的渗透率非常低，导致页岩中和页岩下方超压层中的流体不能流出，从而被有效地圈闭在地层内。因此，盖层不一定是完全不渗透的地层，它通常只是渗透率非常低的地层。

如果密封层是一个厚页岩层，压力增加是渐进的，并且可通过孔隙压力检测技术进行识别。然而，如果密封层是坚硬的结晶岩石（完全没有渗透性），这种转变将是突然的，并且不可能检测到密封层中孔隙压力的增加。

在已知存在超压的区域进行钻井作业时，为了监测过渡带压力的增加，需要钻井作业者监测各种钻井参数、钻井液和钻屑。过渡带的存在，让钻井作业者有机会意识到他们正进入一个超压带。这一作业过程的关键是：尽管过渡带的压力可能很高，但其孔隙空间中的流体不能进入井筒；而当钻头进入过渡带以下的高渗透超压带时，地层流体则会侵入井筒中。在一些地区，为了更容易发现过渡带，作业公司通过降低过平衡（即使这可能引发溢流）钻井的静液柱压力来识别过渡带。需要注意的是，过渡带的超压不会导致流体侵入井筒，因为作为密封层的过渡带，渗透率都非常低。因此，还要用其他方法检测超压。

2.2.7 异常压力的预测和检测

表2.4总结了用于预测（钻前）、监测（随钻）和检测（钻后）

drilling) overpressures are summarised in Table 2.4.

超压的技术。

Table2.4 Methods for predicting and detecting abnormal pressures

Source of data	Parameters	Time of Recording
Geophysical methods	Formation velocity (seismic); Gravity; Magnetics; Electrical prospecting methods	Prior to spudding well
Drilling mud	Gas content; Flowline mudweight; "kicks"; Flowline temperature; Chlorine variation; Drillpipe pressure; Pit volume; Flowrate; Hole fillup	While drilling
Drilling parameters	Drilling rate; $d(dc)$ exponent; Torque; Drag	While drilling; Delayed by the time required for mud return
Drilling cuttings	Shale cuttings; Bulk density; Shale factor; Electrical resistivity; Volume; Shape and Size; Novel geochemical; Physical techniques	While drilling; Delayed by time required for sample return
Well logging	Electrical survey; Resistivity; Conductivity; Shale formation factor; Salinity variations; Interval transit time; Bulk density; Hydrogen index; Thermal neutron; Capture cross section; Nuclear magnetic; Resonance; Downhole gravity data	After drilling
Direct pressure measuring devices	Pressure bombs; Drill stem test; Wire line formation test	When well is tested or completed

2.2.7.1 Predictive Techniques

The predictive techniques are based on measurements that can be made at surface, such a geophysical measurements, or by analysing data from wells that have been drilled in nearby locations (offset wells). Geophysical measurements are generally used to identify geological conditions which might indicate the potential for overpressures such as salt domes which may have associated overpressured zones. Seismic data has been used successfully to identify transition zones and fluid content such as the presence of gas. Offset well histories may contain information on mud weights used, problems with stuck pipe, lost circulation or kicks. Any wireline logs or mudlogging information is also valuable when attempting to predict overpressures.

2.2.7.2 Detection Techniques

Detection techniques are used whilst drilling the well. They are basically used to detect an increase in pressure in the transition zone. They are based on three

2.2.7.1 预测技术

预测技术基于可以在地表上进行的测量（如地球物理测量），或者通过分析邻井（探边井）的数据来实现。地球物理测量通常用于确定可能存在超压的地质条件，如与超压带有关的盐丘。地震数据已成功地用于确定过渡带和流体成分，例如气体的存在。探边井的历史记录包括所使用的钻井液密度及卡钻、井漏或井涌等问题的信息。此外，电缆测井或录井信息对于预测超压也是有价值的。

2.2.7.2 监测技术

通常，监测技术用于钻井时监测超压，其基本原理是监测过渡带压力的增加，主要依据三种

forms of data:

(1) Drilling parameters: observing drilling parameters (e.g.ROP) and applying empirical equations to produce a term which is dependent on pore pressure.

(2) Drilling mud: monitoring the effect of an overpressured zone on the mud (e.g. in temperature, influx of oil or gas).

(3) Drilled cuttings: examining cuttings, trying to identify cuttings from the sealing zone.

1. Detection Based on Drilling Parameters

The theory behind using drilling parameters to detect overpressured zones is based on the fact that:

(1) Compaction of formations increases with depth. ROP will therefore, all other things being constant, decrease with depth

(2) In the transition zone the rock will be more porous (less compacted) than that in a normally compacted formation and this will result in an increase in ROP. Also, as drilling proceeds, the differential pressure between the mud hydrostatic and formation pore pressure in the transition zone will reduce, resulting in a much greater ROP.

The use of the ROP to detect transition and therefore overpressured zones is a simple concept, but difficult to apply in practice. This is due to the fact that many factors affect the ROP, apart from formation pressure (e.g. rotary speed and WOB). Since these other effects cannot be held constant, they must be considered so that a direct relationship between ROP and formation pressure can be established. This is achieved by applying empirical equations to produce a "normalised" ROP, which can then be used as a detection tool.

1) The d exponent

The d exponent technique for detection of overpressures is based on a normalised drilling rate equation developed by Bingham. Bingham proposed the following generalised drilling rate equation:

数据：

（1）钻井参数：观察钻井参数（如机械钻速），并利用经验公式得到一个与孔隙压力有关的项。

（2）钻井液：监测超压层对钻井液的影响（如温度、油气侵入）。

（3）钻屑：为了识别来自密封层的岩屑，要检查钻屑。

1. 基于钻井参数的监测

利用钻井参数监测超压层的理论基础是：

（1）地层的压实程度随深度而增加。因此，在其他条件不变的情况下，机械钻速会随着深度的增加而减小。

（2）过渡带岩石的孔隙度比正常压实地层的孔隙度更大（即欠压实），这将导致机械钻速增加。此外，随着钻井过程的进行，过渡带的静液压力与地层孔隙压力之间的差值将减小，从而使机械钻速大大提高。

利用机械钻速监测过渡带，并由此发现超压带，这仅是一个概念，很难在实践中应用。因为除了地层压力，其他许多因素都会影响机械钻速（如转速和钻压）。由于这些影响因素无法保持恒定，所以建立机械钻速和地层压力之间的直接关系时必须考虑以上因素。利用经验方程可以得到一个"正常的"机械钻速，它可以作为监测超压带的一种工具。

1） d 指数

用于监测超压的 d 指数技术是基于宾汉提出的标准化钻速方程。宾汉提出的标准化钻速方程如下：

$$R = aN^e \left(\frac{W}{B}\right)^d \tag{2.36}$$

where, R= penetration rate (ft/hr); N = rotary speed (r/min); W= WOB (lb); B= bit diameter (in.); a = matrix strength constant; d= formation drillability; e = rotary speed exponent.

式中，R 为钻速，ft/h；N 为转速，r/min；W 为钻压，lb；B 为钻头直径，in；a 为岩石可钻性系数；d 为地层可钻性；e 为转速指数。

Jordan and Shirley re-organised this equation to be explicit in "d". This equation was then simplified by assuming that the rock which was being drilled did not change (a = 1) and that the rotary speed exponent (e) was equal to one. The rotary speed exponent has been found experimentally to be very close to one. This removed the variables which were dependent on lithology and rotary speed. This means however that the resulting equation can only be applied to one type of lithology and theoretically at a single rotary speed. The latter is not too restrictive since the value of e is generally close to 1(one). On the basis of these assumptions and accepting these limitations the following equation was produced:

乔登和雪莉重新修正了这个方程，使其成为 d 的显式。然后通过假设正在钻进的岩性没有变化（a = 1）和转速指数（e）等于 1 来简化这个方程。实验发现，转速指数非常接近于 1。这消除了依赖于岩性和转速的变量。然而，这意味着所得到的方程理论上只适用于岩性和转速都不变的钻进过程。但由于 e 的值通常接近于 1，所以后者的限制并不算严苛。基于这些假设和限制条件，可以得到下列方程：

$$d = \frac{\lg\left(\dfrac{R}{60N}\right)}{\lg\left(\dfrac{12W}{10^6 B}\right)} \tag{2.37}$$

This equation is known as the d-exponent equation. Since the values of R, N, W and B are either known or can be measured at surface the value of the d-exponent can be determined and plotted against depth for the entire well. Values of "d" can be found by using the nomograph in Fig. 2.29. Notice that the value of the d-exponent varies inversely with the drilling rate. As the bit drills into an overpressured zone the compaction and differential pressure will decrease, the ROP will increase, and so the d-exponent should decrease. An overpressured zone will therefore be identified by plotting d-exponent against depth and seeing where the d-exponent reduces (Fig. 2.30)

这个方程被称为 d 指数方程。由于 R、N、W 和 B 的值已知，或者可以在地面测量得到，因此可以确定 d 指数的值，并按整口井的井深绘制出来。通过图 2.29 中的列线图可以查到 d 指数的值。注意，d 指数的值与钻速成反比。当钻头进入超压区时，压实程度和压差将减小，机械钻速将增大，所以 d 指数也将减小。因此，可以通过绘制 d 指数与深度的关系图，并查找图中 d 指数减小的具体位置来确定超压区（图 2.30）。

It should be realised that this equation takes into account variations in the major drilling parameters, but for accurate results the following conditions should be maintained:

该方程考虑了主要钻井参数的变化，但为了得到准确的结果，应保持以下条件：

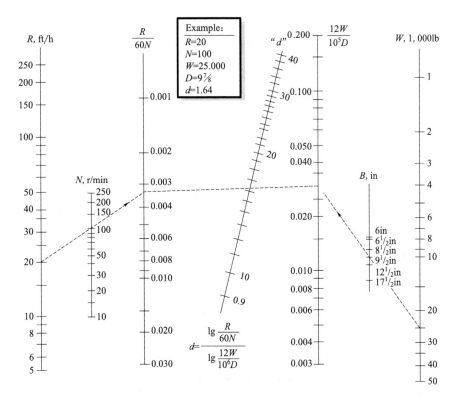

Fig. 2.29 Nomogram for calculating *d* exponent

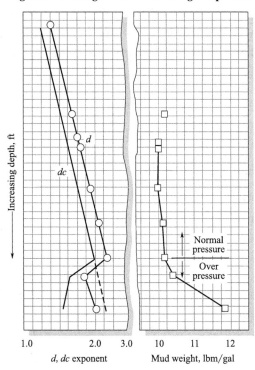

Fig. 2.30 Comparison of *d* and *dc* drilling exponents used in geopressure detection

(1) No abrupt changes in WOB or RPM should occur, i.e. keep WOB and RPM as constant as possible.

（1）钻压和转速不应发生突然变化，比如，尽量保持钻压和

(2) To reduce the dependence on lithology the equation should be applied over small depth increments only (plot every 10ft).

(3) A good thick shale is required to establish a reliable "trend" line.

It can be seen that the *d*-exponent equation takes no account of mudweight. Since mudweight determines the pressure on the bottom of the hole the greater the mudweight the greater the chip hold-down effect and therefore the lower the ROP. A modified *d*-exponent (*dc*) which accounts for variations in mudweight has therefore been derived:

$$dc = d\left(\frac{MW_n}{MW_a}\right) \quad (2.38)$$

where, MW_n = normal mud weight; MW_a = actual mud weight. The *dc* exponent trend gives a better definition of the transition (Fig. 2.30).

The *d* exponent is generally used to simply identify the top of the overpressured zone. The value of the formation pressure can however be derived from the modified *d*-exponent, using a method proposed by Eaton:

$$\frac{P}{D} = \frac{S}{D} - \left[\frac{S}{D} - \left(\frac{P}{D}\right)_n\right]\left[\frac{dc_o}{dc_n}\right]^{1.2} \quad (2.39)$$

where, $\frac{P}{D}$ = fluid pressure gradient (psi/ft); $\frac{S}{D}$ = overburden gradient (psi/ft); dc_o = observed *dc* at given depth; dc_n = *dc* from normal trend (i.e. extrapolated) at given depth.

Eaton claims the relationship is applicable worldwide and is accurate to 0.5lbm/gal.

2) Other Drilling Parameters

Torque can be useful for identifying overpressured zones. An increase in torque may occur of the decrease in overbalance results in the physical breakdown of the borehole wall and more material, than the drilled

转速不变。

（2）为了减少对岩性的依赖，该公式仅适用于较小的深度增量（每10ft绘图一次）。

（3）需要良好的厚页岩层建立可靠的"趋势"线。

可以看出，*d*指数方程没有考虑钻井液密度。由于钻井液密度决定了井底压力，钻井液密度越大，岩屑的压持效应就越大，机械钻速就越低。因此，后来又提出了考虑钻井液密度变化的修正*d*指数（即*dc*指数）：

(2.38)

式中，MW_n为正常钻井液密度；MW_a为实际钻井液密度。*dc*指数更好地定义了的过渡带（图2.30）。

通常只用*d*指数识别超压带的顶部。用伊顿提出的方法，可以通过修正的*d*指数得到地层压力的值：

(2.39)

式中，$\frac{P}{D}$为流体压力梯度，psi/ft；$\frac{S}{D}$为上覆岩层压力梯度，psi/ft；dc_o为给定深度处观测到的*dc*指数；dc_n为正常趋势（外推）线上给定深度处的*dc*指数。

伊顿声称，这一关系式适用于世界各地，且精度可达0.5 lbm/gal。

2) 其他钻井参数

扭矩可用来识别超压带。超压时过平衡程度降低，井壁坍塌，环空中积累的岩石碎屑比正常钻屑多，钻井过程中扭矩增大。也

cuttings is accumulating in the annulus. There is also the suggestion that the walls of the borehole may squeeze into the open hole as a result of the reduction in differential pressure. Drag may also increase as a result of these effects, although increases in drag are more difficult to identify.

2. Drilling Mud Parameters

There will be many changes in the drilling mud as an overpressured zone is entered. The main effects on the mud due to abnormal pressures will be:

(1) Increasing gas cutting of mud;

(2) Decrease in mud weight;

(3) Increase in flowline temperature.

Since these effects can only be measured when the mud is returned to surface they involve a time lag of several hours in the detection of the overpressured zone. During the time it takes to circulate bottoms up, the bit could have penetrated quite far into an overpressured zone.

1) Gas Cutting of Mud

Gas cutting of mud may happen in two ways:

(1) From shale cuttings - if gas is present in the shale being drilled the gas may be released into the annulus from the cuttings.

(2) Direct influx - this can happen if the overbalance is reduced too much, or due to Swabbing when pulling back the drillstring at connections.

Continuous gas monitoring of the mud is done by the mudlogger using gas chromatography. A degasser is usually installed as part of the mud processing equipment so that entrained gas is not re-cycled downhole or allowed to build up in the mud pits.

2) Mud Weight

The mud weight measured at the flowline will be influenced by an influx of formation fluids. The presence of gas is readily identified due to the large decrease in density, but a water influx is more difficult to identify.

有人认为，由于压差减小，裸眼处的井眼可能缩径，这会导致钻井过程中摩阻的增加。尽管摩阻增加识别起来较困难，但它依然可以作为超压的识别参数。

2. 钻井液参数

当进入超压带时，钻井液会发生许多变化。异常压力对钻井液的主要影响是：

（1）增加钻井液气侵；

（2）降低钻井液密度；

（3）回流管线温度升高。

由于这些结果只能在钻井液返回地面时测量，因此在探测超压带时会有几个小时的时间延迟。在钻井液由井底向上循环的过程中，钻头可能已经深入超压带了。

1）钻井液气侵

钻井液气侵有两种方式：

（1）来自页岩岩屑：如果所钻遇的页岩中存在天然气，岩屑中所含的天然气将释放到环空中。

（2）直接侵入：气体通过井壁直接进入井眼中，这种情况发生在过平衡被打破时，或者由起钻时抽汲作用引起。

钻井液记录仪利用气相色谱对钻井液进行连续的气体监测。可以用除气器来处理钻井液中的气体，这样钻井液夹带的气体就不会再次下井重复循环，也不会在钻井液池中聚集。

2）钻井液密度

地层流体的侵入将影响在回流管线测量的钻井液密度。由于密度大幅降低，天然气的存在很容易识别，但地层水的侵入则较

Continuous measurement of mud weight may be done by using a radioactive densometer.

3) Flowline Temperature

Under-compacted clays, with relatively high fluid content, have a higher temperature than other formations. By monitoring the flowline temperature therefore a decrease in temperature will be observed when drilling through normally pressured zones. This will be followed by an increase in temperature when the overpressured zones are encountered (Fig.2.31). The normal geothermal gradient is about 1 degree F/100 ft. It is reported that changes in flowline temperature up to 10 degree F/100 ft. have been detected when drilling overpressured zones.

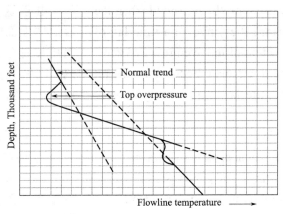

Fig. 2.31 Flowline temperature to detect overpressure

When using this technique it must be remembered that other effects such as circulation rate, mud mixing, etc. can influence the mud temperature.

3. Drilled Cuttings

Since overpressured zones are associated with under-compacted shales with high fluid content, the degree of overpressure can be inferred from the degree of compaction of the cuttings. The methods commonly used are:

(1) Density of shale cuttings;

(2) Shale factor;

(3) Shale slurry resistivity.

Even the shape and size of cuttings may give an indication of overpressures (large cuttings due to low pressure differential). As with the drilling mud parameters these tests can only be done after a lag time of some hours.

1) Density of Shale Cuttings

In normally pressured formations the compaction and therefore the bulk density of shales should increase uniformly with depth (given constant lithology). If the bulk density decreases, this may indicate an undercompacted zone which may be an overpressured zone. The bulk density of shale cuttings can be determined by using a mud balance. A sample of shale cuttings must first be washed and sieved (to remove cavings). These cuttings are then placed in the cup so that it balances at 8.3 lbm/gal (equivalent to a full cup of water). At this point therefore:

$$\rho_s V_s = \rho_w V_t$$

where, ρ_s = bulk density of shale; ρ_w = density of water; V_s = volume of shale cuttings; V_t = total volume of cup.

The cup is then filled up to the top with water, and the reading is taken at the balance point (ρ). At this point:

$$\rho V_t = \rho_s V_s + \rho_w (V_t - V_s)$$

substituting for V_s from the first equation gives:

$$\rho_s = \frac{\rho_w^2}{2\rho_w - \rho}$$

A number of such samples should be taken at each depth to check the density calculated as above and so improve the accuracy. The density at each depth can then be plotted (Fig. 2.32).

2) Shale Factor

This technique measures the reactive clay content in the cuttings. It uses the "methylene blue" dye test to determine the reactive montmorillonite clay present, and thus indicate the degree of compaction. The higher the montmorillonite, the lighter the density - indicating an undercompacted shale.

3) Shale Slurry Resistivity

As compaction increases with depth, water is expelled and so conductivity is reduced. A plot of resistivity against depth should show a uniform increase

进行。

1）页岩岩屑的密度

在正常压力地层中，页岩的压实程度和体积密度应随深度均匀地增加（假设岩性恒定）。如果体积密度减小，则表明可能存在超压的欠压实地层。页岩岩屑的体积密度可以用钻井液密度计来测量。先将页岩岩屑样品进行清洗和筛分（以去除坍塌块），然后放入钻井液密度计中，使游码指示读数为8.3lbm/gal（相当水的密度$1g/cm^3$）。此时：

(2.40)

式中，ρ_s 为 页岩的体积密度；ρ_w 为水的密度；V_s 为页岩岩屑的体积；V_t 为杯子的总体积。

然后将杯子装满水，并取平衡点处读数（ρ）。在这一点上：

(2.41)

代入第一个方程中的 V_s，得到：

(2.42)

应该在每个深度处都采集一些这样的样本，用上述方法计算岩石的体积密度，提高测量的准确性。然后绘制出每个深度对应的密度（图2.32）。

2）页岩因子

该技术测量的是岩屑中的活性黏土含量。它用亚甲蓝试剂来测定岩屑中是否含有活性蒙脱石，并由此说明压实程度。蒙脱石含量越高，密度越轻，表明页岩越欠压实。

3）页岩电阻率

压实程度随深度而增加，孔隙中的水被排出，电导率降低。因此，电阻率随深度的变化曲线应该表现

in resistivity, unless an undercompacted zone occurs where the resistivity will reduce. To measure the resistivity of shale cuttings a known quantity of dried shale is mixed with a known volume of distilled water. The resistivity can then be measured and plotted (Fig. 2.33).

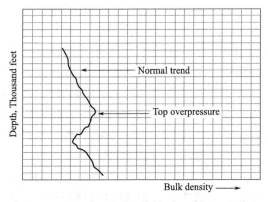

Fig. 2.32   Bulk density to detect overpressure

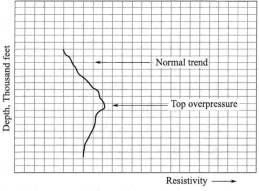

Fig. 2.33   Resistivity to detect overpressure

2.2.7.3   Confirmation Techniques

After the hole has been successfully drilled certain electric wireline logs and pressure surveys may be run to confirm the presence of overpressures. The logs which are particularly sensitive to undercompaction are : the sonic, density and neutron logs. If an overpressured sand interval has been penetrated then the pressure in the sand can be measured directly with a repeat formation tester or by conducting a well test.

2.2.8   Formation Fracture gradient

Audio 2.6

When planning the well, both the formation pore pressure and the formation fracture pressure for all of the formations to be penetrated must be estimated (Fig. 2.34). The well operations can then be designed such that the pressures in the borehole will always lie between the formation pore pressure and the fracture pressure. If the pressure in the borehole falls below the pore pressure, then an influx of formation fluids into the wellbore may occur. If the pressure in the borehole exceeds the fracture pressure, then the formations will fracture and losses of drilling fluid will occur.

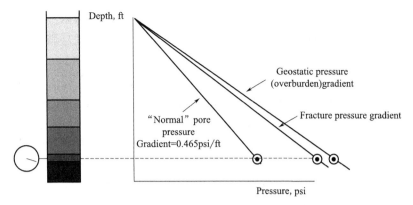

Fig. 2.34 Pore pressure, fracture pressure and overburden pressures for a particular formation

2.2.8.1 Mechanism of Formation Breakdown

The stress within a rock can be resolved into three principal stresses (Fig. 2.35). A formation will fracture when the pressure in the borehole exceeds the least of the stresses within the rock structure. Normally, these fractures will propagate in a direction perpendicular to the least principal stress. The direction of the least principal stress in any particular region can be predicted by investigating the fault activity in the area.

2.2.8.1 地层破裂机理

岩石内部的应力可以分解为三个主应力（图2.35）。当井眼压力超过岩石结构的最小主应力时，地层就会破裂。通常，这些裂缝会沿着垂直于最小主应力的方向传播。通过调查区域内的断层活动，可以预测任何特定区域内的最小主应力方向。

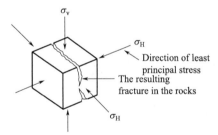

Fig. 2.35 Idealised view of the stresses acting on the block

To initiate a fracture in the wall of the borehole, the pressure in the borehole must be greater than the least principal stress in the formation. To propagate the fracture the pressure must be maintained at a level greater than the least principal stress.

2.2.8.2 The Leak-Off Test, Limit Test and Formation Breakdown Test

The pressure at which formations will fracture when exposed to borehole pressure is determined by conducting one of the following tests:

(1) Leak-off test;

要在井壁上形成裂缝，井眼压力必须大于地层的最小主应力。要使裂缝扩展，该压力必须保持在大于最小主应力的水平上。

2.2.8.2 漏失实验、极限测试和地层破裂实验

当地层暴露于井筒压力下，地层破裂压力可以通过以下测试来确定：

（1）漏失实验；

(2) Limit Test;

(3) Formation Breakdown Test.

The basic principle of these tests is to conduct a pressure test of the entire system in the wellbore (See Fig. 2.36) and to determine the strength of the weakest part of this system on the assumption that this formation will be the weakest formation in the subsequent open hole. The wellbore is comprised of (from bottom to top): the exposed formations in the open hole section of the well (generally only 5-10ft of formation is exposed when these tests are conducted); the casing; the wellhead; and the BOP stack. The procedure used to conduct these tests is basically the same in all cases. The test is conducted immediately after a casing has been set and cemented. The only difference between the tests is the point at which the test is stopped. The procedure is as follows:

（2）极限测试；

（3）地层破裂实验。

这些测试的基本原理是：在假设该地层是后续裸眼井段最薄弱部分的前提下，在井眼中对整个系统进行压力测试（图2.36），并确定该系统最薄弱部分的强度。井筒由以下部分组成（从下到上）：裸眼井段的裸露地层（在进行测试时，通常只有5～10ft的裸露地层）、套管、井口和防喷器组合。这些测试的程序基本上是相同的，即在下入套管并固井后立即进行测试。唯一的区别是测试停止的时间点。其步骤如下：

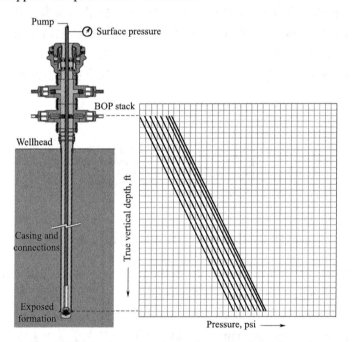

Fig. 2.36 Configuration during formation integrity tests

(1) Run and cement the casing string.

(2) Run in the drillstring and drillbit for the next hole section and drill out of the casing shoe.

(3) Drill 5 ～ 10 ft of new formation below the casing shoe.

(4) Pull the drillbit back into the casing shoe (to

（1）下套管并固井。

（2）为钻下一段井眼下入钻柱和钻头，并钻掉套管鞋。

（3）钻穿套管鞋下方5～10ft的地层。

（4）将钻头上提至套管鞋内

avoid the possibility of becoming stuck in the openhole).

(5) Close the BOPs (generally the pipe ram) at surface.

(6) Apply pressure to the well by pumping a small amount of mud (generally 1/2bbl) into the well at surface. Stop pumping and record the pressure in the well. Pump a second, equal amount of mud into the well and record the pressure at surface. Continue this operation, stopping after each increment in volume and recording the corresponding pressure at surface. Plot the volume of mud pumped and the corresponding pressure at each increment in volume (Fig. 2.37).

Note: the graph shown in Fig. 2.37 represents the pressure all along the wellbore at each increment. This shows that the pressure at the formation at leak off is the sum of the pressure at surface plus the hydrostatic pressure of the mud.

(7) When the test is complete, bleed off the pressure at surface, open the BOP rams and drill ahead.

It is assumed in these tests that the weakest part of the wellbore is the formations which are exposed just below the casing shoe. It can be seen in Fig. 2.36, that when these tests are conducted, the pressure at surface, and throughout the wellbore, initially increases linearly with respect to pressure. At some pressure the exposed formations start to fracture and the pressure no longer increases linearly for each increment in the volume of mud pumped into the well (see point A in Fig.2.37). If the test is conducted until the formations fracture completely (see point B in Fig. 2.37) the pressure at surface will often dop dramatically, in a similar manner to that shown in Fig.2.37.

The precise relationship between pressure and volume in these tests will depend on the type of rock that is exposed below the shoe. If the rock is ductile the behaviour will be as shown in Fig.2.37 and if it is brittle it will behave as shown in Fig. 2.38.

（以免卡在裸眼井段中）。

（5）关闭防喷器组合（通常是半封闸板防喷器）。

（6）通过向井中泵入少量钻井液（一般为1/2bbl）来施加压力。停止泵入并记录井内压力。再泵入等量钻井液，并记录地面压力。连续进行上述操作，并在每次泵入钻井液停止后记录相应的地面压力。作出每次泵送钻井液的体积与相对应的压力曲线图（图2.37）。

注意：图2.37中的曲线代表每次泵送钻井液时沿井筒的压力。这表明，漏失点的地层压力等于地面压力加上静液压力。

（7）测试完成后，在地面释放压力，打开防喷器组合的闸板，并继续钻进。

在这些测试中均假设井筒最薄弱的部分是套管鞋下方的裸露地层。从图2.36可以看出，在进行测试时，地面和整个井筒的压力最初呈线性增长。在一定压力下，裸露地层开始破裂，压力不再随泵入钻井液量的增加而线性增加（图2.37中的A点）。如果试验进行到地层完全破裂（如图2.37中的B点），地面压力通常会急剧下降，就类似于图2.37所示。

在这些测试中，压力和钻井液泵入体积之间的关系取决于套管鞋下方裸露岩石的类型。如果是韧性岩石，如图2.37所示；如果是脆性岩石，如图2.38所示。

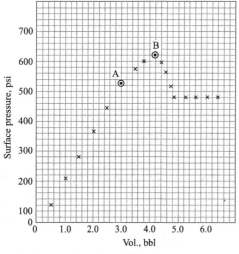

Fig. 2.37　Behavior of a ductile rock　　　Fig.2.38　Behavior of a brittle rock

The Leak-off test is used to determine the pressure at which the rock in the open hole section of the well just starts to break down (or leak off). In this type of test the operation is terminated when the pressure no longer continues to increase linearly as the mud is pumped into the well (Fig. 2.39). In practice the pressure and volume pumped is plotted in real time, as the fluid is pumped into the well. When it is seen that the pressure no longer increases linearly with an increase in volume pumped (Point C) it is assumed that the formation is starting to breakdown. When this happens a second, smaller amount of mud (generally 1/4 bbl) is pumped into the well just to check that the deviation from the line is not simply an error (Point D). If it is confirmed that the formation has started to leak off then the test is stopped and the calculations below are carried out.

The Limit Test is used to determine whether the rock in the open hole section of the well will withstand a specific, predetermined pressure. This pressure represents the maximum pressure that the formation will be exposed to whilst drilling the next wellbore section. The pressure to volume relationship during this test is shown in Fig.2.40. This test is effectively a limited version of the leak-off test.

The Formation Breakdown Test is used to determine the pressure at which the rock in the open hole section

漏失实验用于确定裸眼井段岩石刚开始压裂（或漏失）时的压力。测试时，将钻井液泵入井中直至地面压力不再呈线性增长时终止作业（图2.39）。在实际应用中，将钻井液泵入井中的同时，会实时绘制压力与泵入体积的关系图。当发现压力不再随泵送体积的增加而线性增加时（C点），则假定地层开始破裂。这时，再注入一小部分钻井液（通常为1/4bbl），目的是检查偏离直线是不是由误差（D点）造成的。如果证实地层已经开始漏失，则停止测试并执行后续的计算。

极限测试用来确定裸眼井段的岩石是否能承受一个具体的预定压力。该压力代表在钻下一个井段时，地层将要面临的最大压力。测试期间压力与泵入体积的关系如图2.40所示。这种测试实际上是一种限制压力的漏失实验。

地层破裂实验用于确定裸眼井段岩石完全破裂时的压力。如

of the well completely breaks down. If fluid is continued to be pumped into the well after leak off and breakdown occurs the pressure in the wellbore will behave as shown in Fig. 2.41.

果在漏失和破裂发生后继续向井中泵入流体，井筒内的压力变化将如图2.41所示。

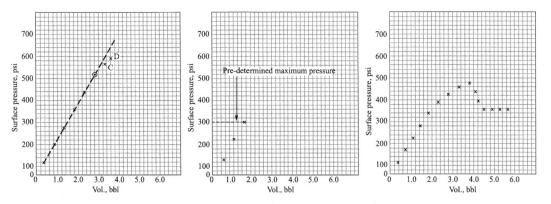

Fig.2.39　P-V behaviour during a leak off test

Fig.2.40　P-V behaviour in a limit test

Fig.2.41　Behaviour in a FBT test

1. Leak Off Test Calculations

In a Leak-Off test the formation below the casing shoe is considered to have started to fracture at point A on Fig. 2.37. The surface pressure at ponit A is known as the leak off pressure and can be used to determine the maximum allowable pressure on the formation below the shoe. The maximum allowable pressure at the shoe can subsequently be used to calculate:

(1) The maximum mudweight which can be used in the subsequent openhole section.

(2) The Maximum Allowable Annular Surface Pressure (MAASP).

The maximum allowable pressure on the formation just below the casing shoe is generally expressed as an equivalent mud gradient (EMG) so that it can be compared with the mud weight to be used in the subsequent hole section.

Given the pressure at surface when leak off occurs (point A in Fig. 2.37) just below the casing shoe, the maximum mudweight that can be used at that depth, and below, can be calculated from:

ρ_{max} = Pressure at the shoe when Leak-off occurs / True Vertical Depth of the shoe;

ρ_{max} = Pressure at surface and hydrostatic pressure of

1. 漏失实验计算

在漏失实验中，人们认为套管鞋下方的地层在图2.37中的点A开始破裂。因此，点A的地面压力称为漏失压力，可用于确定套管鞋以下地层的最大允许压力。最大允许压力后续可用来计算以下数据：

（1）裸眼井段的最大钻井液密度；

（2）最大允许环空地面压力（MAASP）。

套管鞋下方地层的最大允许压力通常用等效钻井液梯度（EMG）表示，以便与后续井段使用的钻井液密度进行比较。

考虑到套管鞋下方发生漏失时的地面压力（图2.37中的A点），在该点及以下深度可使用的最大钻井液密度ρ_{max}可由以下公式计算：

ρ_{max} = 漏失发生时套管鞋处的压力/套管鞋的实际垂直深度；

ρ_{max} =（地面压力+井内液柱

mud in well / True Vertical Depth of the shoe.

Usually a safety factor of 0.5lbm/gal (0.026 psi/ft) is subtracted from the allowable mudweight.

It should be noted that the leak-off test is usually done just after drilling out of the casing shoe, but when drilling the next hole section other, weaker formations may be encountered.

2. The Equivalent Circulating Density

It is clear from all of the preceding discussion that the pressure at the bottom of the borehole must be accurately determined if the leak off or fracture pressure of the formation is not to be exceeded. When the drilling fluid is circulating through the drillstring, the borehole pressure at the bottom of the annulus will be greater than the hydrostatic pressure of the mud. The extra pressure is due to the frictional pressure required to pump the fluid up the annulus. This frictional pressure must be added to the pressure due to the hydrostatic pressure from the colom of mud to get a true representation of the pressure acting against the formation at the bottom of the well. An equivalent circulating density (*ECD*) can then be calculated from the sum of the hydrostatic and frictional pressure divided by the true vertical depth of the well. The *ECD* for a system can be calculated from:

$$ECD = \frac{MW + p_d}{0.052D} \tag{2.43}$$

where, *ECD* = effective circulating density (lbm/gal); *MW*= mud weight (lbm/gal); p_d = annulus frictional pressure drop at a given circulation rate (psi); *D* = depth (ft).

The *ECD* of the fluid should be continuously monitored to ensure that the pressure at the formation below the shoe, due to the *ECD* of the fluid and system, does not exceed the leak off test pressure.

3. MAASP

The Maximum Allowable Annular Surface Pressure-MAASP- when drilling ahead is the maximum closed in (not circulating) pressure that can be applied to the annulus at surface before the formation just below the casing shoe will start to fracture (leak off). The MAASP can be determined from the following equation:

MAASP = Maximum Allowable pressure at the formation just below the shoe minus the Hydrostatic Pressure of mud at the formation just below the shoe.

2.2.8.3 Calculating the Fracture Pressure

The leak-off test pressure described above can only be determined after the formations to be considered have been penetrated. It is however necessary, in order to ensure a safe operation and to optimize the design of the well, to have an estimate of the fracture pressure of the formations to be drilled before the drilling operation has been commenced. In practice the fracture pressure of the formations is estimated from leakoff tests on nearby (offset) wells.

Many attempts have been made to predict fracture pressures. The fracture pressure of a formation drilled through a normally pressured formation can be determined from the following equations.

Vertical well and $\sigma_2 = \sigma_3$:
$$FBP = 2\sigma_3 - p_o \qquad (2.44)$$

Vertical well and $\sigma_2 > \sigma_3$:
$$FBP = 3\sigma_3 - \sigma_2 - p_o \qquad (2.45)$$

Deviated well and $\sigma_2 = \sigma_3$:
$$FBP = 2\sigma_3 - (\sigma_1 - \sigma_3)\sin^2\theta_z - p_o \qquad (2.46)$$

Deviated well in the direction of σ_2 and $\sigma_2 > \sigma_3$:
$$FBP = 3\sigma_3 - \sigma_2 - (\sigma_1 - \sigma_3)\sin^2\theta_z - p_o \qquad (2.47)$$

where, FBP = formation breakdown pressure; σ_1 = overburden stress; σ_2 = horizontal stress; σ_3 = horizontal stress; p_o = pore pressure; θ_z = hole deviation.

If the conservative assumption that the formation is already fractured is made, then the equations used to calculate the fracture pressure of the formations are simplified significantly.

Eaton proposed the following equation for fracture gradients:

$$G_f = (G_o - G_p)\left[\frac{\upsilon}{1-\upsilon}\right] + G_p \qquad (2.48)$$

where, G_f = fracture gradient (psi/ft); G_o = overburden gradient (psi/ft); G_p = pore pressure gradient (psi/ft); v = Poisson's ratio.

Poisson'sratio is a rock property that describes the behaviour of rock stresses (σ_1) in one direction (least principal stress) when pressure (σ_p) is applied in another direction (principal stress).

$$\frac{\sigma_1}{\sigma_p} = \frac{v}{1-v} \tag{2.49}$$

Laboratory tests on unconsolidated rock have shown that generally:

$$\frac{\sigma_1}{\sigma_p} = \frac{1}{3} \tag{2.50}$$

Field tests however show that v may range from 0.25 to 0.5 at which point the rock becomes plastic (stresses equal in all directions). Poisson'sratio varies with depth and degree of compaction (Fig. 2.42).

式中，G_f 为破裂压力梯度，psi/ft；G_o 为上覆岩层压力梯度，psi/ft；G_p 为孔隙压力梯度，psi/ft；v 为泊松比。

泊松比是一个岩石属性，它描述了当压力（σ_p）作用于另一个主应力方向时，岩石最小主应力方向上的应力（σ_1）反应。

通常，松散岩石的室内实验表明：

然而，现场试验表明，v 的取值范围为 0.25～0.5。当 v=0.5 时，岩石变为塑性（各方向应力相等）。此外，泊松比会随着深度和压实程度而变化（图 2.42）。

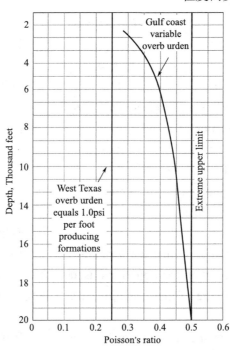

Fig. 2.42 Variation of Poisson'sratio with depth

Matthews and Kelly proposed the following method for determination of fracture pressures in sedimentary

马休斯和凯利提出了以下确定沉积岩破裂压力的方法：

rocks:

$$G_f = G_p + \frac{\sigma K_i}{D} \tag{2.51}$$

where, G_f = fracture gradient (psi/ft); G_p = pore pressure gradient (psi/ft); K_i = matrix stress coefficient; σ = matrix stress (psi); D = depth of interest (ft).

The matrix stress (σ) can be calculated as the difference between overburden pressure, S and pore pressure, p. i.e.

$$\sigma = S - p \tag{2.52}$$

2.2.8.4 Summary of Procedures

When planning a well the formation pore pressures and fracture pressures can be predicted from the following procedure:

(1) Analyse and plot log data or d-exponent data from an offset (nearby) well.

(2) Draw in the normal trend line, and extrapolate below the transition zone.

(3) Calculate a typical overburden gradient using density logs from offset wells.

(4) Calculate formation pore pressure gradients from equations (e.g. Eaton).

(5) Use known formation and fracture gradients and overburden data to calculate a typical Poisson's ratio plot.

(6) Calculate the fracture gradient at any depth.

Basically the three gradients must be estimated to assist in the selection of mud weights and in the casing design. One example is shown in Fig. 2.43. Starting at line A representing 16.5 lbm/gal mud it can be seen that any open hole shallower than 10,200ft will be fractured. Therefore a protective casing or liner must be run to seal off that shallower section before 16.5 lbm/gal mud is used to drill below 10200ft.

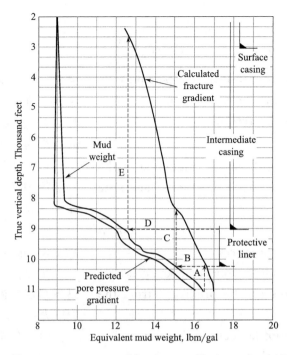

Fig. 2.43 Example of how pore pressure and fracture gradients can be used to select casing seats

To drill to 10,200ft a 15 lbm/gal mud (line B) must be used. This mud will breakdown any open hole above about 8,300ft(line C). This defines the setting depth of the protective casing (and the height of the liner).

To drill to 8,300 ft a 13 lbm/gal mud is required (line E). This mud will breakdown any open hole above 2,500 ft, so this defines the surface casing shoe. Note that casing shoes are usually set below indicated breakdown points as an added safety factor.

2.3 Exercise

(1) ① At a given depth in a sedimentary basin, the overburden stress equivalent density is equal to 1.95 g/cm³ from density logs. From fracturing data, the horizontal stresses equivalent density are defined as 1.75 and 1.77 g/cm³. What type of stress state exists in this

要钻达10200ft，必须使用15lbm/gal的钻井液（线B）。这种钻井液会将8300ft（线C）以上的裸眼井段压裂。这就确定了中间套管的下入深度（和尾管的高度）。

为了钻到8300ft，需要使用13lbm/gal的钻井液（线E）。这种钻井液会将2,500ft以上的裸眼井段压裂，这就决定了表层套管鞋的深度。需要注意的是，考虑到附加的安全因素，套管鞋通常安装在指定的破裂点以下。

2.3 习题

（1）① 在沉积盆地的某一深度处，根据密度测井可以确定上覆岩层压力当量密度为1.95g/cm³。根据压裂数据可以确定水平地应力当量密度分别为1.75g/cm³ 和

field? Is this what you would expect in a sedimentary basin? ② The overburden stress equivalent density is given as 1.81 g/cm³, whereas the two horizontal stresses equivalent density are estimated to be 1.92 and 1.64 g/cm³. What stress state is this? ③ If there is normal pore pressure in question ① and question ②, compute the effective stresses, also called the rock matrix stresses (A normal pore pressure equivalent density is often defined as the density of seawater, 1.03 g/cm³).

(2) The gradient plot in Fig. 2.44 is from a well in the North Sea. Assume that the caprock is located at the 9⅝ in. casing point at 2,350 m. Call this location point B. Furthermore, assume that there is vertical communication down to 2,600 m. Call this location point A.

① Using the pore pressure data, calculate the density of the oil in the reservoir in this interval.

② Assume that the fluid in the depth interval A–B is not oil, but condensate of density 0.5 g/cm³. Compute the pore pressure in this interval for the new values of fluid density.

(3) ① Assume that we have a 1,100-m-deep vertical well. At this depth, the overburden stress equivalent density is 1.9g/cm³, while the two horizontal stresses equivalent density are 1.51g/cm³. There exists a normal pore pressure equivalent density of 1.03g/cm³ in the formation. Determine the fracturing pressure for the borehole. ② Further interpretation of the well data reveals that the two horizontal stresses are actually different. Using the same overburden stress as above, but assuming that the horizontal stresses equivalent density are equal to 1.61 and 1.45 g/cm³, respectively, compute the fracture pressure now. ③ Discuss the effect of anisotropic stresses–that is, equal horizontal stresses vs. different horizontal stresses. Which gives highest fracture pressure?

1.77g/cm³。请问该地区应力状态是何种类型？是否为沉积盆地中的预期应力状态？②上覆岩层压力当量密度为1.81g/cm³，而两个水平地应力当量密度分别为1.92和1.64 g/cm³。这是什么样的压力状态？③如果问题①和问题②中的孔隙压力正常，请计算有效应力，也就是岩石基质应力（正常孔隙压力定义当量密度为海水密度1.03g/cm³）。

（2）图2.44是北海一口井的压力梯度图。假设盖层位于9⅝in套管段，套管下入深度为2350m，称为点B。地层垂向连通至2600 m。2600m处称为点A。

① 利用孔隙压力数据，计算储层中原油的密度。

② 假设层段A–B中的液体不是油，而是密度为0.5g/cm³的凝析油。根据这一密度计算该区间的孔隙压力。

（3）①假设有一口1100m深的直井。在1100m深度处，上覆岩层压力当量密度为1.9g/cm³，而两个水平地应力当量密度均为1.51g/cm³。地层中存在1.03g/cm³的正常孔隙压力当量密度。请确定井筒的破裂压力。②对油井数据的进一步解释表明，两个水平地应力实际上不同，假设水平地应力当量密度分别等于1.61g/cm³和1.45g/cm³，上覆岩层压力当量密度依然为1.9g/cm³，请重新计算破裂压力。③讨论地应力各向异性的影响，即相同水平地应力和不同水平地应力两种情况下，哪个破裂压力更高？

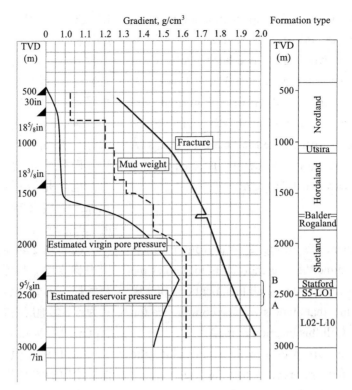

Fig 2.44 Gradient plot from a well in the North Sea

(4) A vertical well is subjected to equal horizontal stresses. The following data apply: Horizontal strsess equivalent density: 1.5 g/cm³, Pore prsesure equivalent density: 1.03 g/cm³, Angle of internal friction: 30°, rock cohesive strength equivalent density: 0.4 g/cm³.

① Compute the critical collapse pressure for the wellbore.

② During drilling, one observes that the rock is not consolidated, because sand particles drop out of the wellbore wall. Compute the critical collapse pressure now.

(5) The following pore pressure information has been supplied for the well you are about to drill.

（4）直井的水平地应力相同，参数如下：水平地应力当量密度：1.5g/cm³，孔隙压力当量密度：1.03g/cm³，内摩擦角：30°，岩石黏结强度当量密度：0.4g/cm³。

① 计算井筒的临界坍塌压力。

② 在钻井过程中，人们观察到砂粒从井壁上落下，岩石没有固结，现在计算临界坍塌压力。

（5）已知一口井的孔隙压力信息如下表所示。

Depth Below(ft)	0	1000	5000	8000	8500	9000	9500
Drill Floor Pressure(psi)	0	465	2325	3720	6800	6850	6900

① Plot the following pore pressure/depth information on a P-Z diagram.

② Calculate the pore pressure gradients in the

① 请绘制孔隙压力—深度曲线。

② 计算从地表到8000ft、

formations from surface; to 8,000ft; to 8,500ft; and to 9,500ft. Plot the overburden gradient (1 psi/ft) on the above plot.Determine the mudweight required to drill the hole section: down to 8,000ft; down to 8,500ft; and down to 9,500ft. Assume that 200 psi overbalance on the formation pore pressure is required.

③ If the mudweight used to drill down to 8000ft were used to drill into the formation pressures at 8500ft what would be the over/underbalance on the formation pore pressure at this depth?

④ Assuming that the correct mudweight is used for drilling at 8,500ft but that the fluid level in the annulus dropped to 500 ft below drillfloor, due to inadequate hole fill up during tripping. What would be the effect on bottom hole pressure at 8,500ft?

(6)　① Whilst drilling the 12¼in hole section of a well the mudloggers were recording the data as shown in the table below. Plot the d and dc exponent and determine whether there are any indications of an overpressured zone.

8500ft、9500ft 处的地层孔隙压力梯度；在图上绘制上覆岩层压力梯度。假设需要大于地层孔隙压力 200psi 的安全余量。确定钻至 8000ft、8500ft 和 9500ft 所需的钻井液密度。

③ 如果用钻 8000ft 的钻井液密度钻至 8500ft，那么该深度处钻井液密度与地层孔隙压力的超/欠压值是多少？

④ 假设在 8500ft 处钻井时使用了正确的钻井液密度，但由于起下钻期间井眼灌浆不足，环空中的液位下降 500ft。请问 8500ft 处的井底压力会受到什么影响？

（6）① 在钻井 12¼in 井段时，录井技术员记录的数据如下表所示。请绘制 d 指数和 dc 指数曲线，并确定是否存在超压带。

Depth (ft)	ROP(ft/h)	RPM (r/min)	WOB(lb)	Mud Weight(lbm/gal)
7500	125	120	38	9.5
7600	103	120	38	9.5
7700	77	110	38	9.5
7800	66	110	38	9.6
7900	45	110	35	9.6
8000	37	110	37	9.8
8100	40	110	35	9.8
8200	42	110	33	9.9
8300	41	100	33	10.0
8400	44	100	38	10.25
8500	34	100	38	10.25
8600	33	100	40	11
8700	32	110	42	11

② If an overpressured zone exists, what is the depth of the top of the transition zone.

③ Use the Eaton equation to estimate the formation pressure at 8,600 ft.

Assume a normal formation pressure of 0.465 psi/ft. an overburden gradient of 1.0 psi/ft and a normal mud weight for this area of 9.5 lbm/gal.

(7) While performing a leak off test the surface pressure at leak off was 940 psi. The casing shoe was at a true vertical depth of 5,010 ft and a mud weight of 10.2 lbm/gal was used to conduct the test. Calculate the Maximum bottom hole pressure during the leakoff test and the maximum allowable mud weight at 5010 ft. Allowing a safety factor of 0.5 lbm/gal, Calculate the maximum allowable mud weight at 5,010 ft.

(8) A leakoff test was carried out just below a 13⅜in casing shoe at 7,000 ft. TVD using 9.0 lbm/gal mud. The results of the tests are shown below. What is the maximum allowable mudweight for the 12¼in hole section？

BBLS Pumped	1.0	1.5	2.0	2.5	3.0	3.5	4.0	4.5	5.0
Surface Pressure (psi)	400	670	880	1100	1350	1600	1800	1900	1920

(9) If the circulating pressure losses in the annulus of the above well is 300 psi when drilling at 7,500ft with 9.5 lbm/gal mud, what would be the *ECD* of the mud at 7,500ft.

(10) The following information has been gathered together in an attempt to predict the fracture pressure of a formation to be drilled at 8,500ft: Vertical Depth of formation = 8,500 ft, Pore Pressure = 5,300 psi, Overburden pressure = 7,800 psi, Poissons Ratio = 0.28. Calculate the fracture pressure of the formation at 8,500 ft.

② 如果存在超压带，过渡区顶部的深度是多少？

③ 使用伊顿方程估算 8600ft 处的地层压力。

假设正常地层压力为 0.465 psi/ft。上覆岩层压力梯度为 1.0 psi/ft，该区域的正常钻井液密度为 9.5lbm/gal。

（7）进行漏失实验时，漏失时的地面压力为940psi。套管鞋位于垂深5010ft处，使用10.2lbm/gal 的钻井液密度进行试验。计算漏失实验期间的最大井底压力和5010ft处的最大允许钻井液密度。如果允许安全系数为0.5lbm/gal，请计算5010ft处的最大允许钻井液密度。

（8）在TVD为7000ft的13⅜in套管鞋下方，使用9.0lbm/gal钻井液进行了漏失实验，测试结果如下表所示。12¼in井段的最大允许钻井液密度是多少？

（9）用9.5lbm/gal钻井液钻井，在7500ft处环空压耗为300psi，那么7500ft处钻井液的 *ECD* 是多少？

（10）为了预测8500ft处待钻地层的破裂压力，收集了以下信息：垂深为8500ft；孔隙压力为5300psi；上覆岩层压力为7800psi；泊松比为0.28。请计算8500ft处地层的破裂压力。

Drilling Tools 3
钻井工具

3.1 Drilling Bits

3.1.1 Introduction

A drilling bit is the major tool that conducts the cutting action located at the end of the drill-string. The bit generates the drilling action by scraping, chipping, gouging, or grinding the rock. Drilling fluid is circulated through the bit to remove the drilled cuttings generated inside the wellbore.

Audio 3.1

There are many variations of bit designs available. The selection of the bit for a particular application will depend on the type of formation to be drilled as well as the expected operating conditions during the drilling process. The performance of a bit is a function of several operating parameters including WOB, RPM, mud properties, and hydraulic efficiency. The drilling engineer must be aware of the design variations, the impact of the operating conditions on the performance of the bit, and the wear generated on the bit in order to be able to select the most appropriate bit for the formation to be drilled.

3.1.2 Historical Development

3.1.2.1 Roller-Cone Bits

Throughout the early 1900s, the performance of roller-cone bits was superior to other bit types and, as a result, their popularity grew as they became the most widely used drill-bit type. The first roller-cone bits were invented by and evolved from a patent by Howard Hughes in 1909 that described a rotary drill bit with two rotating cones. In the early 1930s, the tricone bit

3.1 钻头

3.1.1 简介

位于钻柱尾端的钻头是切削岩石的重要工具。钻头通过刮、削、凿或研磨来完成钻进。钻井液循环流经钻头，清除井眼内产生的岩屑。

钻头设计有许多不同的方法。钻头选型取决于要钻进的地层类型及钻进过程中预期的作业条件。钻头的性能取决于钻压、转速、钻井液性能和水力参数等工作参数。钻井工程师必须了解设计变量、作业条件对钻头性能和钻头磨损的影响，只有这样才能为要钻进的地层选择最合适的钻头。

3.1.2 发展历史

3.1.2.1 牙轮钻头

20世纪早期，牙轮钻头因性能优越成为使用最广泛的钻头类型。第一只牙轮钻头是由霍华德·休斯于1909年发明的一项专利发展而来，该专利描述了一种带有两个旋转锥体的旋转钻头。20世纪30年代初，引入了三牙轮钻

was introduced, with cutters designed for hard and soft formations (Fig. 3.1).

By the late 1940s, the industry was venturing into deep drilling, which means harder rocks such as limestone and chert, slow penetration rates, and reduced bit life. Because conventional milled-tooth bits were simply inadequate for these drilling environments, in 1949, Hughes Tool Company introduced the first three-cone bit using tungsten carbide inserts in the cutting structure. This bit was characterized by short and closely spaced inserts. Failure in the roller cones was mostly from bearing failure, and not to structure failure. Still, developments achieved mostly focused on bearing enhancements, leg and cutter metallurgy, and hydraulics. Today, modern insert bits are used routinely in many areas from top to bottom in low-solids drilling-mud systems.

3.1.2.2 Fixed-Cutter Bits

In the early 1900s, fishtail bits or drag bits, which were the early versions of fixed-cutter bits, were introduced. They were made of steel and were configured with two blades or paddles usually covered with harder alloy coatings or cutting tips to extend life. Their major application was mainly to drill very soft rock formations (Fig. 3.2).

头，其切削齿适合切削硬地层和软地层（图3.1）。

到20世纪40年代后期，石油行业开始涉足深井钻井，这意味着要钻探石灰岩和燧石等更坚硬的岩石，这使得钻头的机械钻速降低且寿命缩短。由于常规的铣齿钻头根本无法应对这些钻井条件，因此，休斯工具公司于1949年推出了首个在切削结构中使用碳化钨镶齿的三牙轮钻头。其特点是镶齿短且齿距密。牙轮钻头失效主要是由于轴承失效而非结构故障，因此技术进展主要集中在改进轴承、牙爪和切削齿材料，以及水力参数方面。现在，结合低固相钻井液体系，镶齿牙轮钻头已广泛应用于由浅到深的诸多地层中。

3.1.2.2 固定切削齿钻头

20世纪初出现的鱼尾钻头或刮刀钻头是早期的固定切削齿钻头。它们由钢制成，具有两个刀翼，为了延长使用寿命，刀翼通常覆盖有硬质合金涂层或切削刃。这种钻头主要用于钻进非常松软的地层（图3.2）。

Fig.3.1 Tricone bit

Fig.3.2 Drag bits

The first bits to use diamond for oil-well drilling were dubbed natural-diamond bits because they used natural diamonds as their cutting elements. These bits

开始用于石油钻采的金刚石钻头是天然金刚石钻头，因为它们通常使用天然金刚石作为切削

were first used in the 1940s and were commonly used through the 1980s. There was a significant increase of interest in the natural-diamond industry in the late 1970s, when thermally stable poly-crystalline (TSP) bits and poly-crystalline-diamond-compact (PDC) technology was developed. Both of these bits use small disks of synthetic diamond to provide the scraping (cutting) surface (Fig. 3.3).

部件。这种钻头在20世纪40年代首次使用并一直沿用到80年代。20世纪70年代末，随着热稳定聚晶金刚石钻头（TSP）和聚晶金刚石复合片（PDC）技术的发展，人们对天然金刚石工业的兴趣显著增加。这两种钻头均用较小的人造金刚石复合片构成刮削（切削）表面（图3.3）。

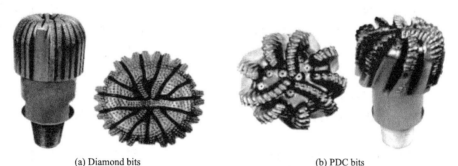

(a) Diamond bits　　　　　　(b) PDC bits

Fig.3.3　Diamond bits and PDC bits

The TSP bit was the first synthetic-diamond component used by the drill-bit industry and represents the evolutionary link to the modern PDC cutter, but it is tolerant of much higher temperatures than a conventional PDC bit.

TSP钻头是钻头行业首次使用人造金刚石组件的钻头，与现代PDC钻头切削刃一样，都是技术革新的产品，但它比传统PDC钻头更耐高温。

By the mid-1980s, the PDC drill bit had evolved. In these years, the PDC bit industry experimented with a wide variety of bit configurations and cutters. Both roller-cone and PDC bits underwent tremendous material and technological advances in the 1990s.

20世纪80年代中期，PDC钻头得到了长足发展，进行了多种PDC钻头结构和切削齿试验。90年代，牙轮钻头、PDC钻头在材料和技术上都有了较大进步。

3.1.3　Bit Types

Rotary drilling uses two types of drill bits: roller-cone bits and fixed-cutter bits. Roller-cone bits have one or more cones containing cutting elements, usually referred to as inserts, which rotate about the axis of the cone as the bit is rotated at the bottom of the hole. Milled-tooth (or steel-tooth) bits are typically used for drilling relatively soft formations. Tungsten-carbide-insert (TCI) bits are used in a wider range of

3.1.3　钻头类型

旋转钻井通常使用两种类型的钻头：牙轮钻头和固定切削齿钻头。牙轮钻头有一个或多个牙轮，牙轮上有通常称为镶齿的切削元件，当钻头在井底旋转时，镶齿绕牙轮轴线旋转。铣齿（或钢齿）钻头通常用于相对松软的地层。碳化钨镶齿（TCI）钻头可

formations, including the hardest and most abrasive drilling applications. Fixed-cutter bits, including PDC, impregnated, and diamond bits, can drill an extensive array of formations at various depths. All fixed-cutter bits consist of fixed blades that are integral with the body of the bit and rotate as a single unit.

3.1.3.1 Roller-Cone Bits

Roller-cone bits are classified as milled-tooth or insert. In milled-tooth bits, the cutting structure is milled from the steel making up the cone. In insert bits, the cutting structure is a series of inserts pressed into the cones. Roller cone bits have a large variety of tooth designs and bearing types, and are suited for a wide variety of formation types and applications.

The drilling action of a roller-cone bit depends to some extent on the offset of its cones. The offset of the bit is a measure of how much the cones are moved so that their axes do not intersect at a common point on the center line of the hole. Offsetting causes the cone to stop rotating periodically as the bit is turned and scrapes the hole bottom much like a drag bit. This action tends to increase drilling speed in most formation types. However, it also pro-motes faster tooth wear in abrasive formations. Cone offsets vary between 0.5 and 0.375 in. for soft-formation roller-cone bits, and are usually between 0.0325 and 0.0 in. for hard-formation bits.

The shape of the bit teeth also has a large effect on the drilling action of a roller-cone bit. Long, widely spaced steel teeth are used for drilling soft formations. As the rock type becomes harder, the tooth length and the cone offset must be reduced to prevent tooth breakage. The drilling action of a bit with zero cone offset is essentially a crushing action. The smaller teeth also allow more room for the construction of stronger bearings (Fig. 3.4).

用于各种地层，包括最坚硬和最耐磨的地层。固定切削齿钻头包括 PDC 钻头、孕镶式钻头和金刚石钻头，可用于不同深度的地层。所有固定切削齿钻头的固定刀翼均与钻头本体固结成一体，并作为一个整体旋转。

3.1.3.1 牙轮钻头

牙轮钻头分为铣齿钻头和镶齿钻头。铣齿钻头的切削齿是由牙轮毛坯铣削加工而成的。镶齿钻头的切削齿是一系列压嵌到牙轮上的牙齿。牙轮钻头具有多种齿形设计和轴承类型，可适用于多种地层类型和不同的应用范围。

牙轮钻头的钻进效果在一定程度上取决于牙轮移轴的大小。牙轮轴线相对于钻头轴线平移一段距离，使它们不会相交于井眼中心线上的一个共同点。移轴导致牙轮旋转时周期性停止旋转，并像刮刀钻头那样刮削井底。在大多数地层类型中，这种作用有助于提高钻速。然而，它也加快了牙齿在研磨性地层中的磨损。对于软地层，牙轮钻头的移轴范围为 0.375～0.5in；对于硬地层，移轴范围通常为 0～0.0325in。

牙齿的形状对牙轮钻头的钻进也有很大影响。长而稀疏分布的钢齿适用于钻进软地层。随着岩石硬度的增加，必须减小齿长和牙轮移轴以防止牙齿断裂。零移轴牙轮钻头的钻进本质上是压碎作用。较小的牙齿还为构造更坚固的轴承提供了更多空间（图 3.4）。

Fig.3.4 Soft- (left) and hard- (right) formation roller-cone bits

Because formations are not homogeneous, sizable variations exist in their drill ability and this has a large impact on cutting-structure geometry. For a given WOB, wide spacing between inserts or teeth results in improved penetration and relatively higher lateral loading on the inserts or teeth. Closely spacing inserts or teeth reduces loading at the expense of reduced penetration. The design of inserts and teeth themselves depends largely on the hardness and drill-ability of the formation. Penetration of inserts and teeth, cuttings-production rate, and hydraulic requirements are interrelated, as shown in Table 3.1.

地层的非均质性使可钻性存在很大差异，这对切削结构的几何形状有很大的影响。对于给定钻压，镶齿或铣齿之间的宽间距能够提高机械钻速，并对牙齿形成相对较高的侧向载荷。相反，镶齿或铣齿之间的窄间距可以减小载荷，但同时机械钻速也会降低。镶齿和铣齿的设计很大程度上取决于地层的硬度和可钻性。镶齿和铣齿钻头的机械钻速、钻屑生成速度和水力参数之间相互关联，见表3.1。

Table3.1 Relationship between inserts, teeth, cuttings-production rate, hydraulic requirements, and the formation

Formation Characteristics	Insert/Tooth Spacing	Insert/Tooth Properties	Penetration and Cuttings Generation	Cleaning Flow-Rate Requirements
Soft	Wide	Long and sharp		
Medium	Relatively wide	Shorter and stubbier	Relatively high	Relatively high
High	Close	Shorter and stubbier	Relatively low	Relatively low

The action of bit cones on a formation is of prime importance in achieving a desirable penetration rate. Soft-formation bits require a gouging action. Hard-formation bits require a crushing action. These actions are governed primarily by the degree to which the cones roll and skid. Maximum gouging (soft-formation) actions

牙轮钻头作用于地层的首要目标是达到预期的机械钻速。对于软地层，钻头需要凿削等作用。对于硬地层，钻头需要压碎等作用。这些作用主要受牙轮滚动和滑动的控制。要获得最大的凿削

require a significant amount of skid.

Conversely, a crushing (hard-formation) action requires that cone roll approach a "true roll" condition with very little skidding. For soft formations, a combination of small journal angle, large offset angle, and significant variation in cone profile is required to develop the cone action that skids more than it rolls.

The journal is the load-carrying surface of the bearing on a bit, and journal angle is the angle subtended between the axis of rotation of the roller cones and a plane perpendicular to the axis of rotation of the drill bit. Hard formations require a combination of large journal angle, no offset, and minimum variation in cone profile. These will result in cone action closely approaching true roll with little skidding.

3.1.3.2 Fixed-Cutter Bits

In general, fixed-cutter bits are categorized under two groups: PDC bits (fail the rock through a shearing process) and diamond bits made up of impregnated, natural-diamond and TSP elements (fail the rock through a grinding process). There does exist a third category, which also fails the rock through a shearing process and which is referred to as a tool steel-bladed bit or a drag bit (Fig. 3.2); however, drag bits are rarely used in the industry today.

The major difference between fixed-cutter bits and roller-cone bits is that fixed-cutter bits do not have any moving parts, which is an advantage, especially with small hole sizes in which space is not available for the cone (bearing) systems with proper teeth structure. The introduction of hard-facing to the surface of the blades and the design of fluid passageways greatly improved the performance of fixed-cutter bits. Because of the dragging (scraping) action of the fixed-cutter bits, high RPM and low WOB are applied.

1. PDC Bits

PDC bits use small disks of synthetic diamond to provide the scraping (cutting) surface. The small discs may be manufactured in any size and shape and

（软地层）作用需要牙轮产生较大的滑动。

反之，压碎（对于硬地层）作用要求牙轮达到几乎不滑动的"真正滚动"状态。对于软地层，为了使牙轮的滑动作用大于滚动作用，需要小轴颈角、大移轴角和牙轮剖面显著变化的结构组合。

轴颈是钻头轴承承受载荷的界面。轴颈角是牙轮旋转轴与垂直于钻头旋转轴的平面之间的夹角。硬地层需要大轴颈角、无移轴和牙轮剖面变化小的组合。这将使牙轮的运动接近真滚动，而几乎没有滑动。

3.1.3.2 固定切削齿钻头

通常固定切削齿钻头分为两类：PDC钻头（通过剪切作用破碎岩石）和孕镶天然金刚石或TSP金刚石钻头（通过研磨作用破碎岩石）。还有第三类钻头，也通过剪切作用破碎岩石，称为钢质刀翼钻头或刮刀钻头（图3.2）。但是，刮刀钻头在当今行业中很少使用。

固定切削齿钻头和牙轮钻头的主要区别在于，固定切削齿钻头没有任何运动部件，这是一个优势，尤其是在小井眼中，牙轮（轴承）系统没有足够的可用空间。刀翼表面的硬化处理和钻井液的流道设计大大提高了固定切削齿钻头的性能。此外，采用高转速和低钻压更宜发挥固定切削齿钻头的切削（刮削）作用。

1. PDC钻头

PDC钻头使用小的人造金刚石复合片进行刮削（切削）。它的小复合片可以制成任何尺寸和形

are not sensitive to failure along cleavage planes as natural-diamond bits are. PDC bits have been run very successfully in many fields all around the world. TSP bits are manufactured similarly to PDC bits, except TSP bits can resist much higher operating temperatures than PDC bits. One commonly used PDC bit is the dual-diameter bit. Dual-diameter bits have a unique geometry that allows them to drill and under-ream. To achieve this, the bits must be capable of passing through the drift diameter (i.e., the smallest inner diameter of a tubular material) of a well casing and then drilling an oversized (larger than casing diameter) hole. State-of-the-art dual-diameter bits are similar to conventional PDC drill bits in the way that they are manufactured. They typically incorporate a steel-body construction and a variety of PDC and/or diamond enhanced cutters. They are unitary and have no moving parts (Fig. 3.5). The maximum benefit of dual-diameter bits is realized in swelling or flowing formations in which the risk of sticking pipe can be reduced by drilling an oversized hole.

2. Impregnated Bits

Impregnated-bit bodies are PDC matrix materials that are similar to those used in cutters. The working portions of impregnated bits are unique, such that they contain matrix impregnated with diamonds (Fig. 3.6).

Both natural and synthetic diamonds are prone to breakage from impact. When embedded in a bit body, they are supported to the greatest extent possible and are less susceptible to breakage. However, because the largest diamonds are relatively small, cut depth must be small and ROP must be achieved through increased rotational speed. They are most frequently run in conjunction with turbo drills and high-speed positive-displacement motors that operate at several times normal rotational velocity for rotary drilling (500 to 1,500r/min). During drilling, individual diamonds in a bit are exposed at different rates. Sharp, fresh diamonds are always being exposed and placed into service (Fig. 3.7). Better bit performance and reduction in the number of required bits have been re-ported in abrasive and heterogeneous

状，不像天然金刚石钻头那样对沿裂解面的破坏十分敏感。PDC钻头已成功应用于世界各地的众多油田。TSP钻头的制作与PDC钻头类似，不同之处在于TSP钻头比PDC钻头更能耐高温。有一种常用的PDC钻头是双径钻头。双径钻头具有独特的几何形状，可以方便地进行钻进和扩眼。要实现这一功能，钻头必须先通过套管的通径（即套管的最小内径），然后再钻出超径（大于套管直径）井眼。最先进的双径钻头在制造方式上类似于传统的PDC钻头。它们通常采用钢制本体结构和各种PDC（金刚石）强化切削齿，是一体式的，没有活动部件（图3.5）。双径钻头的最大优点可在膨胀型或流动型地层中钻一个超大井眼，从而降低卡钻的风险。

2. 孕镶式钻头

孕镶式钻头体是PDC胎体材料，类似于切削齿所用的材料。它们的单一工作部件是孕镶有金刚石的胎体。（图3.6）。

天然金刚石和人造金刚石在冲击作用下都易破裂。但嵌入钻头体后，它们会得到最大程度的支撑且不易损坏。然而，即便是最大的金刚石，其尺寸也相对较小，这导致其切削深度相应较浅，因此必须通过提高转速来提高机械钻速。它们通常与涡轮钻具或高速螺杆钻具搭配使用，其转速可达常规旋转钻井转速的数倍（500～1500r/min）。钻井过程中，钻头上的每一颗金刚石露出的程度不同。锋利且干净的金刚石交替切削岩石（图3.7）。据报道，孕镶式钻头替代牙轮钻头和PDC钻

formations when impregnated bits with turbines are used instead of roller-cone bits and PDCs.

头与涡轮钻具配合使用时，能够在研磨型和非均质地层中取得更好的钻头性能，并减少所需钻头数量。

Fig.3.5 Dual-diameter PDC bit   Fig.3.6 Impregnated bit

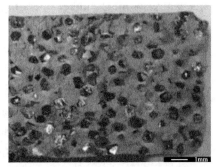

Fig.3.7 Presentation of diamonds impregnated in a cross-sectional view of a matrix body

3. Diamond Bits

The term "diamond bit" normally refers to bits incorporating surface-set natural diamonds as cutters. Diamond bits are used in abrasive formations. The cutting action of a diamond bit is developed by scraping away the rock. Diamond bits drill by a high-speed plowing action that breaks the cementation between rock grains. Fine cuttings are developed in low volumes per rotation. To achieve satisfactory ROPs with diamond bits, they must, accordingly, be rotated at high speeds. Despite its high wear resistance, diamond is sensitive to shocks and vibrations. Thus, caution must be taken when running a diamond bit. Effective fluid circulation across the face of the bit is also very important to prevent overheating of the diamonds and the matrix material, and to prevent bit balling (i.e., cuttings agglomerating on the bit). The cutting elements are typically placed among and along shallow waterways intended to provide some level of cooling and cleaning.

Diamond bits are described in terms of the profile of their crown, the size of diamond stones (stones per carat), total fluid area incorporated into the design, and fluid-course design (radial or cross-flow). Diamonds do

3. 金刚石钻头

金刚石钻头通常是指表面镶嵌天然金刚石作为切削刃的钻头。金刚石钻头用于研磨型地层，其破岩作用是通过刮擦岩石而产生的，即通过高速犁削来破碎岩石颗粒间的胶结物。每次旋转会产生少量细小的岩屑颗粒。所以为了使金刚石钻头达到令人满意的机械钻速，必须相应地提高转速。尽管金刚石具有很高的耐磨性，但它对冲击和振动非常敏感。因此，在使用金刚石钻头时必须加倍注意。覆盖整个钻头表面的有效钻井液循环对于防止金刚石和胎体材料过热，以及钻头泥包（即钻屑在钻头上结块）也非常重要。为了充分冷却和清洁钻头，通常将切削部件沿水槽安装或安装在水槽之间。

金刚石钻头根据其冠部的轮廓、金刚石的大小（每克拉多少颗）、设计总流体面积、水力结构（辐射式或逼压式）来描述。金刚

not bond with other materials. They are held in place by partial encapsulation in a matrix bit body. Diamonds are set in place on the drilling surfaces of bits.

3.1.3.3 Hybrid Bits

Significant advances have been made in PDC-cutter technology, and fixed-blade PDC bits have replaced roller-cone bits in numerous operations. However, in some applications for which the roller-cone bits are uniquely suited—such as drilling hard, abrasive and interbedded formations; complex directional-drilling applications; and, in general, applications in which the torque requirements of a conventional PDC bit exceed the capabilities of a given drilling system—the hybrid bit can substantially enhance the roller-cone bit's performance while generating a lower level of harmful dynamics compared to a conventional PDC bit.

In a hybrid bit, the intermittent crushing of a roller-cone bit is combined with the continuous shearing and scraping of a fixed-blade bit (Fig. 3.8). The central portion of the borehole is cut solely by PDC cutters on the primary blades, while the more-difficult-to-drill outer portion is being disintegrated by the combined action of the cutting elements on the rolling cutters and the fixed blades. The rolling cutters are biased toward the backside of the blades to open up a space (or junk slot) in front of the blades for the return of cuttings and the placement of nozzles. Hydraulically activated expandable hybrid bits having tricones outside and PDC cutters inside perform successfully in hard formations.

3.1.4 Rock-Failure Mechanism During Drilling

The method in which rock fails is important in bit design and selection. Formation failure occurs in two modes: brittle failure and plastic failure. The mode in which a formation fails depends on rock strength, which is a function of composition and downhole conditions such as depth, pressure, and temperature.

Audio 3.2

Video 3.1

Formation failure can be depicted with stress-strain curves (Fig. 3.9). Stress—applied force per unit area—can be tensile, compressive, torsional, or shear. Strain is the deformation caused by the applied force. Under brittle failure, the formation fails with very little or no deformation. For plastic failure, the formation deforms elastically until it yields, followed by plastic deformation until rupture.

地层破坏可以用应力—应变曲线表示（图3.9）。应力（单位面积上施加的力）可以是拉力、压力、扭力或剪切力。应变是由应力引起的形变。在脆性破坏下，地层在很少甚至没有形变的情况下发生破裂。对于塑性破坏，地层会弹性形变直至屈服，然后塑性变形直到破裂。

Fig. 3.8 Hybrid bit

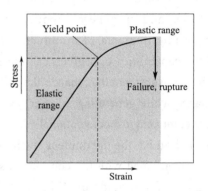

Fig.3.9 Stress-vs.-strain relation of formations

To operate a given bit properly, the drilling engineer needs to understand as much as possible about the basic mechanisms of rock removal that are at work, including: (1) shearing; (2) grinding; (3) erosion by fluid jet action; (4) crushing. To some extent, these mechanisms are interrelated. While one may be dominant for a given bit design, more than one mechanism is usually present. In this discussion, only the two basic rotary-drilling-bit types will be discussed: roller-cone bits and fixed-cutter bits.

3.1.4.1 Roller-Cone Bits

The action of bit cones on a formation is of prime importance in achieving a desirable penetration rate. Soft-formation bits require a scraping action. Hard-formation bits require a crushing action. These actions are governed primarily by the degree to which the cones roll and skid. Maximum scraping (soft-formation) actions require a significant amount of skid. Conversely, a crushing (hard-formation) action requires

为了正确使用钻头，钻井工程师需要尽可能多地了解破岩机理，包括：（1）剪切；（2）研磨；（3）钻井液喷射冲蚀；（4）压碎。这些机理在某种程度上是相互关联的。虽然在具体钻头设计中某种机理起主导作用，但通常多种机理共同作用。下面只讨论两种基本的旋转钻头类型：牙轮钻头和固定切削齿钻头。

3.1.4.1 牙轮钻头

牙轮与地层的相互作用对于实现理想的机械钻速来说至关重要。钻软地层时需要刮削作用，钻硬地层时需要压碎作用。这些作用主要取决于牙轮的滚动和滑动程度。要使刮削（软地层）作用最大化，需要显著的滑动作用；反之，压碎（硬地层）作用需要牙

that cone roll approach a "true roll" condition with very little skidding.

Insert bits designed with a large cone-offset angle for drilling soft formations employ all of the basic mechanisms of rock removal. However, the crushing action is the predominant mechanism present for IADC Series 3, 7, and 8 roller-cone bits. Because these bit types are designed for use in hard, brittle formations in which penetration rates tend to be low and drilling costs tend to be high, the crushing mechanism is of considerable economic interest. Maurer conducted experiments using an original setup of a single-tooth impacting on a rock sample under simulated borehole conditions. He found that the crater mechanism depended to some extent on the pressure differential between the borehole and the rock-pore pressure. At low values of differential pressure, the crushed rock beneath the bit tooth was ejected from the crater, while at high values of differential pressure the crushed rock deformed in a plastic manner and was not ejected completely from the crater. The crater mechanism for both low and high differential fluid pressure is described in Fig. 3.10. The sequence of events shown in this figure is described by Maurer as follows.

用于软地层的大移轴角镶齿钻头在设计上采用了所有的基本破岩机理。IADC 3、7 和 8 系列牙轮钻头的主要破岩机理是压碎作用。因为这些钻头用于坚硬的脆性地层，这些地层机械钻速往往很低，导致钻井成本很高，所以压碎破岩机理具有重要的经济价值。莫勒在模拟钻井条件下，使用原始设置的单齿冲击岩石样品进行实验。他发现齿坑在一定程度上取决于井眼与孔隙压力之间的压力差。在较低的压差下，钻头牙齿下方的碎屑从齿坑崩出，而在较高压差下，碎屑以塑性方式变形而没有完全从凹坑崩出。图 3.10 描述了低压差和高压差下的齿坑机理。莫勒对这一图中所示事件的顺序进行了描述。

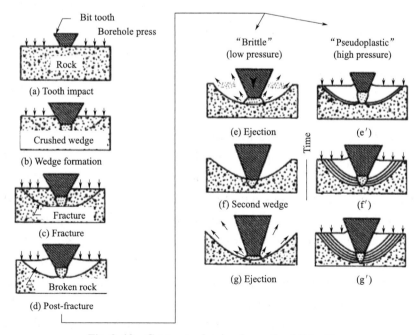

Fig. 3.10 Crater mechanism beneath a bit tooth

As a load is applied to a bit tooth (a), the constant pressure beneath the tooth increases until it exceeds the compressive strength of the rock, and a wedge of finely powdered rock then is formed beneath the tooth (b). As the force on the tooth increases, the material in the wedge compresses and exerts high lateral forces on the solid rock surrounding the wedge until the shear stress exceeds the shear strength of the solid rock and the rock fractures (c). These fractures propagate along a maximum-shear surface, which intersects the direction of the principal stresses at a nearly constant angle as predicted by the Mohr failure criterion. The force at which fracturing begins beneath the tooth is called the threshold force. As the force on the tooth increases above the threshold value, subsequent fracturing occurs in the region above the initial fracture, forming a zone of broken rock (d). At low differential pressures, the cuttings formed in the zone of broken rock are ejected easily from the crater (e). The bit tooth then moves forward until it reaches the bottom of the crater, and the process may be repeated (f, g). At high differential pressures, the downward pressure and frictional forces between the rock fragments prevent ejection of the fragments (e′). As the force on the tooth is increased, displacement takes place along fracture planes parallel to the initial fracture (f′, g′). This gives the appearance of plastic deformation, and craters formed in this manner are called pseudoplastic craters.

3.1.4.2 Fixed-Cutter Bits

High-speed movies (Murray and MacKay 1957) of full-scale bits drilling at atmospheric conditions with air as the circulating fluid have verified that the mechanisms of failure for insert bits with little or no offset is not too different from that observed in single-insert impact experiments. However, the drilling action of insert bits designed with a large offset for drilling soft, plastic formations is considerably more complex than the simple crushing action that results when no offset is used; because each cone alternately rolls and drags, considerable wedging and twisting action is present.

当载荷施加到钻头牙齿时（a），牙齿下方的压力持续增加，直到超过岩石抗压强度时，就会在牙齿下方形成一个岩石碎屑的楔形坑（b）。如果施加于牙齿上的力继续增加，则楔形坑中的岩石受到压缩，并向其周围的岩石施加较高的横向力，直到剪切应力超过岩石的抗剪强度时岩石发生破裂（c）。裂缝将沿着最大剪切面传播，最大剪切面与主应力方向以几乎不变的角度相交，这个角度可由莫尔破坏准则预测。令牙齿下方岩石开始破裂时的压力为门限压力。当牙齿上的力增加并高于门限值时，在原始裂纹上方的区域内再次发生破裂，形成一个岩石破碎区（d）。当压差较小时，岩石破碎区中形成的岩屑易于从齿坑中弹出（e）。然后牙齿向下移动到齿坑的底部，这一过程不断重复进行（f、g）。当压差较大时，向下的压力和岩石碎块之间的摩擦力会阻碍碎块弹出（e′）。随着牙齿上力的增加，沿着平行于初始断裂的断裂平面发生位移（f′、g′）。这看上去像塑性变形，形成的齿坑被称为假塑性齿坑。

3.1.4.2 固定切削齿钻头

在常压下，以空气作为循环流体进行全尺寸钻头钻进的高速影像证实：具有较小移轴或无移轴的镶齿钻头破岩机理与单齿冲击实验中观察到的没有太大区别。然而，用于软塑性地层的大移轴镶齿钻头，其破岩机理要比不移轴钻头产生的简单压碎作用复杂得多。这是因为每个牙轮交替滚动和滑动，存在明显的楔入和扭转作用。

The cutting mechanism of PDC bits drills primarily by shearing such that the cutters have sufficient axial force to penetrate into the rock surface and simultaneously have the available torque for bit rotation. The resultant force defines a plane of thrust for the cutter. Cuttings are then sheared off at an initial angle relative to the plane of thrust, which is dependent on rock strength (Fig. 3.11). In shear, the energy required to reach the plastic limit for rupture is significantly less than that required by compressive stress. PDC bits, thus, require less WOB than roller-cone bits.

Natural-diamond bits are designed to remove rock primarily by a grinding action. Diamond bits with very large diamonds may possess a shallow plowing action. As the bit is rotating, the exposed diamonds grind against and remove the rock with a very shallow depth of cut. The intent is that as these bits wear, either the exposed diamonds are sharpened because of small fractures during drilling, or new diamonds become exposed as the less-abrasion resistant body material is worn away. This is a self-sharpening effect that is consistent with other grinding tools where a new grinding surface is continually exposed.

PDC 钻头的破岩机理主要是通过剪切作用，使切削刃有足够的轴向力楔入岩石表面，同时为钻头旋转提供扭矩。这两种作用的合力确定了切削刃的推力平面。然后，切削刃以一个相对于推力平面的初始角剪切岩屑，剪切面取决于岩石强度（图3.11）。在剪切过程中，达到破裂的塑性极限所需的能量明显小于压缩应力所需的能量。因此，PDC 钻头所需的钻压小于牙轮钻头。

天然金刚石钻头主要通过研磨作用破岩。安装大尺寸金刚石的钻头具有犁削作用。随着钻头的旋转，暴露的金刚石以非常浅的切入深度研磨岩石。随着这些钻头的磨损，要么使已经暴露的金刚石在钻进过程被小裂缝磨尖，要么随着耐磨材料的磨损暴露出新的金刚石。这是一种自磨锐作用，与不断磨出新磨削面的其他磨削工具一样。

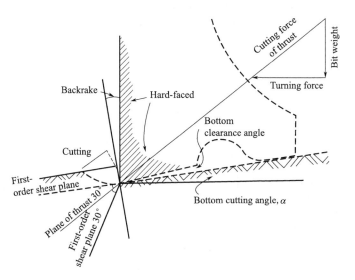

Fig.3.11　Shear and thrust on a cutter

The depth of cut of the bit is determined by the rock strength, the WOB applied, and the dull condition of the bit. The cut geometry at hole bottom is a helix and can be expressed in terms of the bottom-cutting angle or helix angle α. The helix angle α is a function of the cutter penetration per revolution L_p and radius r from the center of the hole. This relation can be defined by

$$\tan\alpha = \frac{L_p}{2\pi r}$$

(3.1)

钻头的切削深度取决于岩石强度、施加的钻压和钻头磨损情况。井底的切削几何形状为螺旋形，可用井底切削角或螺旋角α表示。螺旋角α是每转切削刃吃入量L_p和井眼半径r的函数。其关系式如下：

Rock-mechanics experts have applied several failure criteria in an attempt to relate rock strength measured in simple compression tests to the rotary-drilling process. One such failure criterion often used is the Mohr theory of failure. The Mohr criterion states that yielding or fracturing should occur when the shear stress exceeds the sum of the cohesive resistance of the material and the frictional resistance of the slip planes or fracture plane. The Mohr criterion is stated mathematically by

$$\tau = \pm(c + \sigma_n \tan\theta)$$

(3.2)

岩石力学专家应用了若干个破坏准则，试图将在简单的压缩试验中测量的岩石强度与旋转钻井过程联系起来。其中一个常用的破坏准则就是莫尔破坏准则。莫尔破坏准则规定，当剪应力超过材料的黏聚力和滑移面或断裂面的摩擦阻力之和时，材料发生屈服或断裂。用数学式表示为：

where, τ is the shear stress at failure, c is the cohesive resistance of the material, σ_n is the normal stress at the failure plane, and θ is the angle of internal friction. As shown in Fig. 3.12, this is the equation of a line that is tangent to Mohr's circles drawn for at least two compression tests made at different levels of confining pressure.

式中，τ是破坏时的剪切应力，c是材料的黏聚力，σ_n是在滑移面上的正应力，θ是内摩擦角。如图3.12所示，这是一条与莫尔圆相切的直线方程，它是在不同围压下进行了至少两次压缩实验得到的。

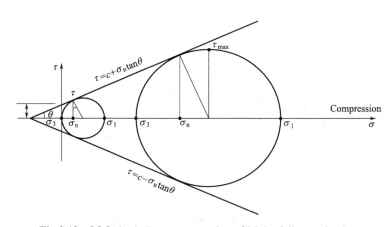

Fig.3.12 Mohr's circle representation of Mohr failure criterion

To understand the use of the Mohr criterion, consider a rock sample to fail along a plane, as shown in Fig. 3.13 (a), when loaded under a compressive force F and a confining pressure p. The compressive stress σ_1 is given by

$$\sigma_1 = \frac{F}{\pi r^2} \tag{3.3}$$

The confining stress is given by

$$\sigma_3 = p \tag{3.4}$$

If we examine a small element on any vertical plane bisecting the sample, the element is in the stress state given in Fig. 3.13(b). Furthermore, we can examine the forces present along the failure plane at failure using the free-body elements shown in Fig. 3.13 (b). The orientation of the failure plane is defined by the angle ϕ between the normal-to-the-failure plane and a horizontal plane. It is also equal to the angle between the failure plane and the direction of the principal stress σ_1. Both a shear stress τ and a normal stress σ_n must be present to balance σ_1 and σ_3. Summing forces normal to the fracture plane [Fig. 3.13(c)] gives

$$\sigma_n dA_n = \sigma_3 dA_3 \cos\phi + \sigma_1 dA_1 \sin\phi \tag{3.5}$$

The unit area along the fracture plane, dA_n, is related to the unit areas dA_1 and dA_3

$$dA_3 = dA_n \cos\phi \tag{3.6}$$

$$dA_1 = dA_n \sin\phi \tag{3.7}$$

Making these substitutions in the force-balance equation gives

$$\sigma_n = \sigma_1 \sin^2\phi + \sigma_3 \cos^2\phi = \frac{1}{2}(\sigma_1 + \sigma_3) - \frac{1}{2}(\sigma_1 - \sigma_3)\cos(2\phi) \tag{3.8}$$

Summing the forces parallel to the fracture plane gives

$$\tau dA_n = \sigma_1 dA_1 \cos\phi - \sigma_3 dA_3 \sin\phi \tag{3.9}$$

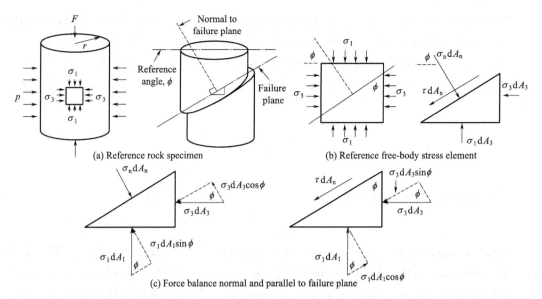

(a) Reference rock specimen
(b) Reference free-body stress element
(c) Force balance normal and parallel to failure plane

Fig.3.13 Mohr's circle graphical analysis

Expressing all unit areas in terms of dA_n and simplifying yields

以 dA_n 表示所有单位面积并简化得：

$$\tau = (\sigma_1 - \sigma_3)\sin\phi\cos\phi = \frac{1}{2}(\sigma_1 - \sigma_3)\sin(2\phi) \qquad (3.10)$$

Note that Eqs.3.8 and 3.10 are represented graphically by the Mohr's circle shown in Fig. 3.12. Note also that the angle of internal friction θ and 2ϕ must sum to 90°. The angle of internal friction for most rocks varies from approximately 30 to 40°.

注意，式（3.9）和式（3.10）由图 3.12 所示的莫尔圆表示。同时还要注意，内摩擦角 θ 和 2ϕ 之和必定为 90°。大多数岩石的内摩擦角为 30°～40°。

The Mohr failure criterion can be used to predict the characteristic angle between the shear plane and the plane of thrust for a fixed-cutter bit. Assuming an angle of internal friction of approximately 30° implies

对固定切削齿钻头来说，可以使用莫尔破坏准则来预测剪切面和推力面之间的特征角。假定内摩擦角约为 30°，则：

$$2\phi = 90° - 30° \qquad (3.11)$$

or

或者：

$$\phi = 30°$$

This value of ϕ has been verified experimentally by Gray etal. (1962) in tests made at atmospheric pressure.

此 ϕ 值已由格雷等在常压下进行的试验所证实。

3.1.5 Wear Mechanism

3.1.5 磨损机理

Audio 3.3

The cutting structures of bits often experience catastrophic failures resulting from dynamic-load variations beyond the capacity of the cutting elements.

钻头的切削结构经常会遭受严重的破坏，这是动载荷的变化超过了切削部件的承受能力所致。通过持续改进切削结构可以逐步

Continuous improvements to cutting structures have steadily increased penetration rates but, at the same time, have elevated their sensitivity to these dynamic effects. Lateral motion of the bit is a major cause of premature cutting-structure wear.

3.1.5.1 Cutter Wear
1. Factors Affecting Wear

The information about the instantaneous rate of bit wear is required for determining the total time interval of bit use. Therefore, it is mandatory to identify the influence of various drilling parameters on the instantaneous rate of bit wear. The rate of cutter wear depends primarily on formation abrasiveness, cutting-element size and geometry, bit weight, rotary speed, and the cleaning and cooling action of the drilling fluid. For roller-cone bits, the major influences on wear on cutters are as follows.

1) Effect of Tooth Height on Rate of Tooth Wear

Campbell and Mitchell showed experimentally that the rate at which the height of a steel tooth can be abraded away by a grinding wheel is inversely proportional to the area of the tooth in contact with the grinding wheel. The shape of steel bit teeth is generally triangular in cross section when viewed from either the front or the side. Thus, almost all milled-tooth bits have teeth that can be described using the geometry shown in Fig.3.14. The bit tooth initially has a contact area given by

$$A_i = w_{x1} w_{y1}$$

where, A_i is bit tooth initial contact area; w_{x1}、w_{y1} are bit tooth initial width in x、y direction.

提高钻头的机械钻速,但同时也增加了它们对动态载荷的敏感性。钻头的横向运动是切削结构磨损的主要原因。

3.1.5.1 切削齿磨损
1. 磨损的影响因素

要确定钻头的寿命就需要获取牙齿磨损速率的信息。因此必须确定各种钻进参数对牙齿磨损速率的影响。牙齿磨损速率主要取决于地层研磨性、切削部件尺寸和几何形状、钻压、转速及钻井液的清洁和冷却作用。对于牙轮钻头,磨损的主要影响因素如下。

1)齿高对牙齿磨损速率的影响

坎贝尔和米契尔的实验表明,钢齿齿高被砂轮磨掉的速率与钢齿和砂轮的接触面积成反比。当从前面或侧面观察时,钢齿的断面形状一般为三角形。因此,几乎所有铣齿钻头的牙齿都能用图3.14中的几何形状来描述。新钻头牙齿的接触面积为:

(3.12)

式中,A_i 为新钻头牙齿的接触面积;w_{x1}、w_{y1} 分别为新钻头牙齿 x、y 方向的宽度。

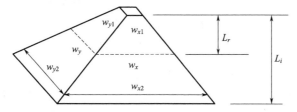

Fig.3.14 Typical shape of a milled tooth as a function of fractional tooth wear

After removal of tooth height L_r of the original tooth height L_i the bit tooth has a contact area given by

将原齿高 L_i 磨损掉 L_r 后,牙齿的接触面积变为:

$$A = w_x w_y = \left[w_{x1} + h(w_{x2} - w_{x1})\right]\left[w_{y1} + h(w_{y2} - w_{y1})\right] \quad (3.13)$$

The ratio L_r/L_i is defined as the fractional tooth wear, h:

将比值 L_r/L_i 定义为牙齿的磨损比 h，则：

$$h = \frac{L_r}{L_i} \quad (3.14)$$

Expressing the contact area A in terms of fractional tooth wear h yields

牙齿的磨损比来表示接触面积 A，为：

$$\begin{aligned} A &= \left[w_{x1} + h(w_{x2} - w_{x1})\right]\left[w_{y1} + h(w_{y2} - w_{y1})\right] \\ &= (w_{x1}w_{y1}) + \left[w_{y1}(w_{x2} - w_{x1}) + w_{x1}(w_{y2} - w_{y1})\right]h + (w_{x2} - w_{x1})(w_{y2} - w_{y1})h^2 \end{aligned} \quad (3.15)$$

If we define the geometry constants G_1 and G_2 by

如果定义几何常数 G_1 和 G_2：

$$G_1 = \left[w_{y1}(w_{x2} - w_{x1}) + w_{x1}(w_{y2} - w_{y1})\right]A_i \quad (3.16)$$

$$G_2 = \left[(w_{x2} - w_{x1})(w_{y2} - w_{y1})\right]A_i \quad (3.17)$$

The contact area A can be expressed by

则接触面积 A 可以表示为：

$$A = A_i(1 + G_1 h + G_2 h^2) \quad (3.18)$$

Because the instantaneous wear rate dh/dt is proportional to the inverse of the contact area A,

由于瞬时磨损速率 dh/dt 与接触面积 A 的倒数成正比，即：

$$\frac{dh}{dt} \propto \frac{1}{A_i(1 + G_1 h + G_2 h^2)} \quad (3.19)$$

the initial wear rate, when $h = 0$, is proportional to A_i. Thus, expressing dh/dt in terms of a standard initial wear rate $(dh/dt)_s$ gives

当 $h=0$ 时的初始磨损速率与 A_i 成正比。因此 dh/dt 用标准初始磨损速率 $(dh/dt)_s$ 表示：

$$\frac{dh}{dt} \propto \left(\frac{dh}{dt}\right)_s \frac{1}{(1 + G_1 h + G_2 h^2)} \quad (3.20)$$

For most bit types, the dimension $w_{x2} - w_{x1}$ is small compared with $w_{y2} - w_{y1}$. This allows a constant H_2 to be chosen such that the wear rate can be approximated using

对于大部分类型的钻头，与 $(w_{y2} - w_{y1})$ 相比 $(w_{x2} - w_{x1})$ 较小。这样就可以选择常量 H_2 近似表示磨损速率：

$$\frac{dh}{dt} \propto \left(\frac{dh}{dt}\right)_s \frac{1}{(1 + H_2 h)} \quad (3.21)$$

The use of Eq. 3.21 in place of Eq. 3.20 greatly simplifies the calculation of tooth wear as a function of rotating time. A case-hardened bit tooth or a tooth with hardfacing on one side often will be self-sharpening as the tooth wears. Even though the mechanism of self-

用式（3.21）替换式（3.20），极大简化了牙齿磨损的计算。例如，硬化处理或一侧表面硬化处理的钻头牙齿通常是随牙齿磨损自锐的，尽管自磨锐齿的机理与

sharpening tooth wear is somewhat different than in the abrasive-wear experiments of Campbell and Mitchell , a constant H_2 usually can be selected such that the instantaneous wear rate can be predicted by use of Eq. 3.21.

Insert teeth used in roller-cone bits usually fail by fracturing of the brittle tungsten carbide. For this tooth type, fractional tooth wear h represents the fraction of the total number of bit teeth that have been broken. The wear rate (dh/dt) does not decrease with increasing fractional tooth wear h. On the contrary, there is some evidence that the tooth breakage accelerates as the number of broken teeth beneath the bit increases. This type of behavior could be modeled with a negative value for H_2 in Eq.3.21. However, this phenomenon has not been studied in detail, and in practice a value of 0 is recommended for H_2 when using insert bits.

2) Effect of Bit Weight on Rate of Tooth Wear

Galle and Woods published one of the first equations for predicting the effect of bit weight on the instantaneous rate of tooth wear. The relation assumed by Galle and Woods is

$$\frac{dh}{dt} \propto \frac{1}{1-\lg\left(\dfrac{W}{d_b}\right)} \tag{3.22}$$

where, W is the bit weight in 1,000lbf units and d_b is the bit diameter in inches. Note that $W/d_b < 10.0$.

The wear rate at various bit weights can be expressed in terms of a standard wear rate that would occur for a bit weight of 4,000 lbf per inch of bit diameter. Thus, the wear rate relative to this standard wear rate is given by

$$\frac{dh}{dt} \propto \frac{0.3979\left(\dfrac{dh}{dt}\right)_s}{1-\lg\left(\dfrac{W}{d_b}\right)} \tag{3.23}$$

坎贝尔和米契尔的磨蚀实验有所不同，但通常可以选择一个常量 H_2，并通过式（3.21）来预测瞬时磨损速率。

牙轮钻头镶齿的失效通常是由于碳化钨的脆性破裂。对于这种牙齿类型，牙齿磨损比 h 代表了崩、断和掉的齿数与新钻头总齿数的比值。磨损率（dh/dt）并不随牙齿磨损比 h 的增加而降低。相反，有证据表明，随着钻头下方断齿数量的增加，断齿的速度会加快。可以用 H_2 为负值的式（3.21）来表征这种现象。尽管如此，人们尚未对这种现象进行详细研究。工程实践中使用镶齿钻头时，一般建议 H_2 取值为0。

2）钻压对牙齿磨损速率的影响

盖尔和伍兹首先提出用方程计算钻压对牙齿磨损速率的影响。他们假定的关系为：

式中，W 是钻压，单位 1000lbf；d_b 是钻头直径，单位为 in。注意，W/d_b <10.0。

不同钻压下的磨损速率可以用每英寸钻头直径在 4000lbf 钻压下的标准磨损速率来表示。因此，相对于这个标准磨损速率的磨损速率为：

Note that d*h*/d*t* becomes infinite for $W/d_b = 10$. Thus, this equation predicts that the teeth will fail instantaneously if 10,000 lbf per inch of bit diameter were applied. Later authors used a simpler relation between the bit weight and tooth wear rate. Perhaps the most commonly used relation is

$$\frac{dh}{dt} \propto \frac{1}{\left(\frac{W}{d_b}\right)_{max} - \frac{W}{d_b}} \tag{3.24}$$

where, $(W/d_b)_{max}$ is the maximum bit weight per inch of bit diameter at which the bit teeth would fail instantaneously, and $W/d_b < (W/d_b)_{max}$. Expressing this relation in terms of a standard wear rate at 4,000 lbf per inch of bit diameter yields

$$\frac{dh}{dt} \propto \left(\frac{dh}{dt}\right)_s \cdot \left[\frac{\left(\frac{W}{d_b}\right)_{max} - 4}{\left(\frac{W}{d_b}\right)_{max} - \frac{W}{d_b}}\right] \tag{3.25}$$

3) Effect of Rotary Speed on Rate of Tooth Wear

The first published relation between the instantaneous rate of tooth wear and the rotary speed N also was presented by Galle and Woods for milled-tooth bits. The Galle and Woods relation is

$$\frac{dh}{dt} \propto N + 4.34 \times 10^{-5} N^3 \tag{3.26}$$

However, several more-recent authors have shown that essentially the same results can be obtained using the simpler relation

$$\frac{dh}{dt} \propto N^{H_1} \tag{3.27}$$

where, H_1 is a constant. Also, H_1 was found to vary with the bit type used. The Galle and Woods relation applied only to milled-tooth bit types designed for use in soft formations. Expressing the tooth wear rate in terms of a standard wear rate that would occur at 60 r/min yields

注意，当 $W/d_b=10$ 时 d*h*/d*t* 为无穷大。因此，该方程预测，如果每英寸钻头直径施加10000lbf钻压时，牙齿将会瞬间失效。后来有些学者在钻压和牙齿磨损速率之间采用一个简单的关系，或许是使用最普遍的关系：

式中，$(W/d_b)_{max}$ 为牙齿即将失效时瞬间每英寸钻头直径的最大钻压，并且 $W/d_b < (W/d_b)_{max}$。用每英寸钻头直径在4000lbf钻压下的标准磨损速率来表示这种关系：

3）转速对牙齿磨损速率的影响

第一个发表铣齿钻头瞬时牙齿磨损速率和转速间关系的是盖尔和伍兹。该关系式为：

但是，后来的一些学者用一种较简单的关系也得到了基本相同的结果：

式中，H_1 是常数。他们同时发现，H_1 随使用钻头类型的不同而不同。盖尔和伍兹的关系仅用于软地层铣齿钻头。用钻速为60r/min的标准磨损速率表示牙齿磨损速率为：

$$\frac{dh}{dt} \propto \left(\frac{dh}{dt}\right)_s \left(\frac{N}{60}\right)^{H_1} \quad (3.28)$$

4) Effect of Hydraulics on Rate of Tooth Wear

The effect of the cooling and cleaning action of the drilling fluid on cutting-element wear rate (dh/dt) is much more important for fixed-cutter bits (diamond and PDC) than it is for roller-cone bits. Cutting elements on fixed-cutter bits must receive sufficient flow to prevent the buildup of excessive heat, which leads to graphitization of the diamond materials, which thus accelerates the wear process. The flow velocities also must be maintained high enough to prevent clogging of fluid passages with drilled cuttings. The design of the fluid-distribution passages on a diamond or PDC bit is extremely important and varies considerably among the various bits available. However, the manufacturer will usually specify the total flow area (TFA) needed for specific drill bits based on rig type and capabilities, flow rates, drilling-fluid type and properties, BHA design, formation characteristics, drilling program, and anticipated ROP. Considering a specific TFA, nozzle sizing must also be based on the need to minimize recirculation of the drilling fluid, cuttings regrinding, and stagnation zones on the bit face. In addition to minimizing cutting-element wear rate, efficient fluid distribution and cleaning also improves ROP. It is generally assumed that as long as the flow is present to clean and to cool the drill bit, the effect of hydraulics on cutting-element wear rate can be ignored or can be assumed to be taken care of.

2. Wear on PDC Cutters

PDC-cutter wear can be divided into two categories, depending on the basic cause of the wear. The first category is steady-state wear that is normally associated with the development of uniform wear flats on the PDC cutter and the gradual degradation in the ROP over the bit life, which is a function of operating parameters applied

4）水力因素对牙齿磨损速率的影响

对于固定切削齿钻头（金刚石钻头和PCD钻头），钻井液的冷却和清洁作用对切削刃磨损速率（dh/dt）的影响要比牙轮钻头更大。固定切削齿钻头的切削部件必须获得足够的液流才能防止过多的热量积累，过热会导致金刚石材料的石墨化，从而加速磨损过程。同时还必须保持高流速，以防止钻屑堵塞流体通道。金刚石或PDC钻头上的液流通道分布设计至关重要，并且在各种可用的钻头之间差异很大。制造商通常会根据钻机类型和能力、排量、钻井液类型和性质、井底钻具组合设计、地层特征、钻进程序和预期的机械钻速来设计特定钻头所需的总流量面积（TFA）。考虑到特定的TFA，喷嘴尺寸还必须基于尽量减小钻井液再循环、钻屑重复研磨和钻头面滞留区的需要。除了最大程度地减小切削部件的磨损速率之外，有效的液流分布和净化作用还可以提高机械钻速。一般认为，只要有液流清洗和冷却钻头，就可以忽略水力因素对切削元件磨损速率的影响，也可以假定要加以处理。

2. PDC钻头切削齿磨损

根据磨损的基本原因，PDC钻头的磨损可分为两类。第一类是稳态磨损，这通常与PDC钻头切削齿上均匀磨损平面的情况及由此引起钻头机械钻速降低有关，机械钻速降低受钻井作业参数、

to the bit and individual cutters, cutter temperature, cutter velocity, formation properties, and cutter properties. The second category of wear is the result of impact loading of the cutters. This type of wear may be caused by dynamic loading of the bit during bit whirl or from drilling through heterogeneous formations.

PDC cutters tend to wear in a manner somewhat similar to that of steel-tooth inserts. Some examples of PDC-cutter wear are presented in Fig. 3.15

切削齿数量、切削齿温度、切削速度、地层特性及切削齿性能等影响。第二类磨损是钻头冲击载荷的结果。这种磨损可能是钻头旋转过程中的动态载荷或钻穿非均质地层所致。

PDC切削齿与钢质镶齿的磨损方式相似。图3.15展示了PDC切削齿磨损的一些图片。

  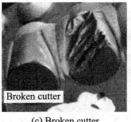 

(a) Worn cutter　　(b) Chipped cutter　　(c) Broken cutter　　(d) Failed cutter because of high impact loads

Fig.3.15　PDC-cutter-wear examples

The shape of the PDC cutter, which is usually circular, provides a different relationship between fractional tooth wear and cutter-contact area. For a zero backrake angle, the cutter-contact area is proportional to the length of the chord defined by the lower surface of the cutter remaining after removal of the cutter height L_r (Fig. 3.16).

PDC切削齿的形状通常是圆形，它的牙齿磨损量和牙齿接触面积之间有不同的关系。对于零后倾角，切削齿接触面积与弦长成正比，而弦长是磨掉切削齿磨损高度L_r后剩下的下表面（图3.16）。

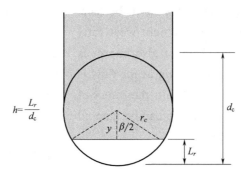

Fig. 3.16　PDC-blank geometry as a function of fractional cutter wear h for a zero backrake angle

Because the fractional tooth wear is given by

因为牙齿磨损量由下式给出：

$$h = \frac{L_r}{d_c} \qquad (3.29)$$

and the dimension y is

且分量y为：

$$y = r_c \cos\frac{\beta}{2} \qquad (3.30)$$

then

$$h = \frac{r_c - y}{d_c} = \frac{r_c - r_c \cos\frac{\beta}{2}}{2r_c} = \frac{1 - \cos\frac{\beta}{2}}{2} \qquad (3.31)$$

where, r_c = radius of PDC cutter; d_c = diameter of PDC cutter.

Solving this expression for the subtended angle β yields

$$\cos\frac{\beta}{2} = 1 - 2h \qquad (3.32)$$

Because the contact area is directly proportional to the chord length subtended by the angle β, then

$$A \propto 2\frac{d_c}{2}\sin\frac{\beta}{2} \qquad (3.33)$$

And the wear rate is inversely proportional to this contact area:

$$\frac{dh}{dt} \propto \left(\frac{dh}{dt}\right)_s \frac{1}{d_c \sin\frac{\beta}{2}} \qquad (3.34)$$

The wear rate decreases with increasing fractional tooth wear h between 0 and 0.5. Above this range, the wear rate increases with increasing h.

For PDC cutters with nonzero backrake angles, the total contact area of the PDC wafer or diamond layer and the tungsten carbide substrate becomes more complex. However, the preceding analysis remains representative of the geometry of the PDC layer, which is believed to be the predominant determinant, in terms of the wear resistance of the PDC cutter. Typical failure modes observed of diamond enhanced inserts are shown in Fig. 3.17. Chipping and fracture resistance can be improved at the expense of the hardness and wear resistance of the materials. This trade off between wear resistance and chipping resistance hinders the development of super hard materials for demanding drilling applications.

Warren and Sinor pointed out the effects of bit whirl on PDC cutters. Bit whirl caused the cutters to

所以有：

式中，r_c 为 PDC 切削齿半径；d_c 为 PDC 切削齿直径。

求解该表达式得到内夹角 β 为：

由于接触面积与夹角 β 所夹的弦长成正比，因此：

磨损速率与该接触面积成反比：

当牙齿磨损比 h 在 0～0.5 范围内时，磨损速率随 h 增加而降低。超过这个范围，磨损速率随 h 增加而增加。

对于具有非零后倾角的 PDC 切削齿，PDC 复合片或金刚石涂层与碳化钨基底的总接触面积求解更加复杂。但是前面的分析仍然代表了 PDC 层的几何形状，就 PDC 切削齿的耐磨性而言，它（几何形状）仍被认为是主要的决定因素。强化金刚石镶齿的典型破坏形式如图 3.17 所示。可以通过牺牲材料硬度和耐磨性为代价来改善抗崩碎性和抗断裂性。耐磨性和抗崩碎性之间的折中取舍阻碍了超硬材料在苛刻钻井情况下的应用。

沃伦和西诺指出了钻头涡动对 PDC 切削齿的影响。钻头涡动

be damaged by impact loading, even in homogeneous rock, by allowing the cutters to move sideways and backward so that the diamond cutting edge was damaged. This damage to the cutting edge allowed the carbide substrate to contact the rock and generate heat. This in turn further damaged the diamond and gave the appearance that the cutter had been damaged by heat, when in fact the root cause of the damage was the initial chipping.

使切削齿承受冲击载荷而损坏，即使在均质岩石中，切削齿横向和向后移动也会导致切削齿边缘损坏。这种损坏会使硬质合金基体接触岩石发热，进一步损坏金刚石，导致切削齿热损坏。实际上损坏的根本原因还是初始的崩碎。

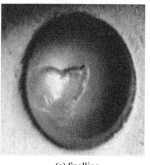 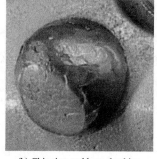 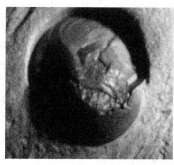

(a) Spalling　　(b) Chipping and heat checking　　(c) Breakage

Fig.3.17　Typical failure modes of PDC inserts

Laboratory results for antiwhirl and conventional PDC bits by Dykstra et al. have shown that vibrations of the former were an order of magnitude less than those of the latter. Backward whirl of the conventional PDC bit was prevalent in both hard and soft formations. Roller-cone-bit tests suggested that they too were subject to backward whirl and that the lateral vibrations that resulted were an order of magnitude worse than the axial vibrations commonly associated with these bits. Johnson noted that proper blade and cutter alignment during the design stage of PDC bits eliminates whirl. Fig. 3.18 shows the evidence of the bit whirl by comparing the bottomhole patterns and presents an alternative bit profile to eliminate this problem (Fig.3.19 and Fig.3.20).

戴克斯特拉等人进行的抗涡动和常规PDC钻头实验结果表明，前者的震动比后者的震动小一个数量级。常规PDC钻头在硬地层和软地层中都普遍存在向后涡动。牙轮钻头实验表明，它们也受到向后涡动的影响，所产生的横向震动通常比轴向震动差一个数量级。约翰逊指出在PDC钻头的设计阶段，合理的刀刃和切削齿排列可以消除涡动。图3.18显示了钻头涡动和不涡动的井眼情况，并给出了消除涡动的可用钻头剖面（图3.19和图3.20）。

Diamond bits also wear by breakage or loss of the diamond-cutter elements. The wear rate of diamond bits is, thus, not sensitive to the fractional cutter wear. The wear rate of diamond bits is far more sensitive to the amount of cooling provided by the flow of drilling fluid across the face of the bit.

金刚石钻头也会因金刚石切削部件的破裂或丢失产生磨损。但金刚石钻头的磨损速率对切削齿磨损量并不敏感，对流过钻头表面的钻井液所能提供的冷却程度更为敏感。

 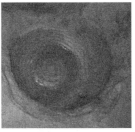 

Fig.3.18 Stable (left) and unstable (right) PDC at high RPM (instability caused continuous fracturing)

Fig.3.19 A ring PDC bit, which is very effective in avoiding whirl and in providing high control on steerability

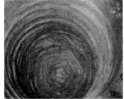 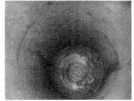

Fig. 3.20 Bit whirl (left) and full-contact gauge ring (right)

Glowka and Stone discussed the wear mechanisms for PDC bits and the dependence of wear on cutter temperature. Above 1,382°F (750℃), wear was shown to accelerate because of thermal deterioration on diamond grain pullout, resulting in catastrophic cutter failure. At temperatures below 1,382°F, the primary mode of wear was described as microchipping abrasive wear. Glowka and Stone have also shown that wear rate increased dramatically above 622°F (350℃). Because of the accelerated wear above 662°F, it is defined as the critical cutter temperature. The following equation is derived for wear-flat temperature (i.e., the temperature at the flat surface of the cutter having direct contact with the formation):

格洛卡和斯通讨论了PDC钻头的磨损机理及磨损与切削齿温度之间的关系。在高于1382°F（750℃）的情况下，由于金刚石晶粒拉拔时的热劣化，磨损加剧，导致切削齿严重失效。在低于1382°F时，磨损的主要方式为微屑磨粒磨损。格洛卡和斯通还指出，在622°F（350℃）以上时磨损速率急剧增加。由于高于662°F时磨损加速，所以662°F被定义为切削齿的临界温度。磨损速平面温度（即与地层直接接触的切削平面上的温度）可由下面方程推导出来：

$$T_w = T_f + \frac{k_f F_n v_f}{A_w}\left[1 + \frac{3\sqrt{\pi}}{4}fk_{hf}\left(\frac{100v}{L_w \alpha_f}\right)^{\frac{1}{2}}\right]^{-1} \tag{3.35}$$

where, T_w is the mean cutter wear-flat temperature, ℃; T_f is the fluid temperature, ℃; v_f is the cutting speed, m/s; α_f is the rock thermal diffusivity, cm²/s; A_w is the cutter wear-flat area, cm²; F_n is the normal force on cutter, N; f is the thermal response function, (cm² ℃)/W; k_{hf} is

式中，T_w是切削齿磨损平面平均温度，℃；T_f是流体温度，℃；v_f是切削速度，m/s；α_f是岩石的热扩散率，cm²/s；A_w是切削齿磨损面的面积，cm²；F_n是切削齿

the rock thermal conductivity, W/(cm ℃); k_f is the friction coefficient between rock and cutter; and L_w is the wear-flat length, cm. Thermal response function f is the effective thermal resistance of the cutter and is a function of cutter configuration, thermal properties, and cooling rates.

3. Wear Equation. Roller-Cone Bits

A composite tooth-wear equation can be obtained by combining the relations approximating the effect of tooth geometry, bit weight, and rotary speed on the rate of tooth wear (Bourgoyne and Young). Thus, the instantaneous rate of tooth wear is given by

$$\frac{dh}{dt} = \frac{1}{\tau_H}\left(\frac{N}{60}\right)^{H_1}\left[\frac{\frac{W}{d_b}-4}{\left(\frac{W}{d_b}\right)_{max}-\frac{W}{d_b}}\right]\left(\frac{1+\frac{H_2}{2}}{1+H_2 h}\right) \qquad (3.36)$$

where, h is the fractional tooth height that has been worn away; t is the time, hours; H_1, H_2, and $(W/d_b)_{max}$ are constants; W is the bit weight, 1,000lbf units; N is the rotary speed, r/min; and τ_H is the formation-abrasiveness constant, hours.

The tooth-wear rate formula given by Eq. 3.36 has been normalized so that the abrasiveness constant τH is numerically equal to the time in hours required to completely dull the bit teeth of the given bit type when operated at a constant bit weight of 4,000lbf/in and a constant rotary speed of 60r/min. The average formation abrasiveness encountered during a bit run can be evaluated using Eq. 3.36, and the final tooth wear h_f can be observed after pulling the bit. If we define a tooth-wear parameter J_2 using

3. 牙轮钻头磨损方程式

通过综合考虑牙齿几何形状、钻压、转速对牙齿磨损速率的影响，就可以得到一个综合的牙齿磨损方程式（布尔戈因和杨）。这时，瞬时牙齿磨损速率为：

式中，h 是已被磨掉的牙齿磨损比；t 是时间，h；H_1、H_2、$(W/d_b)_{max}$ 是常数；W 是钻压，1000lbf；N 是转速，r/min；τ_H 是地层研磨性常数，单位是 h。

由式（3.36）给出的牙齿磨损速率方程已经归一化，这样研磨性常数 τ_H 数值上等于在 4000lbf/in 的恒定钻压和 60r/min 的恒定转速条件下，给定钻头类型的牙齿完全磨钝所需的时间。钻头钻进过程中的平均地层研磨性可用式（3.36）进行估计。起出钻头后可观察到最终牙齿的磨损 h_f。如果使用以下方法定义牙齿磨损参数 J_2：

$$J_2 = \left[\frac{\left(\frac{W}{d_b}\right)_{max}-\frac{W}{d_b}}{\left(\frac{W}{d_b}\right)_{max}-4}\right]\left(\frac{60}{N}\right)^{H_1}\left(\frac{1}{1+\frac{H_2}{2}}\right) \qquad (3.37)$$

Then Eq. 3.36 can be expressed by

则式（3.36）可以表示为：

$$\int_0^{t_b} dt = \tau_H J_2 \int_0^{h_f} (1 + H_2 h) dh \qquad (3.38)$$

Integration of this equation yields

对上式积分后得：

$$t_b = \tau_H J_2 \left(h_f + H_2 \frac{h_f^2}{2} \right) \qquad (3.39)$$

Solving for the abrasiveness constant τ_H gives

求解研磨性常数 τ_H 得到：

$$\tau_H = \frac{t_b}{J_2 \left(h_f + H_2 \dfrac{h_f^2}{2} \right)} \qquad (3.40)$$

Although Eqs. 3.36 through 3.40 were developed for use in modeling the loss of tooth height of a milled-tooth bit, they also have been applied with some degree of success to describe the loss of insert teeth by breakage. Insert bits are generally operated at lower rotary speeds than milled-tooth bits to reduce impact loading on the brittle tungsten carbide inserts. In hard formations, high rotary speeds may quickly shatter the inserts (Estes). Recommended values of H_1、H_2, and $\left(\dfrac{W}{d_b} \right)_{max}$ are shown in Table 3.2 for the various voller-come-bit classes.

虽然式（3.36）至式（3.40）用于建立铣齿钻头齿高损失模型，但它们也已成功地应用于描述由于脱落和折断失掉的镶齿。作业时，为了减少对脆性碳化钨镶齿的冲击载荷，镶齿钻头的转速通常比铣齿钻头更低。在硬地层中，高转速可能会使镶齿快速粉碎（埃斯蒂斯）。针对各种牙轮钻头类型推荐的 H_1、H_2、$\left(\dfrac{W}{d_b} \right)_{max}$ 见表3.2。

Table 3.2 Recommended tooth-wear parameters for roller-cone bits

Bit Class	H_1	H_2	$\left(\dfrac{W}{d_b} \right)_{max}$
1-1 to 1-2	1.9	7	7
1-3 to 1-4	1.84	6	8
1-3 to 1-4	1.8	5	8.5
2-3	1.76	4	9
3-1	1.7	3	10
3-2	1.65	2	10
3-3	1.6	2	10
4-1	1.5	2	10

3.1.5.2 Bearing Wear (Roller Cones Only)

Audio 3.4

The prediction of bearing wear is much more difficult than the prediction of tooth wear. Like tooth wear, the instantaneous rate of bearing wear depends on the current condition of the bit. After the bearing surfaces become damaged, the rate of bearing wear increases greatly. However, because the bearing surfaces cannot be examined readily during the dull-bit evaluation, a linear rate of bearing wear usually is assumed. Also, bearing manufacturers have found that for a given applied force, the bearing life can be expressed in terms of total revolutions as long as the rotary speed is low enough to prevent excessive heat. Thus, bit bearing life usually is assumed to vary linearly with rotary speed.

1. Factors Affecting Bearing Wear

The effect of bit weight and RPM on bearing life depends on the number and type of bearings used and whether the bearings are sealed. When the bearings are not sealed, bearing lubricationis accomplished with the drilling fluid, in which case the mud properties also affect bearing life.

The hydraulic action of the drilling fluid at the bit is also thought to have some effect on bearing life. As flow rate increases, the ability of the fluid to cool the bearings also increases. However, it is generally believed that flow rates sufficient to lift cuttings will also be sufficient to prevent excessive heat in the bearings. Lummus has indicated that a jet velocity that is too high can be detrimental to bearing life. Erosion of bit metal can occur, which leads to failure of the bearing grease seals. In the example discussed by Lummus, this phenomenon was important for bit hydraulic-horsepower values greater than 4.5 hp per square inch of hole bottom. However, a general model for predicting the effect of hydraulics on bearing wear was not presented.

2. Bearing-Wear Equation

One bearing-wear formula used to estimate bearing life is given by Bourgoyne et al. :

3.1.5.2 轴承磨损（仅对于牙轮钻头）

轴承磨损的预测要比牙齿磨损的预测困难得多。像牙齿磨损一样，轴承的瞬时磨损速率取决于钻头的工作状况。在轴承的表面有损伤后，轴承磨损速率急剧增加。然而，由于在钻头磨钝评估过程中不能轻易检查到轴承表面，因此通常假设轴承磨损速率呈线性。此外，轴承制造商已经发现，对于一个给定的力，只要其转速足够低，防止过热，轴承的寿命可以用总转数来表示。因此，钻头轴承的寿命通常假定为随转速线性变化。

1. 影响轴承磨损的因素

钻压和转速对轴承寿命的影响取决于所用轴承的数量、类型及轴承是否密封。当轴承非密封时，轴承润滑通过钻井液实现，这种情况下，钻井液的性质也会影响轴承的寿命。

钻井液在钻头上的水力作用也会对轴承寿命产生一定影响。随着排量增加，钻井液冷却轴承的能力也随之增加。然而，通常认为满足携岩的排量也将满足防止轴承过热。鲁姆斯指出，喷射速度过高会损害轴承的使用寿命，可能冲蚀钻头金属材质从而导致轴承润滑脂密封失效。在鲁姆斯讨论的示例中，这种现象对于钻头水功率在井底超过 4.5hp/in^2 时是很重要的。但是，他没有给出预测水力参数对轴承磨损影响的通用模型。

2. 轴承磨损公式

布尔戈因等人给出了一种用于估算轴承寿命的轴承磨损公式：

$$\frac{db}{dt} = \frac{1}{\tau_B}\left(\frac{N}{60}\right)^{B_1}\left(\frac{W}{4d_b}\right)^{B_2} \tag{3.41}$$

where, b is the fractional bearing life that has been consumed; t is the time, hours; N is the rotary speed, r/min; W is the bit weight, 1,000 lbf; d_b is the bit diameter, in; B_1 and B_2 are the bearing-wear exponents; and τ_B is the bearing constant, hours. Recommended values for the bearing-wear exponents are given in Table 3.3

式中，b 是已消耗的轴承寿命；t 是时间，h；N 是转速，r/min；W 是钻压，1000lbf；d_b 是钻头直径，in；B_1、B_2 是轴承磨损指数；τ_B 是轴承常数，h。表 3.3 给出了轴承磨损指数的推荐值。

Table 3.3 Recommended bearing-wear exponent for roller-cone bit

Bearing Type	Drilling-FluidType	B_1	B_2
Nonsealed	Barite mud	1	1
	Sulfide Mud	1	1
	Water	1	1.2
	Clay/water Mud	1	1.5
	Oil-based Mud	1	2
Sealed roller bearings	—	0.7	0.85
Sealed journal bearings	—	1.6	1

Note that the bearing-wear formula given by Eq. 3.41 is normalized so that the bearing constant τ_B is numerically equal to the life of the bearings if the bit is operated at 4,000 lbf/in. and 60 r/min. The bearing constant can be evaluated using Eq. 3.41 and the results of a dull-bit evaluation. If we define a bearing-wear parameter J_3 using

注意，式（3.41）给出的轴承磨损公式是标准化的，轴承常数 τ_B 在数值上等于钻头在 4000lbf/in 钻压和 60r/min 转速条件下的轴承寿命。可用式（3.41）对轴承常数进行估计并得到钻头磨损评价结果。如果用以下方法定义轴承磨损参数 J_3：

$$J_3 = \left(\frac{60}{N}\right)^{B_1}\left(\frac{4d_b}{W}\right)^{B_2} \tag{3.42}$$

Eq.3.41 can be expressed by

则式（3.41）可以表示为：

$$\int_0^{t_b} dt = J_3 \tau_B \int_0^{b_f} db \tag{3.43}$$

where, b_f is the final bearing wear observed after pulling the bit. Integration of this equation yields

式中，b_f 是起出钻头后观察到的最终轴承磨损。对该方程积分得到：

$$t_b = \tau_B J_3 b_f \tag{3.44}$$

Solving for the bearing constant τ_B gives

$$\tau_B = \frac{t_b}{J_3 b_f} \qquad (3.45)$$

Journal-bearing insert-bit runs without excessive insert breakage or gauge wear typically fail because of seal (bearing) wear.

The bearing wear is proportional to the frictional work, which mainly depends on the travel distance and the contact pressure between two surfaces of cone and journal, which are related to rotary speed of the bit and WOB. Also, bearing wear is dependent on bit type, formation, BHA, and downhole conditions. In addition, the wear is related to bit diameter and to time. The equation for bearing dull grade, is assumed as follows (Hareland et al.):

$$b_f = K(d_b)^a (t)^b (W)^c (N)^d \qquad (3.46)$$

where, K, a, b, c, and d are constants that need to be determined using offset data.

求解轴承常数 τ_B：

(3.45)

有滑动轴承的镶齿钻头如果钻进时没有发生过渡镶齿破裂或保径磨损，则通常的失效原因是密封（轴承）磨损。

轴承磨损与摩擦做功成正比，而做功主要取决于锥体和轴颈两个表面之间的运行距离和接触压力，这与钻头转速和钻压有关。而且，轴承磨损取决于钻头类型、地层、底部钻具组合和井下条件。此外，磨损与钻头直径和时间有关。轴承磨损等级的公式假定如下（哈雷兰等）：

(3.46)

式中，K、a、b、c 和 d 是需要用邻井数据确定的常数。

3.2 The Drillstring

3.2.1 Introduction

3.2 钻柱

3.2.1 简介

Audio 3.5

The drillstring (drillstem) is the major component of a rotary drilling system. The typical drillstring consists of a kelly, a drillpipe with tool joints, drill collars, and stabilizers. The part of the drillstring above the bit is called the bottomhole assembly (BHA). Fig. 3.21 illustrates the usual arrangement of drillstring components. The bit is attached to the drill collars by means of a bit sub.

In conventional rotary drilling, the rotary motion produced by a rotary table is transmitted to the drillpipe by a square or hexagonal pipe called the kelly. For effective rock destruction, the lower part of the drill collars is slacked off onto the drill bit to provide the so-called weight on bit (WOB). Cuttings generated by the

钻柱是旋转钻井系统的主要组成部分。典型的钻柱由方钻杆、带接头的钻杆、钻铤和稳定器组成。钻头上方的钻柱称为井底钻具组合（BHA）。图 3.21 为钻柱常见的装配方式。钻头通过钻头短节安装到钻铤上。

在常规的旋转钻井中，由转盘产生旋转运动，并通过横截面为方形或六边形的方钻杆传递给钻杆。为了有效破岩，将钻铤的重量施加在钻头上形成钻压（WOB）。钻头产生的钻屑被钻井液携带出井底。钻井液从钻柱内

rock bit are removed from the bottom of the hole by the drilling fluid, which is circulated inside the drillstring and through the drill bit into the annular space between the drillstring and the borehole wall. Stabilizers are placed above the bit to control the direction in which the drill bit will penetrate the formation. Downhole motors with bent subs and rotary-steerable tools are also used for controlling the direction in which the bit drills.

向下循环，通过钻头进入钻柱与井壁之间的环空。在钻头上方安装稳定器，控制钻头钻进地层的方向。带有弯接头和旋转导向工具的井下动力钻具也可用于控制钻头的钻进方向。

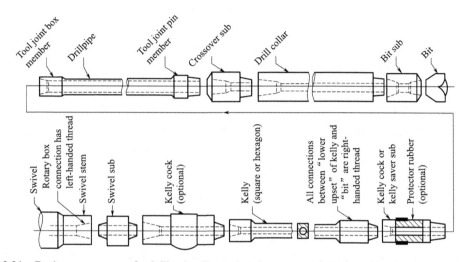

Fig. 3.21 Basic components of a drillstring Reproduced courtesy of the American Petroleum Institute

The major functions of a drillstring in conventional rotary drilling operations are

(1) To transmit rotary motion from the rotary table to a drill bit;

(2) To convey drilling fluid to the working face of the bit;

(3) To produce WOB for effective drilling action.

(4) To provide control of borehole direction.

In addition to the elements already mentioned, the drillstring may include shock absorbers, junk baskets, drilling jars, reamers, and other equipment. The drillpipe itself may serve for drillstem testing, completion, well stimulation (e.g., fracturing, acidizing), and fishing operations.(retrieving the string if some of the elements fail and a portion of the BHA is lost in the hole).

New developments involve using the drillstring

常规旋转钻井作业中，钻柱的主要功能是：

（1）将旋转运动从转盘面传递到钻头；

（2）将钻井液输送到钻头的工作面；

（3）为有效钻进作业提供钻压；

（4）控制井眼方向。

除上述部件外，钻柱还可包括减振器、打捞篮、震击器、扩眼器和其他工具设备。钻杆本身可用于钻杆测试、完井、增产（如压裂、酸化）和落鱼作业等。（落鱼作业是指当底部钻具组合的某些部件失效并掉落井中时，打捞收回落物的作业）

钻柱还可作为传递井下信息

as a vehicle for sending downhole information to the top of the hole. High-speed-telemetry drillpipe can provide high-quality downhole dynamic data along with logging information (gamma scans, density images, etc.) that can be effectively used for real-time drilling optimization.

3.2.2 Drill Collars

The drill collars are the major part of the BHA. The BHA, if properly designed, serves several purposes, including loading of the drill bit with the WOB, as previously mentioned. The mechanical and geometric properties of the BHA affect bit performance (i.e., drilling rate and bit wear), hole problems (doglegs, key seats, differential pressure sticking), drillstring vibrations, and drillpipe service life.

Drill collars are manufactured in various sizes and shapes. Conventional drill collars have a round cross section; however, square and spiral drill collars are also used in the drilling industry. Drill collars with a square cross-sectional outside profile are used to increase the stiffness of a BHA, whereas the spiral type is recommended for drilling in areas where differential pressure sticking is a problem. The spiral grooves on the outside surface of these drill collars reduce the contact area between the borehole wall and the drill collars, which, in turn, decreases the sticking force.

The string of drill collars is formed by connecting individual collars (usually with a length of approximately 30 ft) by means of rotary-shouldered connections. Selection charts for drill-collar connections are available from several manufacturers. The connections must be made up with an appropriate amount of makeup torque so that they will not separate under downhole conditions. The recommended makeup torques for different sizes of drill collars are given by API RP 7G (1998). Basic physical properties of new drill collars are given in Table 3.4.

的工具。高速电子钻杆可向地面传递高质量的井下动态数据及测井信息（伽马扫描、密度成像等），这些信息可有效地用于实时钻井优化。

3.2.2 钻铤

钻铤是底部钻具组合的主要部分。合理设计的底部钻具组合具有多种用途，包括给钻头施加钻压。底部钻具组合的力学和几何性质会影响钻头性能（即钻速和钻头磨损）、井眼问题（狗腿、键槽、压差卡钻）、钻柱振动和钻杆使用寿命。

钻铤可制备成各种尺寸和形状。常规钻铤有一个圆形横截面，方形和螺旋形钻铤也用于钻井行业。横截面外轮廓为方形的钻铤可用于增加底部钻具组合的刚度，螺旋形钻铤则推荐用于存在压差卡钻的井段。钻铤外表面上的螺旋槽减小了井壁与钻铤之间的接触面积，进而减小了黏附力。

通过旋转台肩把各个钻铤（通常长约30ft）连接成钻铤串。多家制造商提供了钻铤连接选择表。连接时必须使用适当的上扣扭矩以保证它们在井底条件下不会断开。API RP 7G（1998）给出了不同尺寸钻铤的推荐上扣扭矩。表3.4给出了新钻铤的基本力学性能。

Table 3.4 Properties of new drill collars

Drill-Collar OD Range, in	Minimum Yield Strength, psi	Minimum Tensile Strength, psi
$3\frac{1}{6} \sim 6\frac{7}{8}$	110,000	145,000
$7 \sim 10$	100,000	135,000

3.2.3 Selection of Drill-Collar Size and Length

Many factors affect selection of the drill-collar shape and unit weight. The most important factors are:

(1) Bit size;
(2) OD of the casing that is expected to be run in the hole;
(3) Formation dip angle and heterogeneity;
(4) Hydraulic program (i.e., drilling-fluid type, properties, flow rate, and nozzle size);
(5) Maximum acceptable dogleg (hole curvature);
(6) Required WOB;
(7) Possibility of fishing operations.

If a near-bit stabilizer is not used, to prevent rapid changes in hole deviation (which may make running a casing string difficult or even impossible), the required OD of the drill collars can be calculated as follows:

$$D_{odc} = 2(D_{occ}) - D_b$$

where, D_{odc} = outside diameter of drill collars; D_{occ} = outside diameter of casing coupling; and D_b = bit diameter.

The required length of drill collars depends mostly on the desired WOB, the unit weight of the drill collars, and the drilling-fluid density. Assuming that only drill collars will be used to create bit loading, the required length is given by the formula

$$L_{dc} = \frac{DFW}{w_{dc} K_b \cos\phi} \qquad (3.48)$$

$$K_b = 1 - \frac{\gamma_m}{\gamma_{st}} \qquad (3.49)$$

3.2.3 钻铤尺寸和长度的选择

影响选择钻铤形状和单位重量的因素有很多。最重要的因素有：

（1）钻头尺寸；
（2）预期下入井眼的套管外径；
（3）地层倾角和各向异性；
（4）水力参数（如钻井液类型、性能、排量和喷嘴尺寸）；
（5）可接受的最大狗腿度（井眼曲率）；
（6）所需的钻压；
（7）落鱼作业的可能性。

如果不使用近钻头稳定器，为了防止井斜变化过快（这会导致下套管困难，甚至无法下入），所需钻铤的外径可以计算如下：

(3.47)

式中，D_{odc} 为钻铤的外径；D_{occ} 为套管接箍外径；D_b 为钻头直径。

所需钻铤长度主要取决于预施加的钻压、钻铤的单位重量和钻井液密度。假设只使用钻铤对钻头施加钻压，则所需钻铤长度由如下公式得到：

where, DF = design factor ($DF \cong 1.1$ to 1.2); W = WOB, lbf; w_{dc} = unit weight of drill collars; K_b = buoyancy factor; γ_m = drilling-fluid specific weight; γ_{st} = drill-collar-material density; and ϕ = hole inclination angle from the vertical.

A design factor of approximately 1.15 to 1.20 is recommended in nearly vertical holes to ensure that the part of the drillstring above the drill collar is under effective tension. The concept of effective tension is explained later in this chapter in the section dealing with axial stress in drillpipe. Maintaining the drillpipe under tension not only prevents it from buckling, but also helps prevent lateral movement of the pipe because of the centrifugal forces that are generated while the pipe is being rotated. Generally, in nearly vertical holes, a higher value of the design factor is recommended for higher rotary speeds of the drillstring and higher flow rates. As the hole inclination angle increases, gravity keeps the drill collars on the lower side of the borehole, which results in drag forces.

The pressure-area method suggested by some authors for calculating the required length of drill collars is not correct and therefore should not be used because it does not consider the tri-axial nature of the stresses that are actually observed when the pipe is immersed in the drilling fluid. It can be shown that hydrostatic forces cannot cause buckling as long as the density of the drilling fluid is less than that of the drill-collar material. In other words, under these conditions, the drill collars will not buckle because of hydrostatic forces, no matter how deep the hole is.

3.2.4 Drill-Collar Buckling

3.2.4.1 Buckling in a Vertical Hole

To understand the phenomenon of drill-collar buckling, for the sake of simplicity, let us first consider a steel vertical rod with a length L=5ft (60 in) and a cross-sectional area of 0.5 in². The lower end is resting on a flat surface, and the upper end is loaded with an axial force. The axial force results in an axial compressive stress equal to the magnitude of the force divided by the cross-sectional area of the rod. By gradually increasing the axial force, it can be observed that at a certain force, the rod buckles. If the hinged types of end conditions (no bending moment) are assumed and the rod weight is neglected, the magnitude of the critical force that causes the rod to buckle can be calculated from the well-known Euler equation as

$$F_{cr} = \frac{\pi^2 EI}{L_r^2}$$

(3.50)

where, EI is a bending stiffness that is a product of modulus of elasticity (E) and the moment of inertia (I), and L is the length of the rod. If the rod is made of steel ($E = 30 \times 10^6$ psi), the critical buckling force from Eq. 3.50 is 1,635 lbf for the case under considerations. Note that in Eq. 3.50, the rod weight is not considered (that is, the rod is as-sumed to be weightless). In other words, the practical usefulness of Eq. 3.50 is limited to those cases in which the buckling force is much greater than the weight of the compressed elements.

A similar phenomenon may occur if the WOB is increased above a certain value called the critical WOB. Because the lateral movement of the drill collar is restricted by the borehole wall, the drill collar will contact the wall, as shown in Fig. 3.22.

The force W is called the WOB, which is the vertical component of the force that the formation exerts on the bit. W is also equal to the weight of the portion of the drill collars in drilling fluid that is below the neutral point, reduced by the frictional force, F_f. In other words, the WOB force, W, plus the frictional force, F_f, is equal to the weight in mud of the portion of the drill collars that is below the neutral point. The force H_0 is called the side force at the bit. The vertical force equilibrium requires that the F_h (hook load) be equal to the total weight of the drill-string in drilling fluid minus the weight of the drill collars that are used to generate bit loading reduced by the friction force (drag force). In other words, the neutral point divides the drillstring into two parts. The upper part, with

的大小除以杆的横截面积。通过逐渐增加轴向力可以观察到，杆会在一定轴向力下发生弯曲。如果假设端部为铰链类型（无弯矩）且忽略杆的重量，则根据欧拉方程，计算导致杆弯曲的临界力大小为：

式中，EI 是弯曲刚度，它是弹性模量（E）和惯性矩（I）的乘积，L 是杆的长度。如果杆是由钢制成的（$E=30\times 10^6$psi），对于所考虑的情况，由式（3.50）可得临界屈曲力为 1635lbf。注意，式（3.50）不考虑杆的重量（即假定杆没有重量）。换句话说，实际上式（3.50）仅适用于屈曲力远大于受压部件重量的情况。

如果钻压增加至临界钻压以上，则可能会发生类似的现象。由于钻铤的横向运动受到井壁的限制，因此钻铤将接触到井壁，如图 3.22 所示。

力 W 称为钻压，是钻头施加给地层压力的垂直分量。钻压也等于中性点以下部分钻铤的浮重减去摩擦力 F_f。换句话说，钻压加上摩擦力等于中性点以下部分钻铤在钻井液中的浮重。力 H_0 称为钻头侧向力。钻柱垂直方向上力的平衡关系要求 F_h（大钩载荷）等于钻柱的总浮重减去用于产生钻压的钻铤重量，再减去摩擦力。换言之，中性点将钻柱分为两部分。上部长度为 X_1，处于有效拉伸状态；下部长度为 X_2，处于有效压缩状态。由于钻铤屈曲

length X_1, is in effective tension, and the lower part, with length X_2, is in effective compression. Because the buckling of drill collars is attributable to the amount of drill-collar weight that is slacked off onto the drill bit, Eq.3.50 cannot be used to calculate the critical buckling force.

是由于部分钻铤重量施加在钻头上的压力所致，因此没办法用式（3.50）来计算临界屈曲力。

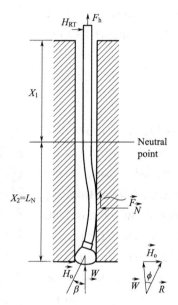

Fig.3.22 Buckling of collars in a vertical hole

A theoretical analysis performed by Lubinski (1951, 1987) revealed that for a friction-less system (that is, friction force is neglected), the critical values of WOB W_{cr} that cause first- and second-order buckling(meaning that the first or the second buckle contacts the borehole wall) can be calculated from the following expressions:

鲁宾斯基进行的理论分析表明，对于无摩擦系统（即忽略摩擦力），临界钻压 W_{cr} 引起一次和二次弯曲（意味着钻柱一次或二次弯曲接触井壁）可以通过以下表达式计算：

$$W_{cr,I} = 1.94 w_{bp} m \tag{3.51}$$

$$W_{cr,II} = 3.75 w_{bp} m \tag{3.52}$$

$$m = \sqrt[3]{\frac{EI}{w_{bp}}} \tag{3.53}$$

where, $W_{cr,I}$=cirtical value of WOB for first-order buckling, $W_{cr,II}$=cirtical value of WOB for second-order buckling, m is a scaling factor or "dimensionless unit", which relates actual to dimensionless length, EI=bending stiffness of the drill collars, and w_{bp}=unit weight of the drill collars in drilling fluid.

式中，$W_{cr,I}$ 为一次屈曲的临界钻压值；$W_{cr,II}$ 为二次屈曲的临界钻压值；m 为比例系数，与实际无量纲长度相关；EI 为钻铤的弯曲刚度；w_{bp} 为钻铤的单位浮重。

Once the drill collars have buckled, the string is no

一旦钻铤发生弯曲，钻柱底

longer vertical at its lower end, and the bit starts to drill an inclined hole.

The direction (inclination) of the force on the bit and the tilt angle are given by:

$$\phi = n\frac{r_c}{m} \tag{3.54}$$

$$\beta = \lambda\frac{r_c}{m} \tag{3.55}$$

$$x_2 = \frac{X_2}{m} \tag{3.56}$$

$$\phi = \arctan(H_0/W) \tag{3.57}$$

where, ϕ = inclination of the resultant bit force; β=tilt angle (the angle between the tangent to the centerline of the drill collars at the bit and the vertical at the bit) (radians); r_c=radial clearance(apparent radius of the hole), r_c=0.5(D_b-OD_{dc}); n、λ=coefficients that depend on the dimensionless distance, x_2, between the bit and the neutral point, as show in Fig.3.23.

Assuming that the drill-bit face-and side-cutting abilities are identical and the formation being drilled is isotropic, the expected instantaneous bit-displacement direction is the same as the direction of the resultant force on the bit. If, however, the side-cutting ability of the drill bit is assumed negligible, the bit will penetrate the formation in the direction in which it is pointed out. In other words, the instantaneous hole angle will be equal to the tilt angle. In general, the direction of bit penetration will be neither that of the tilt angle nor that of the resultant force angle, because formation drillability (formation resistance to drilling) is different in different directions.

The magnitude of the force applied by the buckled drill collars on the hole wall is given by

$$N = fw_{bp}r_c \tag{3.58}$$

$$x_2 = \frac{L_N}{m} \tag{3.59}$$

$$L_N = \frac{W-F}{w_{bp}} \tag{3.60}$$

部就不再垂直，钻头开始钻进一个倾斜的井眼。

钻头受力的方向（井斜）和倾斜的角度由下式给出：

式中，ϕ 为由钻头所受合力引起的井斜；β 为倾斜角（在钻头处，钻铤中心线与垂直方向的夹角）（弧度）；r_c 为径向间隙（井眼视半径），r_c=0.5（D_b-OD_{dc}）；n、λ 为系数，取决于钻头到中性点的无量纲距离 x_2，如图 3.23 示。

假设钻头正面和侧面的切削能力相同且所钻地层为各向同性，则钻头的瞬时位移方向应该与作用在钻头上的合力方向相同。如果假设忽略钻头的侧向切削能力，则钻头将沿指向的方向钻穿地层。换句话说，瞬时井斜角将等于井眼倾角。然而通常的情况是，钻头的钻进方向既不是井眼倾角的方向也不是合力角的方向，因为地层的可钻性（地层抗钻进的能力）在不同的方向上是不同的。

弯曲的钻铤在井壁上施加的力，大小为：

where, f = a coeefficient that depends on the distance from the bit to the neutral point, expressed in dimensionless units, x_2 (Fig. 3.24): $L_N = X_2$ = distance from the bit to the neutral point, ft: $F = \mu N$ = friction force, lbf; and μ = coefficient of friction between the drill collar and the formation at the point of contact.

式中，f 为系数，取决于钻头到中性点的距离，以无量纲单位 x_2 表示（图 3.24）；$L_N = X_2$ = 钻头到中性点的距离，ft：$F(=\mu N)$ 为摩擦力，lbf；μ 为钻铤和地层接触点之间的摩擦系数。

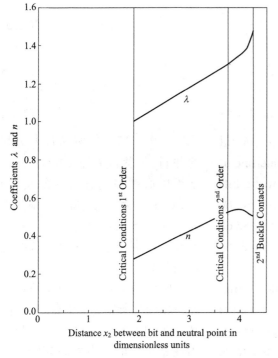

Fig.3.23　Coefficients λ and n as functions of dimensionless distance from the drill bit to the neutral point, x_2

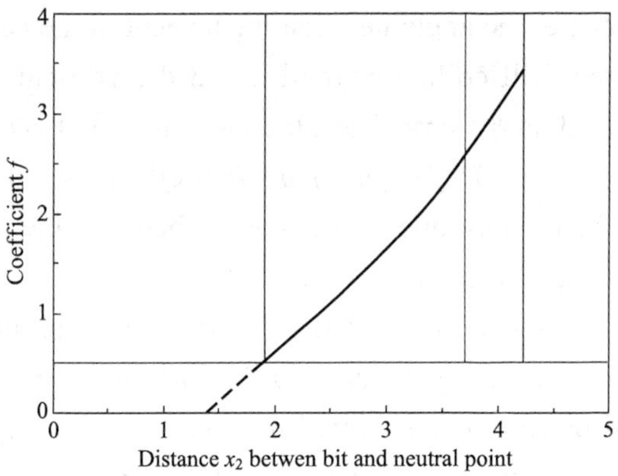

Fig.3.24　Coefficient f for the force that collars exert on the borehole wall upon buckling

Analysis of Eqs. 3.58 and 3.60 indicates that if the coefficient of friction (μ) is known or can be estimated, then the force N can be calculated iteratively. If the drillstring is rotating, the friction force is absorbed by the rotary torque, and the axial fiction can be ignored in Eq. 3.60.

It is evident that buckling generates a bending moment, which in turn produces a bending stress, which is a tension on one side and a compression on the other. This bending stress affects the compound stress in the inner and outer fibers of the drill collars.

The following expression can be used to calculate the bending moment:

式（3.58）和式（3.60）表明，如果已知或可以估计摩擦系数（μ），则可迭代计算力 N。如果钻柱旋转，则摩擦力会被旋转扭矩吸收，则在式（3.60）中可忽略轴向摩擦。

有证据表明，弯曲产生弯矩，并随之产生弯曲应力，使一侧拉伸而另一侧压缩。该弯曲应力影响钻铤内部和外部材质中的复合应力。

以下表达式可用于计算弯矩：

$$M_b = i w_{bp} m r_c \tag{3.61}$$

where, i = the bending-moment coefficient, which is a function of x_2 (Fig.3.25). There are two points in drill collars where the bending moment reaches a maximum.

In Fig. 3.25, the coefficient i_1 corresponds to the point of maximum bending moment that is nearest to the bit, whereas i_2 corresponds to the point above. The dashed lines M_1M_3 and $M_1'M_3'$ represent the dimensionless distance (the ordinate on the right side of Fig. 3.25) to the two points at which the bending moment is a maximum.

式中，i 为弯矩系数，它是 x_2 的函数（图 3.25）。钻铤中有两个点的弯矩达到最大。

在图 3.25 中，系数 i_1 对应于最接近钻头的最大弯矩点，而 i_2 对应于上方的点。虚线 M_1M_3 和 $M_1'M_3'$ 表示到弯矩最大的两个点的无量纲距离（图 3.25 右侧的纵坐标）。

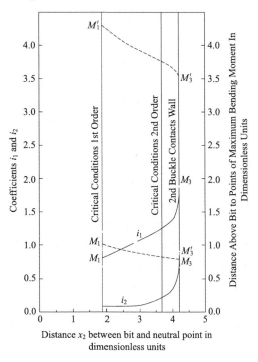

Fig.3.25 Bending coefficients i_1 and i_2, vs. x_2

3.2.4.2 Buckling in an Inclined Hole

If the hole is straight, but not vertical, the critical WOB that induces buckling can be calculated as follows:

3.2.4.2 斜井中的屈曲

如果井眼是直的，但不是垂直的，则引起屈曲的临界钻压可以计算如下：

$$W_{cr} = 2mw_{bp}\sqrt{\frac{2m\sin\varphi}{r_c}} \tag{3.62}$$

Eq.3.62 is based on an experimental study performed by H.B. Woods of Hughes Tool Company. However, Dawson and Pasley's theoretical study of long and frictionless rods provided the following equation:

式 (3.62) 是伍兹基于休斯工具公司进行的实验研究。然而，道森和帕斯利对无摩擦长钻柱的理论研究提供了以下方程：

$$W_{cr} = 2mw_{bp}\sqrt{\frac{m\sin\varphi}{r_c}} \tag{3.63}$$

Clearly, the empirical equation (Eq.3.63) predicts a value of the buckling force greater than that predicted by Eq.3.63 by a factor of $\sqrt{2}$. This is possibly due to the stabilizing effect of the friction force. In other words, Eq. 3.63 provides a fairly conservative estimate of the buckling force in a straight inclined hole and is frequently used by industry at present. This buckling force is called a sinusoidal, snaky, or lateral buckling force. As the compressive force is increased above the lateral buckling force, the pipe eventually develops a 3D helical shape.

Buckling of the drill collars in an inclined hole will require a very high WOB, which is usually outside the range recommended by bit manufacturers for effective drilling. However, because the stiff, heavy drill collars lie on the lower side of the hole, high contact forces are created, which in turn cause increased axial drag, increased rotary torque, and the possibility of differential sticking. High wall friction can also result in whirling (rolling) of drill collars about the well-bore axis rather than rotation about their own axis. This causes very high rotary torque and additional bending loads that are detrimental to the mechanical integrity of the drill-string.

To reduce the problems associated with stiff, heavy drill collars in directional wells, a regular-or heavyweight drillpipe (i.e., a pipe with increased wall thickness) is used to create the desired WOB. If placed in the BHA above the drill collars, a heavyweight drillpipe provides a gradual change in stiffness between the rigid drill collars and the flexible drillpipe. Naturally, a drillpipe that is much more flexible than the drill collars will buckle under much less force than will the drill collars.

Experiments conducted under static conditions (no pipe rotation) in a horizontal well configuration have

显然，经验方程式（3.62）预测的屈曲力比式（3.63）预测的值大$\sqrt{2}$倍。这可能是由于摩擦力的稳定作用。换句话说，式（3.63）为稳斜井段提供了一个相当保守的屈曲力估计，并常用于目前的工业中。这种屈曲力被称为正弦弯曲力、蛇形屈曲力或横向屈曲力。当压缩力超过横向屈曲力时，管串最终发展为三维螺旋形状。

钻铤在斜井中弯曲需要非常高的钻压，这通常超出了钻头制造商建议的有效钻井范围。然而，由于刚度和重量，钻铤位于井眼的低边，形成了较高的接触力，从而导致轴向拉力和旋转扭矩增加，发生压差卡钻的可能性也增大。与井壁过高的摩擦力也会导致钻铤绕井眼轴线旋转（滚动），而不是绕其自身的轴线旋转。这将引起非常高的旋转扭矩和额外的弯曲载荷，它们将损害钻柱的力学完整性。

为了减少定向井中由钻铤的刚性、重量带来的问题，可使用常规或加重钻杆（即壁厚增加的钻杆）来形成所需的钻压。如果将加重钻杆安装在钻铤上方的底部钻具组合中，它会在刚性钻铤和柔性钻杆之间提供一种逐渐变化的刚度。因此，柔性比钻铤高得多的钻杆发生弯曲所需的力将比钻铤小得多。

在水平井的静态条件下（钻柱不旋转）进行的实验表明，受

shown that a pipe subjected to axial loading first buckles into a "snake" shape (as mentioned earlier, this is also called lateral or sinusoidal buckling), and as the force is increased, the pipe eventually assumes a helical shape. This is the so-called helical buckling phenomenon. For pipe with a known bending stiffness (EI), the relationship between the axial force F and the helix pitch p is given by Lubinski as

$$\wp^2 = \frac{8\pi EI}{F} \qquad (3.64)$$

Once the helix pitch has been obtained, the corresponding pipe curvature (k_p) can be calculated from Eq. 3.26 and the bending moment (M_b) from Eq. 3.27:

$$k_p = \frac{4\pi^2 r_c}{\wp^2 + 4\pi^2 r_c^2} \qquad (3.65)$$

$$M_b = EI k_p \qquad (3.66)$$

It can be shown that in a helical pipe configuration, the unit contact force between the pipe and the wellbore is given by Mitchell's equation:

$$w_c = \frac{r_c F^2}{4EI} \qquad (3.67)$$

Clearly, for a pipe with a known bending stiffness (EI), the radial clearance r_c and the axial force (F) are the major factors controlling the bending moment (bending stress) and the unit contact force.

3.2.5 Drillpipe and Tool Joints

The major portion of a drill-string is composed of drillpipe. Drillpipe in common use is made out of steel [steel drill-pipe (SDP)]. In some applications (e.g., drilling long extended-reach wells), it may be better to use aluminum drill-pipe (ADP) or perhaps titanium drillpipe (TDP). To evaluate the usefulness of ADP and TDP compared with SDP, one would need to consider the well-bore path configuration, the down-hole temperatures, the working environment

Audio 3.6

(presence of H_2S and CO_2), and the drag and torque issues. Hot-rolled, pierced seamless tubing, sometimes with the end threaded (eight threads per inch of thread length), is used for tool-joint attachments. Tool joints provide a means of fastening the individual lengths of pipe together. Currently, threaded connections between the drillpipe and tool joints have been almost completely replaced by butt welds. To reinforce the ends of the pipe, the pipe is upset at both ends. The pipe can be internal-upset (IU), external-upset (EU), or internal-and-external-upset (IEU). These three different designs are shown in Fig. 3.26.

端螺纹（每英寸螺纹长度8个螺纹），用于钻杆接头的连接。钻杆接头能够将各种长度的钻杆连接在一起。现在，钻杆和钻杆接头之间的螺纹连接几乎已被对接焊缝完全取代。为了加固钻杆的两端，对钻杆两端都加厚。钻杆可以是内加厚（IU）、外加厚（EU）或内外加厚（IEU）。这三种不同的设计如图3.26所示。

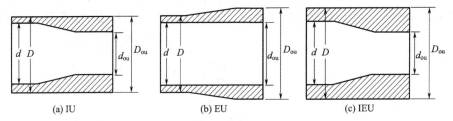

Fig.3.26 Drillpipe upsets

d—ID of drillpipe; D—OD of drillpipe; d_{ou}—ID of joint; D_{ou}—OD of joint

For identification purposes, drillpipe can be classified according to:

(1) Size (nominal OD);
(2) Wall thickness (or nominal unit weight);
(3) Steel grade;
(4) Length ranges.

The API standard drillpipe sizes include 2⅜in, 2⅞in, 3½in, 4in, 4½in, 5in, 5½in, and 6⅝in. The wall thicknesses and corresponding unit weights for the standardized drillpipe sizes are listed in Table 3.5.

钻杆可根据以下方式分类：

（1）尺寸（名义外径）；
（2）壁厚（名义单位质量）；
（3）钢级；
（4）长度范围。

API标准钻杆尺寸包括2⅜in、2⅞in、3½in、4in、4½in、5in、5½in和6⅝in。表3.5列出了标准钻杆尺寸、壁厚和相应的单位重量。

Table 3.5 New drillpipe dimensional data

OD in	Nominal Weight of Thread Couplings lbf/ft	Plain-End Weight lbf/ft	Wall Thickness in	ID in	Cross-Sectional Area, Body of Pipe in^2	Polar Sectional Modulus Z in^3
2⅜	4.85 6.65	4.43 6.26	0.190 0.280	1.995 1.815	1.3042 1.8429	1.320 1.734
2⅞	6.85 10.40	6.16 9.72	0.217 0.362	2.441 2.151	1.8120 2.8579	2.242 3.204
3½	9.50 13.30 15.50	8.81 12.31 14.63	0.254 0.368 0.449	2.992 2.764 2.602	2.5902 3.6209 4.3037	3.922 5.144 5.846

Continued

OD in	Nominal Weight of Thread Couplings lbf/ft	Plain-End Weight lbf/ft	Wall Thickness in	ID in	Cross-Sectional Area, Body of Pipe in^2	Polar Sectional Modulus Z in^3
4	11.85	10.46	0.262	3.476	3.0767	5.400
	14.00	12.93	0.330	3.340	3.8048	6.458
	15.70	14.69	0.380	3.240	4.3216	7.156
4½	13.75	12.24	0.271	3.958	3.6004	7.184
	16.60	14.98	0.337	3.826	4.4074	8.542
	20.00	18.69	0.430	3.640	5.4981	10.232
	22.82	21.36	0.500	3.500	6.2832	11.345
	24.66	23.20	0.550	3.400	6.8251	12.062
5	16.25	14.87	0.296	4.408	4.3743	9.718
	19.50	17.93	0.362	4.276	5.2746	11.416
	25.60	24.03	0.500	4.000	7.0686	14.490
5½	19.20	16.87	0.304	4.892	4.9624	12.222
	21.90	19.81	0.361	4.778	5.8282	14.062
	24.70	22.54	0.415	4.670	6.6296	15.688
6⅝	25.20	22.19	0.330	5.965	6.5262	19.572
	27.70	24.22	0.362	5.901	7.1227	21.156

[from API RP 7G (1998)] Reproduced courtesy of the American Petroleum Institute

The steel grades used and the corresponding minimum tensile yield strength for each are given in Table 3.6.

表3.6列出了每种钻杆的钢级和相应的最小抗拉屈服强度。

Table 3.6 Minimum tensile yield strength for new drillpipe

Steel Grade	D	E	X—95	G—105	S—135	V—150
Minimum Yield Strength, psi	55,000	75,000	95,000	105,000	135,000	150,000

Usually, drillpipe is available in three length ranges:

(1) Range 1: 16～25 ft;
(2) Range 2: 27～30 ft;
(3) Range 3: 38～45 ft.

In regular rotary-drilling operations, the drillpipe most commonly used is Range 2. The minimum mechanical-performance properties of various kinds of drillpipe are given in Table 3.7

通常，钻杆的长度有以下三种：

（1）长度范围1：16～25ft；
（2）长度范围2：27～30ft；
（3）长度范围3：38～45ft。

在常规的旋转钻井作业中，最常用的钻杆是长度范围2。表3.7给出了各种钻杆的最低力学性能。

Table 3.7 Minimum performance properties of new drillpipe

OD in	Nominal Weight lbf/ft	Grade	Wall Thickness in	ID in	Collapse Resistance psi	Internal Yield Pressure psi	Pipe-Body Yield Strength 1,000 lb
2⅜	6.65	E	0.280	1.815	15,600	15,470	138
	6.65	X	0.280	1.815	19,760	19,600	175
	6.65	G	0.280	1.815	21,840	21,660	194
	6.65	S	0.280	1.815	28,080	27,850	249

Contiued

OD in	Nominal Weight lbf/ft	Grade	Wall Thickness in	ID in	Collapse Resistance psi	Internal Yield Pressure psi	Pipe-Body Yield Strength 1,000 lb
2⅞	10.40	E	0.362	2.151	16,510	16,530	214
	10.40	X	0.362	2.151	20,910	20,930	272
	10.40	G	0.362	2.151	23,110	23,140	300
	10.40	S	0.362	2.151	29,720	29,750	386
3½	9.50	E	0.254	2.992	10,040	9,520	194
	13.30	E	0.368	2.764	14,110	13,800	276
	15.50	E	0.449	2.602	16,770	16,840	323
	13.30	X	0.368	2.764	17,880	17,480	344
	15.50	X	0.449	2.602	21,250	21,330	409
	13.30	G	0.368	2.764	19,760	19,320	380
	15.50	G	0.449	2.602	23,480	23,570	452
	13.30	S	0.368	2.764	25,400	24,840	480
	15.50	S	0.449	2.602	30,190	30,310	581
4	11.85	E	0.262	3.476	8,410	8,600	231
	14.00	E	0.330	3.340	11,350	10,830	285
	14.00	X	0.330	3.340	14,380	13,720	361
	14.00	G	0.330	3.340	15,900	15,160	400
	14.00	S	0.330	3.340	20,170	19,490	514

[from API RP 7G (1998)] Reproduced courtesy of the American Petroleum Institute

A tool joint is usually welded onto the drillpipe and consists of a pin and a box, as illustrated in Fig. 3.27. It is made from high-alloy steel with a wide thread (4 to 5 threads per inch) on the pin and the box.

Tool joints are classified by style as: (1)Extra-hole; (2)Wide-open; (3)Slimhole; (4)Full-hole; (5)Internal-flush.

通常将钻杆接头焊接在钻杆上，包括外螺纹和内螺纹，如图3.27所示。它由高级合金钢制成，外螺纹和内螺纹具有宽螺纹（每英寸4至5个螺纹）。

钻杆接头按样式分为以下几类：（1）孔外；（2）全开；（3）细孔；（4）贯眼型；（5）内平型。

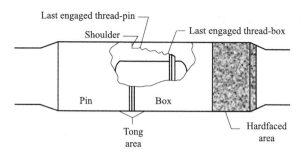

Fig.3.27　Tool joint: rotary-shouldered connection

Table 3.8 shows the corresponding API numbered connections as used in API RP 7G (1988). According to the API standard, the mechanical properties of tool joints shall not be lower than the minimum values as specified below:

(1) Minimum yield strength: 120,000 psi;
(2) Minimum tensile strength: 140,000 psi.

表 3.8 显示了 API RP 7G 标准中 API 数字型接头与旧接头的对应关系。根据 API 标准，钻杆接头的力学性能不得低于以下规定的最小值：

（1）最小屈服强度：120000psi；
（2）最小抗拉强度：140000psi。

Table 3.8　API numbered connections

API Numbered Connection	NC26	NC31	NC38	NC40	NC46	NC50
Internal flush	2⅜	2⅞	3½	4		4½
Full hole					4	
Extra hole					4½	5
Wide open			3½		4	4½
Slimhole	2⅞	3½	4½			
(after API RP 7G(1998))Reproduced courtesy of the American Petroleum Institute						

For applications that require high rotary torque, double-shoulder tool joints are recommended. A double-shoulder connection involves a primary and secondary shoulder. In the hand-tight position, only the primary shoulder makes contact. As more makeup torque is applied to the joint in the power-tight position, the box compresses, the pin elongates, and the secondary shoulder engages. The secondary shoulder provides the desired increase in torsional capacity compared with a conventional API rotary-shoulder connection. More technical information on double-shoulder tool joints is

对于高旋转扭矩的应用，建议使用双台肩钻杆接头。双台肩连接包括主台肩和副台肩。在手紧的位置，只有主台肩会接触。随着力紧位置的接头处上扣扭矩的增大，内螺纹缩紧，外螺纹延长并与副台肩啮合在一起。与传统的 API 旋转台肩式连接相比，副台肩连接提供了所需的更大扭转能力。可以从各制造商处获得有关双台肩钻杆接头的更多技术

available from various manufacturers on the Internet.

3.2.6 Forces Acting on the Drillstring

A drillstring operating in the borehole is subjected to a number of loads, including tension, compression, torsion, bending, and collapse or burst pressure. These forces can be either static or dynamic. The loads can repeat a number of times (cyclic loads) or can be applied over a relatively short period of time (impact loads). As a result, the stress state of a drillpipe is very complex and difficult to describe analytically. Here, for design purposes, a static stress state is assumed, and an appropriate design factor [safety factor (SF)] is used to arrive at a solution that is acceptable for field conditions.

3.2.6.1 Axial Tension/Compression Stresses

The largest tension load exists at the top of the drillstring because of the weight of the drill collars, stabilizers, drillpipe, and other string components, and because of forces attributable to fluid pressure acting on surfaces perpendicular to the drillstring axis. The bottom of the string (immersed in a fluid) is subjected to axial compressive force because of the hydrostatic pressure acting at the bottom of the pipe. The stress produced by an axial load, F, on the cross section, A, of a drillstring can be expressed as:

$$\sigma_a = \frac{F}{A}$$

(3.68)

where, σ_a = the average axial stress, or simply axial stress (tension or compression), psi; F = axial force, lbf; A = cross-sectional area $A = 0.785 (D_{op}^2 - D_{ip}^2)$, in²; D_{op} = pipe OD in; D_{ip} = pipe ID, in. The axial force, F, which is perpendicular to the cross-sectional area, A, can be determined by generating a free-body diagram. Sometimes the stress calculated from Eq. 3.68 is called the actual (true) axial stress because it can be measured by strain gauges.

[Example 3.1] Suppose that a drillstring is

composed of 9,500 ft of 4½in. drillpipe with a unit weight of 18.3 lbf/ft and a cross-sectional area of 4.4074 in², and of 600 ft of drill collars with D_{odc} = 6½ in, D_{idc} = 2¼ in (A_{dc} = 29.1922 in²), and a unit weight of 99 lbf/ft. The WOB is 28,000 lbf, and the drilling-fluid SG is 1.2 (that of water is 1.0). Calculate the axial stress on a cross section of drillpipe located at a depth of 9,000 ft (500 ft from the top of the drill collars).

Solution. A free-body diagram for this situation is shown in Fig. 3.28. Note that only axial forces are considered. System static equilibrium requires that

4½in 的钻杆（单位重量 18.3lbf/ft，横截面积 4.4074in²）和 600ft 的钻铤（D_{odc} = 6½in, D_{idc}=2¼in， 横截面积29.1922in²，单位质量 99lbf/ft）组成。钻压为 28000 lbf，钻井液相对密度为 1.2。计算深度位于 9000ft（距离钻铤顶部 500ft）处的钻杆横截面上所受到的轴向应力。

解：钻杆受力图如图 3.28 所示。注意只考虑轴向力，由系统静态平衡可知：

(a) $\qquad F = W_1 + W_2 + Fp_1 - Fp_2 - W$
(b) $\qquad W_1 = 500 \times 18.3 = 9{,}150 \text{(lbf)}$
(c) $\qquad W_2 = 600 \times 99 = 59{,}400 \text{(lbf)}$
(d) $\qquad Fp_1 = 0.052 \times 1.2 \times 8.34 \times 9{,}500 \times (29.1922 - 4.4074) = 122{,}534 \text{(lbf)}$
(e) $\qquad Fp_2 = 0.052 \times 1.2 \times 8.34 \times 10{,}100 \times 29.1922 = 153{,}440 \text{(lbf)}$

where, F = axial force on the cross section under consideration, lbf; W_1 = weight of drillpipe below the cross section under consideration, lbf; W_2 = weight of drill collars, lbf; Fp_1 = pressure force acting at the top of the drill collars, lbf; Fp_2 = pressure force acting at the bottom of the drill collars, lbf; and W = WOB, lbf.

式中，F 为待求横截面上的轴向力，lbf；W_1 为待求横截面以下钻杆的重量，lbf；W_2 为钻铤的重量，lbf；Fp_1 为作用于钻铤顶部的压力，lbf；Fp_2 为作用于钻铤底部的压力，lbf；W 为钻压，lbf。

Hence,

(f) $\qquad F = 9{,}150 + 59{,}400 + 122{,}534 - 153{,}440 - 28{,}000 = 9{,}644 \text{(lbf)}$

and the axial stress is a tensile stress ($F > 0$):

且轴向应力为拉力（$F>0$）：

(g) $\qquad \sigma_a = \dfrac{9644}{4.4074} = 2188 \text{(psi)}$

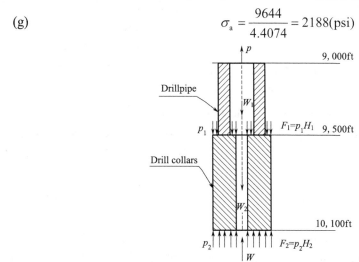

Fig.3.28 Free-body diagram for Example 3.1

To determine whether or not a string fails at a given cross section, one must evaluate the 3D stress state that actually exists at the cross section under consideration. Assuming that the drillpipe is a thin-walled pipe, the tangential and radial stresses are the same and equal to the negative of the fluid pressure at a depth of 9,000 ft:

(h) $\quad\sigma_t=\sigma_r=-0.052\times1.2\times8.34\times9000=-4683\text{(psi)}$

为了确定钻柱是否在给定的横截面上失效，必须评估待求横截面上实际存在的三维应力状态。假设钻杆是薄壁管，切向和径向应力相等，都等于9000ft深度处流体压力：

The drilling industry has adopted the maximum-energy-of-distortion theory to assess the strength of tubulars. The theory was originally proposed by Hubert and was further developed by von Mises. According to the maximum-energy-of-distortion theory, yielding begins when the distortion energy reaches the value of the distortion energy at the yield point in a simple tension test.

钻井行业采用最大畸变能理论来评价管柱强度。理论最初由休伯特提出，并由冯·米塞斯进一步发展。根据该理论，当变形能量值达到简单拉伸试验中的屈服点时开始屈服。

Assuming that there is no torsion, the following equation can be derived by use of the maximum energy-of-distortion theory, which is frequently referred to as the Huber-von Mises criterion for yielding:

假设不存在扭转，可以利用最大畸变能理论推导出以下方程，它通常被称为休伯特—冯·米塞斯屈服准则：

$$\sigma_Y = \sigma_{vm} = \frac{1}{\sqrt{2}}\left[(\sigma_z - \sigma_t)^2 + (\sigma_t - \sigma_r)^2 + (\sigma_r - \sigma_z)^2\right]^{1/2} \qquad (3.69)$$

where, σ_Y = the yield point (limit of elasticity) as determined in a simple tensile stress test; σ_{vm} = von Mises stress; and σ_z, σ_r, σ_t = axial, radial, and tangential stresses, respectively. The right side of Eq.3.69 is frequently called a von Mises stress. Eq.3.69 states that yielding occurs if the equivalent stress equals the yield point of the material. If the equivalent stress is less than the yield point, then, theoretically, yielding does not occur. In other words, deformation occurs in the elastic range and disappears when loading is removed. For the data of Example 3.1, the equivalent stress is

式中，σ_Y 为简单拉伸应力试验中确定的屈服点（弹性极限）；σ_{vm} 为冯·米塞斯应力；σ_z、σ_r、σ_t 分别是轴向、径向和切向应力。公式（3.69）的右边通常被称为冯·米塞斯应力。式（3.69）指出，当等效应力等于材料的屈服点时，就会发生屈服。如果等效应力小于屈服点的应力，则理论上不会发生屈服。换句话说，形变发生在弹性范围内，当载荷消除时形变消失。对于例3.1的数据，等效应力为：

$$\sigma_{vm} = \frac{1}{\sqrt{2}}\left[(2188-4683)^2 + (4683-4683)^2 + (4683-2188)^2\right]^{1/2} = 6871\text{(psi)}$$

In the case under consideration, if the pipe wall is assumed to be thin, the radial and tangential stresses are the same ($\sigma_t=\sigma_r=-p_h$) and equal to the hydrostatic

在所考虑的情况下，如果假设管壁较薄，则径向和切向应力相同（$\sigma_t=\sigma_r=-p_h$），且等于钻井液

pressure (p_h) of the drilling fluid. Hence, the equivalent stress is

$$\sigma_{vm} = \sigma_z + p_h \qquad (3.70)$$

In general, it should be noted that if there is no WOB, then the axial stress at the bottom of the string is equal to $-p_h$, and consequently the effective stress is nil. At the top of the hole, the hydrostatic pressure term p_h is zero, and the effective stress is equal to the axial stress.

It can also be shown that, for the case under consideration, the effective stress can be calculated by dividing the buoyant weight of the string below the cross section under consideration by the corresponding cross-sectional area:

$$\sigma_e = \frac{(L_{dp} w_{dp} + L_{dc} w_{dc}) K_b}{A} \qquad (3.71)$$

where, σ_e = effective axial stress (buoyant stress), psi; L_{dp} = length of drillpipe, ft; w_{dp} = unit weight of drillpipe, lbf/ft; L_{dc} = length of drill collars, ft; w_{dc} = unit weight of drill collars, lbf/ft; and K_b = buoyancy factor.

If WOB, W, is used, then the effective axial stress is

$$\sigma_e = \frac{(L_{dp} w_{dp} + L_{dc} w_{dc}) K_b - W}{A} \qquad (3.72)$$

Eq. 3.71 and Eq. 3.72 are valid for vertical wells. Appropriate adjustments of the weight components are required for holes that are not vertical.

The quantity calculated using Eq.3.71 or Eq.3.72 is called the effective tensile (compressive) stress. API RP 7G (1988) calls this quantity the buoyant tensile stress. Note that this stress does not actually exist because it cannot be measured.

Analysis of Eq.3.72 reveals that there exists a cross section in the drillstring at which the effective axial stress is nil. Above this cross section, the effective stress

的静水压力（p_h）。因此，等效应力为：

通常应该注意的是，如果没有钻压，钻柱底部的轴向应力等于$-p_h$，因此有效应力为零。在井口，静水压力p_h为零，有效应力等于轴向应力。

对于所考虑的情况，还可以通过将所求横截面以下的钻柱浮重除以相应的横截面积来计算有效应力：

式中，σ_e有效轴向应力（浮力应力），psi；L_{dp}为钻杆长度，ft；w_{dp}为钻杆单位重量，lbf/ft；L_{dc}为钻铤长度，ft；w_{dc}为钻铤单位重量，lbf/ft；K_b为浮力系数。

如果钻压W已知，则有效轴向应力为：

式（3.71）和式（3.72）适用于垂直井。对于井眼不垂直的情况，需要适当的调整组件重量。

使用式（3.71）或式（3.72）计算出的量称为有效拉伸（压缩）应力。API RP 7G（1988）将这个量称为浮力拉应力。请注意，实际上这个应力并不存在，因为它无法测量。

分析式（3.72）可知，在钻柱中存在一个有效轴向应力为零的横截面。在该截面上方，有效应

is positive, while below it, the effective axial stress is negative. This neutral cross section can be found from the following equation:

$$(L_{dp}w_{dp} + L_{dc}w_{dc})K_b - W = 0 \quad (3.73)$$

If only the drill collars are used to create WOB, then

$$L_{dc}w_{dc}K_b = W \quad (3.74)$$

$$L_{dc} = \frac{W}{w_{dc}K_b} \quad (3.75)$$

It is immediately apparent that Eq.3.75 may also be obtained by setting $DF = 1$ and $\cos \phi = 1$ in Eq.3.48.

If the length of drill collars is calculated from Eq.3.75, then the neutral point resides right at the top of the drill collars. In other words, the drill collars are under effective compression (or just compression), while the drillpipe is under effective tension (or just tension). For the sake of simplicity, in the following discussion, the term "tension/ compression" refers to effective tension/compression.

Knowledge of the actual axial stress along the drillstring is useful for calculating the length of the drillstring, while the effective stress indicates whether yielding (loading above yield stress) may occur.

In fact, the top cross section of the drillpipe is the only point in the drillstring where the value of the axial stress is equal to the effective axial stress. Once again, it must be remembered that while the axial stress is a stress that actually exists in the pipe (and can be measured), the effective axial stress is only a computational device. The magnitude of the effective (equivalent) stress determines whether or not plastic deformation (yielding) occurs.

3.2.6.2 Torsional Stresses

Torsion in a drillstring is caused by a twisting moment (*T*) called torque and results in a shear or torsional stress (*τ*) and an angle of twist (*φ*). The shear stress and the differential angle of twist can be calculated as

力为正，而在该截面下方，有效轴向应力为负。此中性横截面可以从以下公式中找到：

如果只用钻铤产生钻压，那么：

很明显，如果令式（3.48）中 *DF*=1、cos *φ* =1，也可以得到式（3.75）。

如果根据式（3.75）计算钻铤长度，则中性点刚好位于钻铤的顶部。换句话说，钻铤处于有效压缩（或只受压力），而钻杆处于有效拉伸（或只受拉力）。为了简单起见，在下面的讨论中，拉力（压力）指有效拉力（压力）。

了解沿钻柱的实际轴向应力对于计算钻柱的长度很有用，而有效应力表明了钻柱是否会发生屈曲（载荷超过屈服应力）。

事实上，钻杆顶部的横截面是钻柱轴向应力值等于有效轴向应力的唯一点。同样，必须记住，虽然轴向应力是一个实际存在于钻柱中的应力（并且可以被测量），但有效轴向应力只是一个计算值。有效（等效）应力的大小决定了是否发生塑性变形（屈服）。

3.2.6.2 扭转应力

钻柱的扭转是称为扭矩的扭转惯性量（*T*）引起的，并产生剪切应力或扭转应力（*τ*）及扭转角（*φ*）。剪切应力和扭转角可计算如下：

$$\tau = \frac{Tr}{J} \qquad (3.76)$$

$$\frac{d\phi}{dz} = \frac{T}{GJ} \qquad (3.77)$$

where, τ = shear stress, psi; $d\phi/dz$ = differential angle of twist, in^{-1}; T = torque, in.-lbf; r = distance from the center of the pipe to the point under consideration, in. $D_{ip} \leq 2r \leq D_{op}$; G = shear modulus of elasticity; J = polar moment of inertia;

$$G = \frac{E}{2(1+v)}$$

$$J = \frac{\pi}{32}\left(D_{op}^4 - D_{ip}^4\right)$$

The maximum shear stress occurs at the outer fiber of the pipe, and for this case, Eq. 3.76 can be written as

$$\tau_{max} = \frac{16 D_{op} T}{\pi\left(D_{op}^4 - D_{ip}^4\right)} = \frac{T}{Z} \qquad (3.78)$$

where, Z is the polar sectional modulus. For field engineering calculations, if the horsepower, to rotate the string is known, the corresponding torque is given by Eq. 3.79:

$$T = \frac{5250 HP}{RPM} \qquad (3.79)$$

where, HP = horsepower required to turn the rock bit and drillstring; RPM = drillstring rotary speed, r/min; and T = rotary torque, ft·lbf. The horsepower to rotate the drillpipe is

$$HP_p = C_d D_{op}^2 RPM L_{dp} SG \qquad (3.80)$$

where, C_d = an empirical factor that depends on hole inclination angle = $[(4.8(10^{-5})) - (66.5)(10^{-7})]$ for hole angles ranging from 3° to 5°; L_{dp} = length of drillpipe, ft; and D_{op} = OD of drillpipe, in.; SG = drilling fluid SG (water = 1.0). For hole inclination angles greater than approximately 5°, it is necessary to calculate the torque needed to overcome drag forces and then eventually the corresponding horsepower.

The horsepower to rotate a rotary-roller rock bit in a vertical hole can be estimated from Eq. 3.81:

$$HP_p = C_f W^{1.5} D_b^{2.5} RPM \tag{3.81}$$

where, C_f = an empirical factor ranging from 4×10^{-6} for very hard formations to approximately 14×10^{-5} for very soft formations; D_b = bit diameter, in.; and W = WOB, 10^3 lbf. For practical design applications, the coefficient C_f should be obtained from wells drilled under similar drilling conditions.

In addition to the shear stress given by Eq. 3.76, there is a transverse shear stress produced by the shearing force due to changes in bending moment along the pipe. This stress, however, is usually small during regular rotary drilling operations, and for this reason is not usually included in drillstring design calculations.

3.2.6.3 Bending Stress

In drilling operations, a drillstring frequently undergoes bending because of hole curvature (doglegs), transverse loads, and other disturbances. The bending stress can be calculated from the following equation:

$$\sigma_b = \frac{M_b D_{op}}{2I} \tag{3.82}$$

The bending moment in Eq. 3.82 is a product of pipe bending stiffness and curvature k_p:

$$M_b = EI k_p \tag{3.83}$$

If, as a first approximation, we assume that the pipe shape parallels the wellbore path (i.e., the pipe is in continuous contact with the wellbore—no tool joints), the magnitude of pipe curvature $k_p = \dfrac{1}{R}$ and Eq. 3.82 can be written as

$$\sigma_b = \frac{E D_{op}}{2R} \tag{3.84}$$

where, R is the wellbore radius of curvature.

The hole curvature is usually called the dogleg severity (*DLS*) and is expressed in degrees/100 ft. If, for example, the DLS is 5°/100 ft, this implies that over every foot of hole, the hole curves by 0.05°. The relationship between the radius of curvature and the DLS is

$$R = \frac{5729.6}{DLS} \quad (3.85)$$

where, *DLS* = dogleg severity

3.2.6.4 Pressure-Induced Stresses

Generally, because of drilling-fluid flow, the pressure inside the drillstring is different from the pressure outside. If the drilling fluid is circulated down the drillpipe and up the annular space, the pressure inside the drillstring is greater than the pressure outside. If, however, the drilling-fluid circulation is reversed, the opposite holds true. Consider a certain pipe cross section at which the pressures are p_i inside the pipe and p_o outside the pipe. The pressures p_i and p_o exist because of flow, not because of hydrostatic pressure. These pressures induce axial, tangential, and radial stresses, which can be calculated as

$$\sigma_{an} = \frac{r_i^2 p_i - r_o^2 p_o}{r_o^2 - r_i^2} \quad (3.86)$$

$$\sigma_t = \frac{(p_i - p_o) r_i^2 r_o^2}{(r_o^2 - r_i^2) r^2} + \sigma_{an} \quad (3.87)$$

$$\sigma_r = -\frac{(p_i - p_o) r_i^2 r_o^2}{(r_o^2 - r_i^2) r^2} + \sigma_{an} \quad (3.88)$$

where, σ_{an} = axial neutral stress, psi; σ_t = tangential stress, psi; σ_r = radial stress, psi; r_i, and r_o = inner and outer radii of the pipe, in; and r = radial distance to a point in the cross section under consideration, in (Fig. 3.29).

Eq. 3.87 and Eq. 3.88 are known as Lamé's equations. It should be well understood that the neutral axial stress as given by Eq. 9.46 has no effect on bending. It can be shown, that an increase in the pressure p_i, which results in an increase in the axial stress, does not decrease the pipe tendency to bending. Even a very high value of the inside pressure, p_i, does not prevent bending or buckling of the pipe. On the other hand, an increase in the outside pressure, p_o, although it may produce high compressive stress, does not induce buckling. For this reason, the stress given by Eq. 3.86 is called the neutral axial stress.

降低钻柱弯曲的趋势。即使是很高的内部压力 p_i 也不能阻止钻柱弯曲或屈曲。另外,外部压力 p_o 的增加虽然可能产生高压缩应力,但不会导致弯曲。因此,式(3.86)给出的应力称为中性轴向应力。

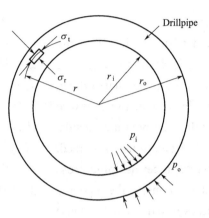

Fig. 3.29 Pipe cross section

3.3 Drillstring Design

3.3 钻柱设计

Audio 3.7

The purpose of drillstring design is to determine the optimal size, grade, and length of drillstring components so that they will be sufficiently strong, yet entail minimum cost. Because of the complexity of the problem, an iterative approach is normally used. A design model is initially assumed, the components of the drillstring are selected,and then the design is refined by incorporating factors that were ignored during the first step. The designer must possess a good knowledge of drillstring performance properties (available sizes, grades, etc.), data from wells already drilled in similar conditions to those of the well under consideration, and also current prices of the drillstring components.

The final design should satisfy the following major criteria:

(1) The load capacity of any drillstring member (divided by an SF) should be greater than or equal to the maximum permissible load.

钻柱设计的目的是确定钻柱组件的最佳尺寸、钢级和长度,以确保在满足钻柱强度要求的同时成本最低。由于问题复杂,通常采用迭代法进行设计。首先假设一个设计模型,选择钻柱的组件,然后通过合并在第一步中被忽略的因素来细化设计。设计人员必须充分了解钻柱的性能特性(如可用尺寸、钢级等)、来自类似条件下已钻井的数据,以及钻柱组件的当前价格。

最终设计应满足以下主要标准:

(1)任何钻柱组件的承载能力(除以安全系数)都应大于或等于最大允许载荷。

(2) Neighboring elements must be compatible. This is accomplished by selecting elements with an appropriate bending-stress ratio (BSR).

(3) The drillstring geometric properties should be selected in conjunction with an optimal hydraulic and casing program.

(4) In deviated wells, drillstring rotation should not produce excessive casing damage.

(5) The total cost of the string should be kept to a minimum.

Keeping in mind the criteria just mentioned and the fact that the drillstring is subjected to loads that may be cyclic and dynamic, the problem of drillstring design is rather complex. Because of the number and complexity of the calculations involved, obtaining a good design requires the use of high-speed computers. In this chapter, a simplified approach is presented to illustrate the concepts involved and their practical importance.

First, the designer selects the drill-collar size based on the hole and casing diameters. Typically, the components of the BHA are selected on the basis of deviation-control requirements. Next, the drillpipe size is selected. To avoid a rapid change in bending stiffness, an intermediate member may be placed between the drill collars and the drillpipe. For best performance, the *BSR* of adjacent members should be less than approximately 5.0 ～ 5.5. The *BSR* is defined as the ratio of the section moduli of two adjacent members of the drillstring. To satisfy this requirement, more than one size of drill collar may be needed. In addition, a few joints of heavy walled pipe are often placed above the collars below the regular drillpipe to obtain a *BSR* < 5.

The maximum allowable length of drillpipe (Section 1) that should be placed above the heavyweight pipe can be calculated as

$$\frac{F_1}{SF} = (L_{dc}w_{dc} + L_{hw}w_{hw} + L_{dpl}w_{dpl})K_b \quad (3.89)$$

（2）相邻组件必须兼容。这可以通过选择具有适当弯曲应力比（BSR）的组件来实现。

（3）钻柱的几何特性应与最优水力参数和井身结构一起选择。

（4）斜井中，钻柱旋转不会造成过度的套管磨损。

（5）钻柱的总成本应该保持最低。

鉴于前面提到的设计标准及钻柱可能受到循环和动态载荷的事实，钻柱设计的问题相当复杂。由于所涉及的计算量大且十分繁琐，因此要获得一个良好的设计，就需要使用高速计算机。在本章中，我们提出了一种简化的方法来说明所涉及的概念及其重要性。

首先，根据井眼和套管直径来选择钻铤尺寸。通常，BHA的组件是根据井斜控制要求来选择的。然后，选择钻杆尺寸。为了避免弯曲刚度的快速变化，可以在钻铤和钻杆之间安装一个中间构件。为了获得最佳性能，相邻组件的 *BSR* 应小于 5.0 ～ 5.5。*BSR* 是钻柱中两个相邻组件的截面模量之比。为了满足这一要求，可能需要多个尺寸的钻铤。此外，为了使 *BSR*<5，会将一些加重钻杆连接在钻铤之上、常规钻杆以下。

安装在加重钻杆上方的第1段钻杆的最大允许长度计算如下：

where L_{hw} = length of heavyweight drillpipe (if used in the drillstring), lbf/ft; w_{hw} = unit weight of heavyweight drillpipe, lbf/ft; L_{dp1} = length of Section 1 of the drillpipe, ft; w_{dp1} = unit weight of Section 1 of the drillpipe, lbf/ft; F_1 = tension load capacity of Section 1 of the drillpipe, lbf; and SF = safety factor.

Eq.3.89 is valid for a vertical hole; however, a corresponding equation can be written for a directional well if the hole shape is known. Here, for the sake of simplicity, only axial and torsional loads are assumed, so

$$F_1 = \sqrt{F_{t1}^2 - 3\left(\frac{A_1}{Z_1}T\right)^2} \qquad (3.90)$$

where, F_{t1} = tensional axial load capacity of Section 1 of the drillpipe $(F_{t1} = S_y A_1)$; A_1 = cross-sectional area of Section 1, in²; Z_1 = sectional modulus of Section 1, in³; and T = rotary torque applied to the drillstring, in·lbf.

The magnitude of the SF depends on the quality of the drillstring, the drilling practices, the working environment, and drilling experience from offset wells. Substituting Eq.3.90 into Eq.3.89 and solving for the length of Section 1 of the drillpipe leads to

$$L_{dp1} = \frac{F_1}{SF K_b w_{dp1}} - \frac{L_{dc} w_{dc}}{w_{dp1}} - \frac{L_{hw} w_{hw}}{w_{dp1}} \qquad (3.91)$$

If the sum $L_{dc} + L_{hw} + L_{dp1}$ is less than the planned depth of the hole, a stronger pipe must be selected for the portion above Section 1. This stronger pipe will be called Section 2 of the drillpipe.

The maximum length of Section 2 of the drillpipe can be calculated as

$$L_{dp2} = \frac{F_2}{SF K_b w_{dp2}} - \frac{L_{dc} w_{dc}}{w_{dp2}} - \frac{L_{hw} w_{hw}}{w_{dp2}} - \frac{L_{dp1} w_{dp1}}{w_{dp2}} \qquad (3.92)$$

$$F_2 = \sqrt{F_{t2}^2 - 3\left(\frac{A_2}{Z_2}T\right)^2} \qquad (3.93)$$

where, F_{t2} = tension load capacity of Section 2 of the drillpipe; L_{dp2} = length of Section 2, ft; w_{dp2} = unit weight of Section 2, lbf/ft; A_2 = cross-sectional area of Section 2, in^2 ; and Z_2 = sectional modulus of Section 2, in^3.

Again, the sum $L_{dc} + L_{hw} + L_{dp1} + L_{dp2}$ is calculated and checked against the total hole depth. If necessary, a third section must be used. Normally, however, to avoid excessive logistical problems, no more than two sections are recommended. A drillstring consisting of more than one size of drillpipe is called a tapered drillstring.

Once the drillpipe sizes and lengths have been determined, tool joints should be selected so that their load capacities are greater than that of the drillpipe. The drillstring configuration as obtained must be checked for margin of overpull, bending loads, drilling-fluid pressure, and drillpipe crushing in slips. Usually there are several different drillstring combinations that satisfy the desired strength requirements. To arrive at the final selection, the cost of the various drillstring configurations is calculated, and the design that yields the lowest cost is chosen. In other words, the designer must consider string mechanical integrity, operational difficulties, and economics to select the optimal set of drillstring components.

Additional calculations should be performed to verify that the design meets the slip-crushing, bending-stress, and drilling-fluid pressure requirements stated in preceding sections of this chapter. Appropriate refinements should be introduced if the predicted stress levels become higher than the maximum acceptable values. In addition, other drillpipe sizes and steel grades should be considered to arrive at other designs. The configuration that meets the safety criteria and yields the least overall cost will become the final drillstring design.

式中，F_{t2}为第2段钻杆的拉力承载能力；L_{dp2}为第2段钻杆的长度，ft；w_{dp2}为第2段钻杆的单位重量，lbf/ft；A_2为第2段钻杆的横截面积，in^2；Z_2为第2段钻杆的截面模量，in^3。

再次计算$L_{dc} + L_{hw} + L_{dp1} + L_{dp2}$，并与总井深进行比较。如有必要，必须使用第三段钻杆。然而，为了避免后勤储备的负担过重，通常情况下建议不超过两段。由一种以上尺寸的钻杆组成的钻柱称为塔式钻柱。

一旦确定了钻杆的尺寸和长度，就应选择负载能力大于钻杆的接头。必须检查所得钻柱结构的拉力极限、弯曲载荷、钻井液压力和钻杆破裂情况。通常有几种不同的钻柱组合可以满足期望的强度要求。为了得到最终的选择，将计算各种钻柱配置的成本，并选择成本最低的设计。换句话说，为了选择最优的钻柱组合，设计者必须综合考虑钻柱的机械完整性、作业难度和经济效益。

为了验证设计结果是否满足本章前面提到的压力、弯曲应力和钻井液压力要求，应进行额外的计算。如果预测的应力水平高于最大允许值，则应进行适当的改进。此外，为了达到其他设计要求，应考虑其他钻柱尺寸和钢级。符合安全标准且总成本最小的配置将成为最终的钻柱设计。

3.4　Exercise

(1) Consider 7½ in by 2½ in steel drill collars in a 12¼ in vertical hole. If the WOB W = 20,000 lbf and the mud weight = 12 lbm/gal, determine:

① Whether drill-collar buckling occurs and, if it does, the distance to the tangency point;

② The side force at the bit;

③ The force that the buckled drill collar applies to the borehole wall;

④ The distance from the bit to the point of maximum bending stress and the corresponding bending stress.

(2) Suppose that a drillstring is composed of 9,500 ft of 4½ in drillpipe with a unit weight of 18.3 lbf/ft and a cross-sectional area of 4.4074 in², and of 600 ft of drill collars with D_{odc} = 6½ in, D_{idc} = 2¼ in (A_{dc} = 29.1922 in²), and a unit weight of 99 lbf/ft. The drilling-fluid SG is 1.2 (that of water is 1.0). Calculate the total elongation of the drillpipe and the equivalent stress at the top of the hole. Assume that there is no WOB.

(3) Determine the maximum tensile load capacity of 3½ in drillpipe with unit weight 15.5 lbf/ft, steel grade E with a yield strength of 75,000 psi, if a rotary torque of 6,000 ft·lbf is applied to the pipe.

(4) Calculate the maximum torque that can be applied to 4-in. grade E drillpipe with a unit weight of 10.46lbf/ft (ID = 3.476in, A = 3.0767in², Z = 5.400in³) if the DLS is 10.0°/100ft. The tensile load applied to the pipe is F = 140,000lbf. To solve this problem, assume that the pipe curvature is the same as the hole curvature.

(5) Name the basic components of a drillstring.

(6) Discuss the major functions of a drillstring.

(7) What is the purpose of using drill collars?

(8) Explain why spiral and square drill collars are sometimes used in a BHA.

(9) List the major factors that affect selection of drill collars.

Drilling Fluids 4
钻井液

Drilling fluid is a critical component in the rotary drilling process. Its primary functions are to remove the drilled cuttings from the borehole whilst drilling and to prevent fluids from flowing from the formations being drilled, into the borehole. Since it is such an integral part of the drilling process, many of the problems encountered during the drilling of a well can be directly, or indirectly, attributed to the drilling fluids and therefore these fluids must be carefully selected and/or designed to fulfil their role in the drilling process.

Audio 4.1

The success of the rotary-drilling process (completion of an oil or gas well) and its cost depend substantially on three important factors: (1) The bit penetrating the rock; (2) The cleaning the bit face and transport of the cuttings to surface; (3) The support of the borehole.

The drilling fluid used affects all of these critical items. The drilling-fluid density and ability to penetrate rock have an effect on the rate of penetration. The hydraulic energy expended on the bottom of the hole and the viscos-ity and flow rate of the fluid affect the cuttings transport. And the density of the fluid and its ability to form a layer on the wellbore (wall cake) affects the wellbore stability and support. It is often said that the majority of the prob-lems in drilling are related in some manner or another to the drilling fluid. The penetration rate of the rotary bit and opera-tional delays caused by circulation loss, stuck drillpipe, caving shale, and the like are significantly affected by the drilling-fluid properties.

The cost of the drilling fluid, is comparatively small as compared to the rig or casing costs; but, the selection of the proper fluid and the testing and control

钻井液是旋转钻井的关键组成部分。其主要功能是在钻进时清除井眼中的钻屑，并防止流体从地层中侵入井筒。由于钻井液是钻井过程中不可或缺的一部分，钻井过程中遇到的许多问题可能直接或间接地和它有关，因此必须细心选择和精心设计钻井液，从而实现钻井液在钻井过程中的重要作用。

旋转钻井作业的成功（油井或气井的完井）及其成本基本上取决于三个重要因素：（1）钻头破岩；（2）钻头清洗和岩屑携带；（3）支撑井壁。

钻井液会影响每一个关键步骤。钻井液的密度和辅助破岩的能力对机械钻速有影响，钻井液的黏度、流速及施加在井底的压力会影响岩屑的运移，而流体的密度及其在井壁上形成滤饼的能力影响井筒的稳定性。通常，钻井中的大多数问题或多或少都与钻井液有关。钻井液的性质会显著影响机械钻速及井漏、卡钻、泥页岩坍塌等造成作业延迟的复杂情况。

钻井液的成本与钻机或套管成本相比相对较小；但是，选择合适的钻井液体系并测试和控制

of its properties has considerable effect on the total well cost. The additives needed to create and maintain the fluid properties can be expensive. The cost of the mud can be as high as 10% ～ 15% of the total cost of the well. Although this may seem expensive, the consequences of not maintaining good mud properties may result in drilling problems which will take a great deal of time and therefore cost to resolve. In view of the high cost of not maintaining good mud properties an operating company will usually hire a service company to provide a drilling fluid specialist (mud engineer) on the rig to formulate, continuously monitor and, if necessary, treat the mud.

4.1 Functions of Drilling Fluids

The functions of a drilling fluid can be categorized as follows:

(1) Cuttings transport: ① Clean under the bit; ② Transport the cuttings up the borehole; ③ Release the cuttings at the surface without losing other beneficial materials; ④ Hold cuttings and weighting materials when circulation is interrupted.

(2) Physicochemical functions.

(3) Cooling and lubricating the rotating bit and drill string.

(4) Fluid-loss control: ① Wall the newly drilled wellbore with an impermeable cake for borehole support; ② Reduce adverse and damaging effects on the formation around the wellbore.

(5) Control subsurface pressure.

(6) Support part of the drillstring and casing weight.

(7) Ensure maximum logging information.

(8) Transmit hydraulic horsepower to the rotating bit.

4.1.1 Cuttings Transport

The drilling fluid should be able to remove rock

钻井液性能，对油井总成本有相当大的影响，这是因为形成和保持钻井液性能所需的添加剂可能非常昂贵。钻井液成本可能高达油井总成本的10%～15%。尽管看起来很昂贵，但如果钻井液不能保持良好的性能则可能会导致钻井复杂情况，并需要大量时间和成本来解决。鉴于钻井液性能不好会导致成本升高，作业者通常会聘请专业服务公司的钻井液专家（钻井液工程师），来配制、持续监测、必要时处理钻井液。

4.1 钻井液功能

钻井液的功能可分为以下几类：

（1）岩屑运移：① 清洁钻头下方岩屑；② 将岩屑携带出井眼；③ 在不损失其他有用材料的情况下在地面排出岩屑；④停止循环时，悬持岩屑和加重材料。

（2）物理化学功能。

（3）冷却润滑旋转的钻头及钻枪。

（4）滤失控制：① 在新钻的井壁上形成防渗滤饼，用于支撑井壁；② 减少对井眼周围地层的伤害。

（5）控制井底压力。

（6）支撑钻柱和套管的重量。

（7）尽可能获取测井信息。

（8）向旋转的钻头传递水功率。

4.1.1 岩屑运移

钻井液能够从钻头下方清洗

fragments or cuttings from beneath the drilling bit, transport them up the wellbore drillstring annulus, and permit their separation at the surface using solids-control equipment. The density and viscosity of the drilling fluid are the properties that control the process of lifting particles that fall down through the flowing fluid by the effect of gravity. The fluid must also have the ability to form a gel-like structure to hold cuttings and weighting materials when circulation is interrupted. Horizontal and high-angle wells require specialized fluid formulations and "sweep" protocols to minimize the risk of barite sag and of low-side cuttings settling, both of which can lead to stuck pipe and loss of the well. The lubricity of the drilling fluid is also a key factor in controlling torque and drag in these types of operations.

4.1.2 Physicochemical Functions

The drilling-fluid system should remain stable when exposed to contaminants and hostile downhole conditions. Among the common natural contaminants are reactive drill solids, corrosive acid gases (e.g., H$_2$S), saltwater flows, and evaporites (e.g., gypsum). The cement used in setting casing and liner strings is also a contaminant to some water-based muds. Wells in certain areas have extremely high bottomhole temperatures, at times approaching 500°F, and, likewise, arctic locations may expose the drilling fluid to subzero temperatures at surface.

4.1.3 Cooling and Lubricating the Rotating Bit and Drillstring

The drilling mud cools and lubricates when the rotating bit drills into the bottom of the hole and when the drillstring rotates against the wellbore walls. The fluid should have the ability to absorb the heat generated by the friction between metallic surfaces and formation. In addition, the fluid should not adversely affect the bit life nor increase the torque and drag between the drillstring and the borehole.

井底岩石碎块或岩屑并携岩至钻柱与套管形成的环形空间内，在地面使用固控设备将其处理分离。钻井液的密度和黏度是控制颗粒在重力作用下从流动钻井液中下落的特性。当停止循环时，钻井液必须能够形成凝胶状结构，以保持岩屑和加重材料悬浮。水平井和大斜度井需要专门的配方和波及特征来尽可能降低重晶石下落和井眼低边产生岩屑沉降的风险，这两种情况都会导致卡钻和井漏。钻井液的润滑性也是控制钻井作业中扭矩和摩阻的关键因素。

4.1.2 物理化学功能

当处在污染和恶劣的井下条件时，钻井液体系应该保持稳定。常见的天然污染物有活性钻屑、腐蚀酸性气体（如H$_2$S）、盐水和蒸发岩（如石膏）。用于坐封套管和尾管的水泥浆对某些水基钻井液来说也是一种污染物。有些地区的井底温度非常高，可能接近500°F（260℃），相反，北极地区的地面钻井液温度可能会低于0℃。

4.1.3 冷却润滑旋转的钻头及钻柱

当钻头钻入井底，钻柱靠着井壁旋转时，需要钻井液进行冷却和润滑。钻井液应具有吸收金属表面与地层摩擦所产生热量的能力。同时，钻井液不会影响钻头的寿命，也不会增加钻柱与井眼之间的扭矩和摩阻。

4.1.4 Fluid-Loss Control

The bit removes lateral support of the drilled wellbore and is immediately replaced by the drilling fluid until the casing is set with cement. The stability of uncased sections of the borehole is achieved by a thin, low-permeability filter cake formed by the mud on the walls of the hole. Also, the cake seals pores and other openings in formations caused by the bit, thereby minimizing liquid loss into permeable formations. Poor fluid-loss control can cause surge (an increase in wellbore pressure under a bit from running into the hole), swab (a decrease in wellbore pressure under a bit from running out of the hole), and circulation-pressure problems. Loss of circulation increases the drilling-fluid cost and the potential for the inflow of fluids from formations. The drilling fluid should not negatively affect the production of the fluid-bearing formation. That is, the drilling mud is designed to reduce adverse effects on formation around the wellbore. In contrast, the fluid must assist in the collection and interpretation of electrical-log information.

4.1.5 Control Subsurface Pressure

A drilling fluid is the first line of defense against well-control problems. The drilling fluid balances or overcomes formation pressures in the wellbore. Typically, this is accomplished with weighting agents such as barite, although there are other chemicals that can be used. In addition, surface pressure can be exerted to give the equivalent pressure needed to balance a formation pressure. An overbalanced condition occurs when the drilling fluid exerts a higher pressure than the formation pressure. An underbalanced condition occurs when the drilling fluid exerts a lower pressure than the formation pressure. Therefore, in underbalanced drilling operations, the borehole is deliberately drilled with a fluid/pressure combina-tion lower than the formation pressure. A balanced condition exists if the pressure exerted in the wellbore is equal to the formation pressure.

4.1.4 滤失控制

钻头钻掉了井眼中原本横向支撑井壁的岩石,在下套管注水泥之前由钻井液来提供横向支撑。钻井液能够在井壁上形成薄且渗透率低的滤饼,进而实现裸眼井段的井壁稳定。同时,滤饼还可以封堵孔隙及由钻头等原因造成的地层裂隙,从而最大限度地减少钻井液滤失到可渗透地层中。滤失控制不好会导致激动(钻头下入井内时井筒压力增加)、抽汲(钻头提出井内时井筒压力降低)和循环压力问题。井漏增加了钻井液的成本,也增加了流体从地层溢出的可能性。钻井液不应对含液地层的生产产生负面影响。也就是说,设计使用的钻井液一定要减少对井筒周围地层的不利影响。同时,钻井液必须便于电测井资料的收集和解释。

4.1.5 控制井底压力

钻井液是井控问题的第一道防线。钻井液可以平衡井眼内地层压力。通常情况下可通过使用重晶石等加重剂,或其他化学材料来实现。此外,也可以通过在地面施加压力,来提供平衡地层压力所需的等效压力。当钻井液施加的压力高于地层压力时,就会出现过平衡状态。当钻井液施加的压力比地层压力低时,就会出现欠平衡状态。在欠平衡钻井作业中,钻井液产生的压力与地面回压之和必须低于地层压力。如果井筒压力等于地层压力,则为平衡状态。

4.1.6 Support the Drillstring and Casing Weight

Any time a material is submerged in a fluid in a gravita-tional field, there is a reaction that offsets the force that gravity exerts. This is often called buoyancy. Nonetheless, in heavily weighted situations, this "buoyancy force" can assist by offsetting some of the weight of a drillstring or casing. This offset is dependent upon the density of the fluid, with higher-density fluids giving more of an offset and lower density "fluids" (e.g., air) not helping much, if at all.

4.1.7 Ensure Maximum Logging Information

The drilling fluid has a profound impact on the electrical and acoustical properties of a rock. Because these properties are what logging tools measure, it is imperative that the correct selection of wireline logging tool or logging-while-drilling (LWD) tool for a given drilling fluid be made. Or, lacking that, then the correct drilling fluid must be used for a given logging tool. In addition, the drilling fluid should facilitate retrieval of information by means of cuttings analysis.

4.1.8 Transmit Hydraulic Horsepower to the Rotating Bit

The hydraulic force is transmitted to the rotating bit when the fluid is ejected through the bit nozzles at a very high velocity. This force moves the rock fragments or cuttings away from the drilled formation beneath the bit. In directional-drilling operations, the hydraulic force powers the downhole hydraulic motor and turns electric-power generators (turbines) for measurement-while-drilling (MWD) and LWD drillstring equipment.

4.2 Drilling-Fluid Categories

According to the World Oil annual classification of fluid systems, there are nine distinct categories of

4.1.6 支撑钻柱和套管的重量

任何时候，一种材料浸没在重力场的流体中时，都会产生一种抵消重力的作用力，这通常被称为浮力。在加重钻井液中，这种浮力可以抵消钻柱或套管的部分重量。抵消的大小取决于流体的密度，密度较高的流体会产生更多的抵消，而密度较低的流体（如空气）则没有多大帮助，甚至根本没有帮助。

4.1.7 尽可能获取测井信息

钻井液对岩石的电学和声学特性有很大的影响。由于这些特性是测井工具测量的内容，所以必须针对给定的钻井液，正确选择电缆测井工具或随钻测井（LWD）工具。或者说，针对给定的测井工具选择正确的钻井液体系。此外，钻井液应便于通过岩屑分析获取信息。

4.1.8 向旋转的钻头传递水功率

当钻井液以极高的速度通过钻头喷嘴喷射时，水力就传递到了钻头。水力会清洗井底并携带岩屑。在定向钻井作业中，钻井液为井下动力钻具提供动力，并驱动涡轮旋转为随钻测量（MWD）和随钻测井（LWD）供电。

4.2 钻井液类型

根据世界石油年会对钻井液体系分类，目前使用的钻井液有

drilling fluids in use today (World Oil 2002). Five categories include freshwater systems, one category covers saltwater systems, two categories include oil- or synthetic-based systems, and the last category covers pneumatic (air, mist, foam, gas) "fluid" systems.

4.2.1 Classification of Drilling Fluids

The principal factors governing the selection of type (or types) of drilling fluids to be used on a particular well are

(1) The characteristics and properties of the formation to be drilled;

(2) The quality and source of the water to be used in building the fluid;

(3) The ecological and environmental considerations.

4.2.1.1 Continuous-Phase Classification

Drilling fluids are categorized according to their continuous phase so that there are (1) Water-based fluids; (2)OBFs; (3)Pneumatic (gas) fluids.

Consider a drop of a drilling fluid. If one could go from a point in one phase to any other point in that same phase, then it is said to be continuous. If one had to cross one phase to get back to the previous phase, then that phase is discontinuous. Solids are always a discontinuous phase. Therefore, drilling fluids are designated by their continu-ous phase.

Water-based drilling muds are the most commonly used fluids, while oil-based muds are more expensive and require more environmental considerations. The use of pneumatic drilling fluids (i.e., air, gas, and foam) is limited to depleted zones or areas where the formations are low pressured. In water-based fluids, the solid particles are suspended in water or brine, while in oil-based muds the particles are suspended in oil. When pneumatic drilling fluids are used, the rock fragments or drill cuttings are removed by a high-velocity stream of air or natural gas. Foaming agents are added to remove minor inflows of water (Darley and Gray 1988).

4.2.1.2 Water-Based Fluids

The majority of wells are drilled with water-based drilling fluids. The base fluid may be fresh water,

九个不同的类别（世界石油2002年）：五类清水体系，一类盐水体系，两类油基或合成体系，一类气体型（空气、雾气、泡沫、天然气）"流体"体系。

4.2.1 钻井液分类

针对具体的井，钻井液类型选择的主要因素有：

（1）待钻地层的特征和性质；

（2）钻井液配制所用水的质量和来源；

（3）生态和环境因素。

4.2.1.1 连续相分类

钻井液按连续相分类，可分为：（1）水基钻井液；（2）油基钻井液；（3）气体钻井液。

如果一滴钻井液可以从单相中的一点移动到同一相中的任何一点，那么这个相就是连续相。如果必须穿过一个相才能回到前一个相，那么这个相是不连续相。钻井液中的固体总是不连续相。所以钻井液用它的连续相来表示。

水基钻井液是最常用的钻井液，油基钻井液则比较昂贵，需要考虑更多的环境因素。气体钻井液（即空气、天然气和泡沫）的使用仅限于枯竭带或地层压力较低的地区。水基钻井液，固体颗粒悬浮在水或盐水中；而油基钻井液，固体颗粒悬浮在油中。当使用气体钻井液时，岩石碎片或钻屑被高速气流或天然气携出。如果要去除少量侵入的水，需要加入起泡剂。

4.2.1.2 水基钻井液

大多数钻井作业采用水基钻井液。基液可以是清水、盐水、

saltwater, brine, or saturated brine. A typical water-based mud composition is illustrated in Fig.4.1.

卤水或饱和盐水。典型的水基钻井液成分如图 4.1 所示。

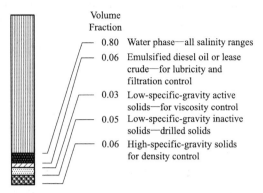

Fig.4.1 Composition of water-based mud

Water-based drilling fluids range from native muds to lightly treated fluids to the more heavily treated, inhibi-tive fluids. They are divided into three major subclassifications:

(1) Inhibitive;
(2) Noninhibitive;
(3) Polymer.

Inhibitive fluids retard clay swelling (i.e., the ability of active clays to hydrate is greatly reduced). For this reason, inhibitive fluids are used for drilling hydratable-clay zones. The ability of the formation to absorb water is not in-hibited when noninhibitive fluids are used. The term noninhibitive refers to the lack or absence of those specific ions (sodium, calcium, and potassium) that are present in inhibitive fluids. Inhibitive fluid systems do not use chemical dispersants (thinners), but native waters.

1. Saltwater Drilling Fluids

Saltwater drilling fluids are used for shale inhibition and for drilling salt formations. They are also known to inhibit hydrates (ice-like formations of gas and water) from forming, which can accumulate around subsea wellheads and well-control equipment, blocking lines and impeding critical operations.

2. Inhibitive Drilling Fluids

Inhibitive drilling fluids are designed to reduce

水基钻井液的范围从原始钻井液到低密度钻井液，再到加重处理的抑制性钻井液。它们主要分为三个亚类：

（1）抑制性；
（2）非抑制性；
（3）高分子聚合物。

抑制性钻井液可延缓黏土膨胀（即活性黏土的水化能力大大降低），因此，抑制性钻井液用于可水化黏土区的钻井。当使用非抑制性钻井液时，地层吸水的能力不会受到抑制。非抑制性是指钻井液中缺乏或不存在抑制性的特定离子（钠、钙和钾）。抑制性钻井液体系不使用化学分散剂（稀释剂），而是使用原生水。

1. 盐水钻井液

盐水钻井液用于抑制页岩和盐层钻进。盐水钻井液还可以抑制水合物（天然气和水形成的冰状物）的形成，水合物会在海底井口和井控设备周围聚集，堵塞管线，阻碍关键作业。

2. 抑制性钻井液

抑制性钻井液旨在减少钻井

chemical reactions between the drilling fluid and the formation. Fluid formulations containing sodium, calcium, or potassium ions minimize shale hydration and swelling. Gypsum ("gyp") drilling fluids are used for drilling anhydrite and gypsum formations. In known H_2S environments, a high pH water-based fluid treated with scavengers, or an OBF with 4～6 ppb of excess lime, usually is the recommended drilling-fluid system.

Incorporating up to 10% oil in a water-based fluid to improve lubricity and inhibition was a long-standing drilling practice; however, the availability of glycol additives and high performance SBFs makes oil additions unnecessary.

3. Organic and Synthetic Polymers

Organic and synthetic polymers are used to provide viscosity, fluid-loss control, shale inhibition, and prevention of clay dispersion in freshwater-or saltwater-based drilling fluids. Most polymers are very effective even at low concentrations and can be run by themselves or added in small quantities to enhance or extend bentonite performance. Specially developed high-temperature polymers are available to help overcome gelation issues.

Polymers function in several ways. Some polymers actually hydrate and swell in much the same manner as con-ventional clay materials. By doing this they thicken the water phase, making the escape of this water into the forma-tion or into the clay structure more difficult, thereby preventing swelling. Large, high molecular-weight polymers will bond onto clay surfaces and literally surround and isolate the clay/shale particle. This is referred to as encapsulation.

4. Membrane-Efficient

Water-Based Drilling Fluids. Wellbore instability predominantly occurs in shales that overlay reservoirs and is an industrywide problem that affects exploration through to development drilling. Shales are fine-grained with high clay content, low permeability, and are

液和地层之间的化学反应。它含有钠离子、钙离子或钾离子，可以最大限度地减少页岩的水化和膨胀。石膏钻井液用于钻探酸性地层和石膏层。在已知的硫化氢环境中，通常推荐使用经过除氧剂处理的高pH值的水基钻井液，或含有4～6 lb/bbl过量石灰的油基钻井液体系。

长期的钻井实践表明，在水基液中加入10%的油可以提高润滑性和抑制性；但乙二醇添加剂和高性能合成基钻井液的出现，使得上述做法没那么必要了。

3. 有机聚合物和合成聚合物钻井液

在淡水或盐水基钻井液中，常用有机聚合物和合成聚合物来提高黏度、控制钻井液漏失、抑制页岩和防止黏土的分散。大多数聚合物即使在低浓度下也非常有效，可以单独使用或将其少量地添加到膨润土中以拓展其性能。

聚合物有多种功能。一些聚合物实际上以与常规黏土材料非常相似的方式水化和膨胀。通过这种方式，它们会使水相稠化，使得水更难逸出到地层或黏土结构中，从而防止膨胀。大分子量的聚合物会黏结在黏土表面，并包围和隔离黏土（页岩）颗粒，这就是所谓的包封。

4. 高效膜水基钻井液

井眼不稳定主要发生在盖层的页岩中，这是一个从勘探到开发阶段都会影响钻井作业的问题。页岩颗粒细，黏土含量高，渗透率低，易与不配伍的钻井液发生

chemically reactive with incompatible drilling fluids. The significant drilling fluid/ shale interaction mechanisms are hydraulic or mud-pressure penetration, chemical potential.

When drilling with improperly designed drilling fluids, shales without effective osmotic membranes progres-sively imbibe water, which leads to mud-pressure penetration. Consequently, the net radial mud support changes over a period of time and near-wellbore formation pore pressure increases. This reduces formation strength and leads to borehole instability.

Past efforts to develop improved water-based fluids for shale drilling have been hampered by a limited under-standing of the drilling fluid/shale interaction phenomenon. Recent studies of fluid/shale interactions have pro-duced new insights into the underlying causes of borehole instability, and these studies suggest new and innovative approaches to the design of water-based drilling fluids for drilling shales (Tare 2002).

In most cases, the two most relevant mechanisms for water transport into and out of shale are the hydraulic-pressure difference between the wellbore pressure (drilling-fluid density) and the shale pore pressure; and the chemical potential difference (i.e., the water activity) between the drilling fluid and the shale.

The fine pore size and negative charge of clay on pore surfaces cause argillaceous materials to exhibit mem-brane behavior. The efficiency is a measure of the capacity of the membrane to sustain osmotic pressure between the drilling fluid and shale formation. If the water activity of the drilling fluid is lower than the formation water activity, an osmotic outflow of pore fluid from the formation, because of the chemical potential mechanism, will lessen the increase in pore pressure resulting from mud-pressure penetration. If the osmotic outflow is greater than the inflow as a result of mud-pressure penetration, there will be a net flow of water out of the formation into the wellbore. This will result in the lowering of the pore fluid pressure below the in-situ value. The associated increase in the effective mud support will lead to an improvement in the stability of the wellbore.

化学反应。钻井液与页岩相互作用机制有水或钻井液压力渗透、化学势。

当使用设计不当的钻井液进行钻井时，没有有效渗透膜的页岩会逐渐渗入水分，产生钻井液压力渗透。结果，在一段时间后，近井眼地层孔隙压力增加，钻井液产生的径向净支撑发生变化。这就降低了地层强度，导致井眼不稳定。

过去努力改进页岩钻井水基钻井液时，由于对钻井液与页岩相互作用现象的认识不足而受到阻碍。最近对钻井液与页岩相互作用的研究，使人们对井眼不稳定的根本原因有了新的认识，为页岩钻井水基钻井液的设计提出了新的创新方法。

在大多数情况下，水进入和离开页岩的两个最相关的机理是在井内（钻井液密度）和页岩孔隙之间形成的流体压差及化学势差（即水活度）。

黏土的孔隙非常细小且表面分布负电荷，这使黏土表现出薄膜的性质。该性质是衡量膜在钻井液和页岩层之间维持渗透压能力的指标。如果钻井液的水活度低于地层水活度，则由于化学势机制，地层孔隙流体从地层中渗透到钻井液，减少由于钻井液压力渗透造成的孔隙压力增加。如果渗透流出量大于钻井液压力导致的流入量，则会有水从地层净流入井筒中。这将导致孔隙流体压力降低到原始值以下。由此，有效钻井液支撑的增加将提高井筒的稳定性。

One of the key parameters that can be manipulated to increase the osmotic outflow is membrane efficiency. The osmotic outflow increases with increasing membrane efficiency. In most conventional water-based fluids, the membrane efficiency is low. Therefore, even if the water activity of the drilling fluid is maintained significantly lower than the shale-water activity, the osmotic outflow may be negligible because of the low membrane efficiency.

As a result of extensive testing, three new generations of water-based drilling fluid systems with high membrane efficiencies (greater than 80%) have been developed, including a 12% NaCl system that generated a membrane efficiency of approximately 85%. Such systems hold promise for operations where OBFs and SBFs are unsuitable or prohibited because of drilling conditions and/or regulations.

4.2.1.3　OBFs

Normally, the high salinity water phase of an invert emulsion or SBF helps stabilize reactive shale and prevent swelling. However, drilling fluids formulated with diesel- or synthetic-based oil and no water phase are used to drill long shale intervals where the salinity of the formation water is highly variable. By eliminating the water phase altogether, the all-oil drilling fluid preserves shale stability throughout the interval.

Diesel and mineral oil OBFs—also called "invert emulsions"—are inhibitive, resistant to contaminants, stable at high temperatures and pressures, lubricious, and noncorrosive. Onshore, they are the flfl uids of choice for drilling troublesome shale sections, extended-reach wells that would be otherwise prone to pipe-sticking problems, and dangerous HP/HT H_2S wells. The drilling effifi ciency of an OBF system can save days, perhaps weeks, on the time required to drill the well. These fluid systems are also subject to stringent disposal regulations because of their toxicity. Mineraloil formulations are considered less toxic than diesel-based fluids, but not a suitable alternative where "greener" SBFs are available. The use of diesel- or mineral-oil-based fluids is absolutely prohibited

控制地层渗透流出量的关键参数之一是膜效率。渗透流出量随着膜效率的提高而增加。在大多数水基钻井液中膜效率很低。因此，即使保持钻井液的水活度显著低于页岩水活度，由于膜效率较低，渗透流出量也可以忽略不计。

经过大量的测试，已开发出三种高膜效率（大于80%）的水基钻井液体系，其中12%的氯化钠含量体系，其膜效率约为85%。当钻井条件和（或）法规不适合或禁止使用油基和合成基钻井液体系时，这类钻井液体系有望用于钻井作业中。

4.2.1.3　油基钻井液

通常，水包油乳化钻井液中的高矿化度水或合成基有助于稳定活性页岩并防止膨胀。然而，对于地层水矿化度变化较大的长页岩层，通常要使用柴油或合成油基配制的无水相钻井液，通过剔除水相，全油型钻井液可在全井段维持页岩稳定性。

柴油和矿物油的油基钻井液也被称为"油包水乳化钻井液"，具有抑制性、抗污染性、高温高压稳定性、润滑性和非腐蚀性。在陆地钻井中，常用来钻复杂页岩段、易发生卡钻问题的大位移井及危险的高压（高温）硫化氢井。油基钻井液的钻井效率较高，可节省几天甚至几周的钻井时间。但由于其具有毒性，在处理时需遵守严格的规定。矿物油配方虽被认为比柴油基钻井液的毒性更低，但在可以使用"更环保"的合成基钻井液时，它依然不是一个合适的选择。在某些地区，严

in some areas. The oil/water ratios typically range from 90∶10 to 60∶40, though both oil and water percentages can be increased beyond these ranges. Generally, the higher the percentage of water, the thicker the drilling fluid. High salinity levels in the water phase dehydrate and harden reactive shales by imposing osmotic pressures.

The basic components of an invert-emulsion fluid include diesel or mineral oil, brine (usually calcium chloride), emulsifiers, oil-wetting agents, organophilic clay, filtration-control additives, and slaked lime.

1. SBFs

Synthetic-based drilling fluids were developed to provide the highly regarded drilling-performance characteristics of conventional OBFs while significantly reducing the toxicity of the base fluid. Consequently, SBFs are used almost universally offshore and they continue to meet the increasingly rigorous toxicity standards im-posed by regulatory agencies. The cost-per-barrel for an SBF is considerably higher than that of an equivalent-density water-based fluid, but because synthetics facilitate high ROPs and minimize wellbore-instability problems, the overall well-construction costs are generally less, unless there is a catastrophic lost-circulation occurrence.

2. Emulsion-Based Drilling Fluids (EBFs)

In early 2002, an SBF formulated with an ester/IO blend became widely used, especially in deepwater operations where temperatures range from 40 to 350°F on a given well. The fluid contains no commercial clays or treated lignites; rheological and fluid-loss-control properties are maintained with specially designed fatty acids and surfactants. The system provides stable viscosity and flat rheological prop-erties over a wide range of temperatures. Gel strengths develop quickly, but are extremely shear-sensitive. As a result, pressures related to breaking cir-culation, tripping or running casing, cementing, and ECD are significantly lower than pressures that occur with conventional invert emulsion fluids. Lost circulation incidents appear to occur less frequently and with a lesser degree of severity where

禁使用柴油或矿物油基钻井液。油基钻井液油水比一般在9∶1到6∶4之间，有时也会超过这个范围。一般来说，水的比例越高，钻井液越稠。水相中的高矿化度可通过渗透压力使活性页岩脱水和硬化。

油基钻井液的基本成分包括柴油或矿物油、盐水（通常是氯化钙）、乳化剂、润油剂、有机黏土、过滤控制剂和熟石灰。

1. 合成基钻井液

开发合成基钻井液的目的是为了在提供传统油基钻井液良好性能特性的基础上，大幅降低钻井液的毒性。因此，合成基钻井液广泛用于海上，以满足监管部门提出的越来越严格的毒性标准。合成基钻井液的成本远高于等密度的水基钻井液，但由于合成材料能够提高机械钻速，并最大限度地减少井壁失稳问题，因此，在不发生严重井漏的情况下，总体建井成本会更低。

2. 酯基钻井液（EBFs）

2002年年初，一种含有酯（IO混合剂）的合成基钻井液得到了广泛的应用，特别是温度范围在40～350℉的深水钻井作业中。该钻井液不含商业黏土或处理过的褐煤；特殊设计的脂肪酸和表面活性剂保持了流变性和滤失性的控制。该体系能够在广泛的温度范围内提供稳定的黏度和流变性能。其静切力提高很快，但对剪切力极为敏感。因此，停泵、起下钻、固井时当量循环密度明显低于常规油基钻井液。在使用酯基钻井液的情况下，井漏发生的次数较少，严重

EBFs are used.

The EBF performs well from an environmental perspective and has met or surpassed stringent oil-retained-on-cuttings regulations governing cuttings discharge in the GOM. The EBF is highly water-and solids-tolerant and responds rapidly to treatment.

4.2.1.4 Pneumatic Drilling Fluids

Pneumatic drilling fluids are most commonly used in dry, hard formations such as limestone or dolomite. In pneumatic drilling-fluid systems, air compressors circulate air through the drillstring and up the annulus to a rotating head. The return "fluid" is then diverted by the rotating head to a flowline leading some distance from the rig to protect personnel from the risk of explosion. Gas from a pressured natural gas source nearby may be substituted for air. Both air and gas drilling are subject to downhole ignition and explosion risks. Sometimes, nitrogen—either from cryogenic sources or generated using membrane systems—is substituted for the pneumatic fluid. Pneumatic drilling fluids are considered to be nondamaging to productive formations.

4.2.1.5 Silicate-Based Drilling Fluids

Field applications of sodium silicate drilling fluids indicate that they appear to provide a sealing effect within shale pore throats and may also increase membrane efficiency (the mobility of solutes through a shale pore network) (Mody 1993). Contact with calcium or magnesium ions, or the decrease in pH caused by dilution of fluid filtrate with pore fluid, may cause the electrolytes to precipitate (Bland 2002). Shale permeability is therefore reduced. However, the excellent shale inhibition characteristics of silicate drilling-fluid systems may be outweighed by other perceived deficiencies in lubricity, thermal stability, and the necessity for a high pH.

Typically, the silicate-based fluid system creates a physical membrane, which, in conjunction with the soluble sodium silicate or potassium silicate, provides primary shale inhibition. The wellbore becomes pressure-

程度也较轻。

从环保角度来看，酯基钻井液性能良好，已经达到或超过了墨西哥湾严格的钻屑含油处理规定。酯基钻井液对水和固体有很强的耐受性，并可快速处理。

4.2.1.4　气体钻井液

气体钻井液最常用于干硬地层，如石灰岩或白云岩。在气体钻井液体系中，空气压缩机将空气循环通过钻柱后，沿环空向上到达旋转头。返回的钻井液随后被旋转控制头导流到离钻机一定距离的管线，以保护人员免受爆炸风险。也可以用来自附近加压天然气源的气体来替代空气。空气钻井和天然气钻井都存在井下着火和爆炸的风险。有时，氮气——要么来自低温源，要么使用膜系统产生——也可以替代气体钻井液。通常认为气体钻井液对产层没有伤害。

4.2.1.5　硅酸盐基钻井液

硅酸盐钻井液的现场应用表明，它们可以在页岩孔喉内提供密封效果，还可提高膜效率（溶质通过页岩孔隙网络的流动性）。通过与钙或镁离子接触，或因流体滤液与孔隙液的稀释而导致的pH值下降，可能会导致电解质沉淀，页岩渗透性因此而降低。然而，硅酸盐钻井液体系优异的页岩抑制特性可能会被润滑性、热稳定性和高pH值等明显的缺陷所掩盖。

通常情况下，硅酸盐钻井液体系会形成一层物理膜，与可溶性硅酸钠或硅酸钾联合提供一级页岩抑制作用。井筒压力独立，

isolated so that filtrate invasion is minimal. Polyglycols may provide secondary inhibition, improve lubricity and filtration control, and stabilize the drilling fluid's physical properties. Polyglycols can withstand the high pH environment of silicate fluids better than conventional lubricants that may hydrolyze at high pH.

Silicate-based fluids have been used successfully in drilling highly reactive gumbo clays in the top hole interval and as an alternative inhibitive fluid in areas where invert-emulsion fluids are prohibited. They may also be useful for drilling highly dispersible formation such as chalk and have been used to stabilize unconsolidated sands.

4.2.2 Drilling Fluid Additives

Audio 4.2

Water-based drilling fluids consist of a mixture of solids, liquids, and chemicals, with water being the continuous phase. Solids may be active or inactive. The active (hydrophilic) solids such as hydratable clays react with the water phase, dissolving chemicals and making the mud viscous. The inert (hydrophobic) solids such as sand and shale do not react with the water and chemicals to any significant degree. Basically, the inert solids, which vary in specific gravity, make it difficult to analyze and control the solids in the drilling fluid (i.e., inert solids produce undesirable effects). Broad classes of water-based drilling-fluid additives are in use today. Clays, polymers, weighting agents, fluid-loss-control additives, dispersants or thinners, inorganic chemicals, lost-circulation materials, and surfactants are the most common types of additives used in water-based muds.

4.2.2.1 Weighting Agents

The most important weighting additive in drilling fluids is barium sulfate ($BaSO_4$). Barite is a dense mineral comprising barium sulfate. The specific gravity of barite is at least 4.20g/cm³ to meet API specifica-tions for producing mud densities from 9 to

19 lbm/gal. However, a variety of materials have been used as weight-ing agents for drilling fluids including siderite (3.08g/cm³), calcium carbonate (2.7～2.8g/cm³), hematite (5.05 g/cm³), ilmetite (4.6g/cm³), and galena (7.5g/cm³).

4.2.2.2 Fluid-Loss-Control Additives

Clays, dispersants, and polymers are widely used as fluid-loss-control additives. Sodium montmorillonite (bentonite) is the primary fluid-loss-control additive in most water-based drilling fluids. The colloidal-sized sodium-bentonite particles are very thin and sheetlike or platelike with a large surface area, and they form a compressible filter cake. Inhibitive mud systems inhibit the hydration of ben-tonite and greatly diminish its effectiveness.

Therefore, bentonite should be prehydrated in fresh water before being added to these systems. The larger and thicker particles of sodium montmorillonite do not exhibit the same fluid-loss-control characteristics.

4.2.2.3 Thinners or Dispersants

Although the original purpose in applying certain substances called thinners was to reduce flow resistance and gel development (related to viscosity reduction), the modern use of dispersants or thin-ners is to improve fluid-loss control and reduce filter cake thickness. The term dispersant is frequently used incor-rectly to refer to deflocculants. Dispersants are chemical materials that reduce the tendency of the mud to coagulate into a mass of particles or "floc cells". In addition, some dispersants contribute to fluid-loss control by plug-ging or bridging tiny openings in the filter cake. For this reason, some dispersants such as lignosulfonate (a highly anionic polymer) are more effective than others as fluid-loss reducers.

Quebracho is a type of tannin that is extracted from certain hardwood trees and used as a mud thinner. It also can be added to mud to counteract cement contamination. High pH is required for quebracho to dissolve readily in cold water. Therefore, quebracho should be added

度9～19lbm/gal的要求。然而，还有许多材料可用作钻井液的加重剂，包括方解石(3.08g/cm³)、碳酸钙(2.7～2.8g/cm³)、铁矿粉(5.05g/cm³)、钛铁矿粉(4.6g/cm³)和方铅矿粉(7.5g/cm³)。

4.2.2.2 滤失控制剂

黏土、分散剂和聚合物被广泛用作滤失控制剂。钠蒙脱石（膨润土）是大多数水基钻井液的主要降滤控制剂。胶体大小的钠—膨润土颗粒很薄，呈片状或板状，表面积很大，它们形成一个可压缩的滤饼。抑制性钻井液体系抑制了膨润土的水化作用，大大降低了其有效性。

因此，膨润土在加入这些体系之前，应先在清水中进行预水化。钠蒙脱石的颗粒较大、较粗，却没有表现出相同的滤失控制特性。

4.2.2.3 降黏剂或分散剂

虽然最初使用降黏剂的目的是为了减少流动阻力和凝胶形成（与降低黏度有关），但分散剂或降黏剂现在的用途是为了提高滤失控制和降低滤饼厚度。"分散剂"一词经常被错误地用来指反凝剂。实际上分散剂是一种化学材料，它能降低钻井液凝结成颗粒或"絮状细胞"的趋势。此外，一些分散剂通过堵塞或桥接滤饼中的微小开口来控制流体滤失。正是由于这个原因，一些分散剂，如木质素磺酸盐（一种高阴离子聚合物）作为降滤失剂比其他分散剂更有效。

栲胶是一种从某些硬木树中提取的单宁，可用作钻井液降黏剂。加到钻井液中用以抵消钻井液的污染。栲胶需要较高的pH值才能在冷水中溶解。因

with caustic soda in equal proportions by weight of 1 part of caustic soda to 5 parts quebracho. Concentration of quebracho varies between 0.5 and 2 lbm/bbl. Safety considerations for mixing these fluids must be observed.

4.2.2.4 Lost-Circulation Materials

In mud parlance, losses of whole drilling fluid to subsurface formation are called lost circulation. Circulation in a drilling well can be lost into highly permeable sandstones, natural or induced for-mation fractures, and cavernous zones; such a loss is generally induced by excessive drilling-fluid pressures. Drill-ing mud flowing into the formation implies a lack of mud returning to the surface after being pumped down a well.

An immense diversity of lost-circulation materials have been used. Commonly used materials include:

(1) Fibrous materials such as wood fiber, cotton fiber, mineral fiber, shredded automobile tires, ground-up cur-rency, and paper pulp

(2) Granular material such as nutshell (fine, medium, and coarse), calcium carbonate (fine, medium, and coarse), expanded perlite, marble, formica, and cottonseed hulls

(3) Flakelike materials such as mica flakes, shredded cellophane, and pieces of plastic laminate

Darley and Gray include an additional group of lost-circulation materials—slurries. Hydraulic cement, diesel oil-bentonite-mud mixes, and high-filter-loss drilling muds harden (increase strength) with time after placement.

4.2.2.5 Surfactants or Surface-Active Agents

A surface-active agent is a soluble organic compound that concen-trates on the surface boundary between two dissimilar substances and diminishes the surface tension between them. The molecular structure of surfactants is made of dissimilar groups having opposing solubility tendencies such as hydrophobic and hydrophilic. They are commonly used in the oil industry as additives to water-based drilling fluid to change the colloidal state of the clay from that of complete dispersion to one of controlled floc-culation. They may be cationic

此，栲胶要与烧碱按1∶5的质量比例加入。栲胶的浓度一般在0.5～2lbm/bbl之间，混合时必须注意安全问题。

4.2.2.4 堵漏剂

各种工作液在压差作用下漏入地层的现象称为井漏。钻井时，循环钻井液可能会漏失到高渗透性砂岩、天然或诱导裂缝及洞穴中；这种漏失通常是由钻井液压力过大引起的。钻井液漏入地层意味着入井钻井液不会再返回地面了。

目前已经使用了大量的堵漏材料。常用的材料包括：

（1）纤维材料，如木纤维、棉纤维、矿物纤维、汽车轮胎碎屑、碎浆和纸浆。

（2）坚果壳（细、中、粗）、碳酸钙（细、中、粗）、膨胀珍珠岩、大理石、胶木、棉籽壳等颗粒状材料。

（3）片状材料，如云母片、碎玻璃纸和塑料板碎片。

达利和格雷指出堵漏材料还包括水泥浆。水凝水泥、柴油—膨润土—泥浆混合物和高滤失量钻井液在充填后会随着时间而硬化（增加强度）。

4.2.2.5 表面活性剂

表面活性剂是一种可溶性的有机化合物，它能在两种不同物质的表面边界上产生作用，降低它们之间的表面张力。表面活性剂的分子结构是由具有相反溶解度的异性基团组成的，如疏水性和亲水性。在石油工业中，它们通常被用作水基钻井液的添加剂，以改变黏土的胶体状态，从完全分散状态变为可控的絮凝状态。它们可以是阳离

(dissociating into a large organic cation and a simple inorganic anion), anionic (dissociating into a large organic anion and a simple inorganic cation), or nonionic (long chains of polymer that do not dissociate).

Surfactants are used in drilling fluids as emulsifiers, dispersants, wetting agents, foamers and defoamers, and to decrease the hydration of the clay surface. The type of surfactant behavior depends on the structural groups of the molecules.

4.2.2.6 Various Other Additives

There are a plethora of other additives for drilling fluids. Some are used for pH control—that is, for chemical-reaction control (inhibit or enhance) and drill-string-corrosion mitigation. There are bactericides used in starch-laden fluids (salt muds in particular) to kill bacteria. There are various contaminate reducers such as sodium acid polyphosphate (SAPP) used while drilling cement to bind up calcium from the cement cuttings. There are corrosion inhibitors, especially H_2S scavengers. There are defoamers to knock out foaming and foaming agents to enhance foaming. There are lubricants for torque-and-drag reduction as well as pipe-freeing agents for when a drill string is stuck.

4.3 Rheological Models of Drilling Fluids

Audio 4.3

The frictional pressure loss term in the pressure balance equation is the most difficult to eval- uate. However, this term can be quite important because extremely large viscoelastic forces must be overcome to move drilling fluid through the long, slender conduits used in the drilling process. Generally, the elastic properties of drilling fluids and cement slurries and their effects during the flow in the hydraulic circuit of a drilling well are negligible. It is common practice in the computation of the frictional losses to consider only the effects of the viscous forces. However, with the new generations of drilling fluids, with polymers intro- duced on a regular basis, tests should be conducted to verify the elastic recovery from deformation that occurs during flow. A mathematical description of the viscous forces present in a fluid is required for the development of friction- loss equations. The rheological models generally used by drilling engineers to approximate fluid behavior are the Newtonian model, the Bingham plastic model, the power-law or Ostwald-de Waele model, and the Herschel- Bulkley model.

4.3.1 Overview of Rheological Models

The viscous forces present in a fluid are characterized by the fluid vis- cosity. To understand the nature of viscosity, consider a fluid contained between two large parallel plates of area A, which are separated by a small distance L (Fig. 4.2). The upper plate, which is initially at rest, is set in motion in the x direction at a constant velocity v. After suffi- cient time has passed for steady motion to be achieved, a constant force F is required to keep the upper plate moving at a constant velocity. The magnitude of the force F was found experimentally to be given by

引入聚合物，应该定期进行测试，从而验证流动过程中发生变形的弹性恢复。建立摩擦损失方程需要对流体中存在的黏性力进行数学描述。钻井工程师通常用来近似描述流体特性的流变模型有牛顿模型、宾汉模型、幂律或奥斯特瓦尔德—德韦勒模型和赫—巴模型。

4.3.1 流变模型概述

流体中存在的黏滞力用流体黏性来表征。为了理解黏度的本质，假设流体在两个面积为 A 的大平行板之间，两个板之间的距离 L 很小（图4.2）。最初静止的上板以恒定的速度 v 在 x 方向上运动。经过足够长的时间实现稳定运动后，需要恒定的力 F 来保持上板以恒定的速度运动。实验发现力的大小由下式给出：

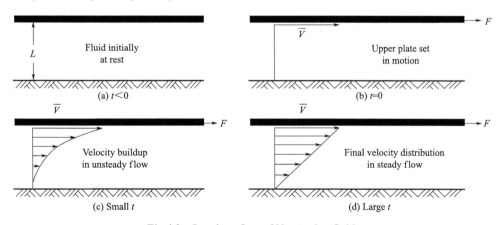

Fig.4.2 Laminar flow of Newtonian fluids

$$\frac{F}{A} = \mu \frac{v}{L} \tag{4.1a}$$

where, the term F/A is called the shear stress exerted on the fluid. The constant of proportionality μ is called the appar- ent viscosity of the fluid. Thus, shear stress is defined by

式中，F/A 称为施加在流体上的剪应力。比例常数 μ 称为流体的表观黏度。因此，剪应力定义为：

$$\tau = \frac{F}{A} \tag{4.1b}$$

Note that the area of the plate, A, is the area in contact with the fluid. The velocity gradient v/L is an expression of the shear rate:

请注意,板的面积是与流体接触的面积。速度梯度 v/L 是剪切速率的表达式:

$$\dot{\gamma} = \frac{dv}{dL} \approx \frac{v}{L} \tag{4.2}$$

The apparent viscosity is defined as the ratio of the shear stress to the shear rate and depends on the shear rate at which the measurement is made and the prior shear-rate history of the fluid. The viscous forces present in a simple Newtonian fluid are characterized by a constant fluid viscosity. However, most drilling fluids are too complex to be characterized by a single value for viscosity. Fluids that do not exhibit a direct proportionality between shear stress and shear rate are classified as non-Newtonian. Non-Newtonian fluids that are shear-dependent are pseudo- plastic or yield-pseudoplastic if the apparent viscosity decreases with increasing shear rate(Fig. 4.3a) and are dilatant if the apparent viscosity increases with increasing shear rate (Fig.4.3b). Many drilling fluids and cement slurries are generally pseudoplastic in nature.

表观黏度定义为剪切应力与剪切速率的比值,取决于测量时的剪切速率和流体之前的剪切速率。简单牛顿流体的黏性力用常数黏度来表征。然而,大多数钻井液都比较复杂,无法用单一的黏度值来表征。剪切应力和剪切速率之间非线性关系的流体,称为非牛顿流体。非牛顿流体表观黏度随剪切速率增加而减小时为假塑性或屈服假塑性流体[图4.3(a)],表观黏度随剪切速率增加而增大时为膨胀性流体[图4.3(b)]。通常很多钻井液和水泥浆都是假塑性流体。

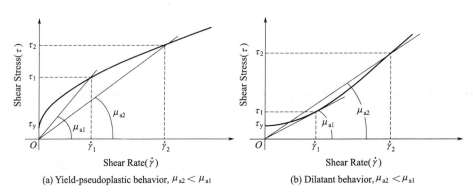

(a) Yield-pseudoplastic behavior, $\mu_{a2} < \mu_{a1}$ (b) Dilatant behavior, $\mu_{a2} < \mu_{a1}$

Fig.4.3 Shear stress vs. shear rate for yield-pseudoplastic and dilatant fluids

The Bingham plastic and the power-law rheological models were used in the past to approximate the pseudo-plastic behavior of drilling fluids and cement slurries. The Bingham model was fairly simple, but the power-law model could handle the behavior of pseudoplastic

过去采用宾汉模型及幂律模型来近似钻井液和水泥浆的假塑性行为。宾汉模型相当简单,但幂律模型比宾汉模型更能反映假塑性钻井液和水泥浆的行为,特

drilling fluids and cement slurries better than the Bingham plastic model, particularly at low shear rates.

However, a typical behavior of the majority of the drilling fluids and of the cement slurries used today includes a yield stress. The behavior of these fluids, called yield-pseudoplastic, is characterized by a trend similar to that of pseudoplastic fluids and by the presence of a finite shear stress at zero shear rate, which is referred to as the yield stress. One of the rheological models that fits better this kind of behavior, both at low and high shear rates, is the Herschel-Bulkley model.

Non-Newtonian fluids that are dependent on shear time (Fig. 4.4) are thixotropic if the apparent viscosity de- creases with time after the shear rate is increased to a new constant value and are rheopectic if the apparent viscosity increases with time after the shear rate is increased to a new constant value. Drilling fluids and cement slurries are generally thixotropic.

别是在低剪切速率情况下。

然而，目前使用的大多数钻井液和水泥浆的典型行为都包括屈服应力。这些流体称为屈服假塑性流体，其特征是存在类似于假塑性流体的趋势，零剪切速率下还存在一定的剪切应力，称为屈服应力。赫—巴模型更适合描述这种高低剪切速率下的流体行为。

触变性和流变性是非牛顿流体与剪切时间相关的属性（图4.4），如果剪切速率增加到一个新的恒定值后保持不变，表观黏度随剪切时间降低，则为触变性；如果剪切速率增加到一个新的恒定值后保持不变，表观黏度随剪切时间增加，则为流变性。钻井液和水泥浆通常具有触变性。

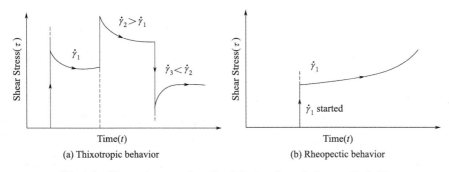

Fig.4.4　Shear stress vs. time for thixotropic and rheopectic fluids

At present, the thixotropic behavior of drilling fluids and cement slurries is not modelled mathematically. How- ever, drilling fluids and cement slurries generally are stirred before measuring the apparent viscosities at various shear rates so that steady-state conditions are obtained. Not accounting for thixotropy is satisfactory for most cases, but significant errors can result when a large number of direction changes and diameter changes are present in the hydraulic circuit of the well

目前，钻井液和水泥浆的触变性没有数学模型。然而，在测量不同剪切速率下的表观黏度之前，通常先搅拌钻井液和水泥浆，以便获得稳态条件。在大多数情况下，可以不考虑触变性，但是当井中水力通道存在大量的方向变化或直径变化较大时，不考虑触变性会导致显著的误差。

4.3.2 Newtonian Model

The Newtonian model states that the shear stress τ is directly proportional to the shear rate γ as follows:

$$\tau = \mu \dot{\gamma} \quad (4.3)$$

where, μ, the constant of proportionality, is known as the apparent viscosity or simply the viscosity of the fluid. In terms of the moving plates (Fig. 4.2), this means that if the force F is doubled, the plate velocity v also will double. Examples of Newtonian fluids are water, gases, high-gravity oils, and brines.

Viscosity is expressed in poise (P). A poise corresponds to 1 dyne·s/cm² or 1 g/(cm·s). In the drilling industry, viscosity generally is expressed in centipoise (cP), where 1cP= 0.01 poise. In SI units, viscosity is expressed in pascal-seconds (Pa·s), which corresponds to 10 poise. Sometimes, viscosity is expressed in units of lbf-sec/ft². The units of viscosity can be related by

$$1\frac{\text{lbf} \cdot \text{s}}{\text{ft}^2} \times \frac{(453.6\text{g}) \times (980.7\text{cm}/\text{s}^2)/\text{lbf}}{(30.48\text{cm}/\text{ft})^2} = 478.83 \text{dyn} \cdot \text{s}/\text{cm}^2$$

$$=47.883\text{Pa} \cdot \text{s}=47,883\text{cP} \quad (4.4)$$

The linear relation between shear stress and shear rate described by Eq. 4.3 is valid only as long as the fluid moves in layers, or lamina. A fluid that flows in this manner is said to be in laminar flow. This is true only at relatively low shear rate, and the pressure-velocity relationship is a function of the viscous properties of the fluid. At high shear rates, the flow pattern changes from laminar flow to turbulent flow, in which the fluid particles move downstream in a tumbling, chaotic motion so that vortices and eddies are formed in the fluid. Dye injected into the flow stream thus would be dispersed quickly throughout the entire cross section of the fluid. The turbulent flow of fluids has not been described mathematically. Thus, when turbulent flow occurs, frictional pressure drops must be determined by empirical correlations.

4.3.2 牛顿模型

牛顿模型指出剪切应力 τ 与剪切速率 γ 成正比，如下所示：

(4.3)

式中，比例常数 μ 被称为表观黏度或简称为流体黏度。这表明在图 4.2 中移动的平板如果力 F 加倍，板速度 v 也将加倍。典型的牛顿流体有水、气体、高比重油和盐水。

黏度以泊（P）为单位。1P= 1dyn·s/cm²=1g/(cm·s)。在钻井行业，黏度通常以厘泊（cP）表示，其中 1 cP= 0.01P。在国际单位制中，黏度用 Pa·s 来表示，相当于 10P。有时，黏度以 lbf·s/ft² 为单位表示，可通过下式联系起来：

(4.4)

式（4.3）描述的剪切应力和剪切速率之间的线性关系仅在流体层流运动时有效。流体按此方式流动时被称为层流。层流仅在相对较低的剪切速率下才成立，此时压力一速度关系是流体黏度的函数。高剪切速率下，流动模式从层流变为紊流，其中流体颗粒以翻滚、混乱的运动向下游移动，从而在流体中形成涡流和漩涡。注入液流中的染料将快速分散在流体的整个横截面上。流体的紊流没有数学方法描述。因此，当发生紊流时，摩擦压降必须由经验关系确定。

4.3.3 Bingham Plastic Model

The Bingham plastic model is defined by

4.3.3 宾汉模型

宾汉模型定义为：

$$\tau = \tau_y + \mu_p \dot{\gamma}, \tau > \tau_y \tag{4.5}$$

A graphical representation of this behavior is shown in Fig. 4.5b.

这种模型的图形表示如图4.5（b）所示。

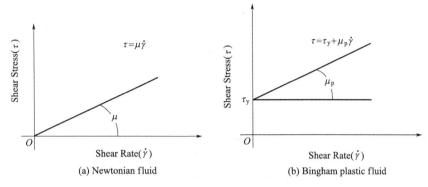

Fig.4.5 Shear stress vs.shear rate for Newtonian fluid and Bingham plastic fluid

A Bingham plastic fluid will not flow until the applied shear stress τ exceeds a certain minimum value τ_y known as the yield stress. After the yield stress has been exceeded, changes in shear stress are proportional to changes in shear rate, and the constant of proportionality is called the plastic viscosity, μ_p. Eq. 4.5 is valid only for laminar flow. Note that the units of plastic viscosity are the same as the units of Newtonian or apparent viscosity. To be consistent, the units of the yield stress τ_y must be the same as the units for shear stress τ. Thus, the yield stress has consistent units of dyn/cm². In SI units, the yield stress is expressed in N/m² or Pa. However, yield stress usually is expressed in field units of lbf/100 ft². The units can be related (at sea level) by

宾汉流体，当剪切应力 τ 超过某个最小值 τ_y（称为屈服应力）时才会流动。超过屈服应力后，剪切应力与剪切速率的变化成正比，比例常数称为塑性黏度 μ_p，式（4.5）只对层流有效。注意，塑性黏度的单位与牛顿流体的黏度或表观黏度单位相同。为保持一致，屈服应力 τ_y 的单位必须与剪切应力 τ 的单位相同。因此，屈服应力的单位为 dyn/cm²。在国际单位制中，屈服应力用 N/m² 或 Pa 表示。现场通常用 lbf/100 ft² 来表示屈服应力。单位转换关系为（在海平面上）：

$$\frac{1\text{lbf}}{100\text{ft}^2} \times \frac{453.6\text{g} \times 980.7\text{cm}/\text{s}^2/\text{lbf}}{100(30.48\text{cm}/\text{ft})^2} = 4.788\text{dyn}/\text{cm}^2 = 0.4788\text{Pa} \tag{4.6}$$

4.3.4 Power-Law Model

The power-law model is defined by

4.3.4 幂律模型

幂律模型的定义如下：

$$\tau = K\dot{\gamma}^n \tag{4.7}$$

A graphical representation of the model is shown in Fig. 4.6.

Like the Bingham plastic model, the power-law model requires two parameters for fluid characterization. However, the power-law model can be used to represent a pseudoplastic fluid ($n < 1$), a Newtonian fluid ($n = 1$), or a dilatant fluid ($n > 1$). Eq. 4.7 is valid only for laminar flow.

The parameter K usually is called the consistency index of the fluid, and the parameter n usually is called either the power-law exponent or the flow behavior index. The deviation of the dimensionless flow behavior index from unity characterizes the degree to which the fluid behavior is non-Newtonian. The units of the consistency index K depend on the value of n. K has units of dyn·s^n/cm² or g/(cm·s^{2-n}). In SI units, the consistency index is expressed in N·s^n/m² or Pa·s^n. In this text, a unit called the equivalent centipoise (eq cP) will be used to represent 0.01 dyn·s^n/cm². Occasionally, the consistency index is expressed in units of lbf·sec^n/ft². The units of the consistency index can be related (at sea level) by

该模型的图形表示如图 4.6 所示。

像宾汉模型一样，幂律模型需要两个参数来表征流体。但幂律模型可用来表示假塑性流体 ($n<1$)、牛顿流体 ($n=1$) 或膨胀流体 ($n>1$)。式（4.7）只对层流有效。

参数 K 通常称为流体的稠度系数；参数 n 通常称为幂律指数或流性指数；表征了流体的非牛顿程度。稠度系数 K 的单位取决于 n 的值。K 可用 dyn·s^n/cm² 或 g/(cm·s^{2-n}) 作为单位。在国际单位制中，稠度系数以 N·s^n/m² 或 Pa·s^n 表示。在本文中，用一个称为等效厘泊（eq cP）的单位来表示 0.01dyn·s^n/cm²。偶尔，稠度系数以 lbf·sec^n/ft² 为单位。稠度系数的单位转换关系为（在海平面上）：

$$1\frac{\text{lbf}\cdot\text{sec}^n}{\text{ft}^2}\times\frac{453.6\text{g}\times980.7\text{cm/s}^2/\text{lbf}}{(30.48\text{cm/ft})^2}=478.83\text{dyn}\cdot s^n/\text{cm}^2=47.883\text{Pa}\cdot s^n=47.883\text{eq cP}$$

(4.8)

4.3.5 Herschel-Bulkley Model

The Herschel-Bulkley model is defined by

$$\tau=\tau_y+K\dot{\gamma}^n, \tau>\tau_y$$

4.3.5 赫—巴模型

赫—巴模型由下式定义：

(4.9)

A graphical representation of the model is shown in Fig. 4.7. The model combines the characteristics of the Bingham and power-law models and requires three parameters for fluid characterization. The Herschel-Bulkley model can be used to represent a yield-pseudoplastic fluid ($n<1$), a dilatant fluid ($n>1$), a pseudoplastic fluid ($\tau_y=0$, $n<1$), a plastic fluid ($n=1$), or a Newtonian fluid ($\tau_y=0$, $n=1$). Eq. 4.50 is valid only for laminar flow.

Like the Bingham plastic model, a fluid represented by this model will not flow until the applied shear stress

赫—巴模型的图形表示如图 4.7 所示。赫—巴模型结合了宾汉模型和幂律模型的特点，需要三个参数来表征。赫—巴模型可用于表示屈服假塑性流体 ($n<1$)、膨胀流体 ($n>1$)、假塑性流体 ($\tau_y=0$, $n<1$)、塑性流体 ($n=1$) 或牛顿流体 ($\tau_y=0$, $n=1$)。式（4.9）只对层流有效。

与宾汉塑性模型一样，赫—巴模型表示当剪切应力 τ 超过最

exceeds a minimum value τ_y, which is called the yield stress. The fluid behaves like a solid until the applied force is high enough to exceed the yield stress. Thus, the yield stress has consistent units of dyn/cm^2 In SI units, the yield stress is expressed in N/m^2 or Pa. However, yield stress usually is expressed in field units of lbf/100 ft^2. The parameter K is called the consistency index of the fluid, and the parameter n usually is called the flow behavior index.

小值 τ_y（称为屈服应力）之前流体像固体一样不会流动。当剪切应力超过屈服应力时，流体才会流动。屈服应力单位为 dyn/cm^2。在国际单位制中，屈服应力以 N/m^2 或 Pa 为单位。然而，现场通常用 lbf/100 ft^2 作为屈服应力的单位。参数 K 称为稠度系数，参数 n 称为流性指数。

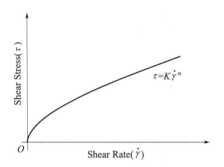

Fig.4.6　Shear stress vs.shear rate for a pseudoplastic power-law fluid

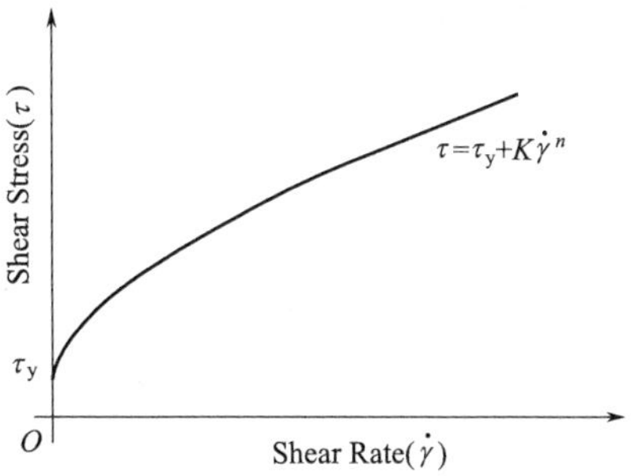

Fig.4.7　Shear stress vs.shear rate for a yield-pseudoplastic Herschel-Bulkley fluid

Generally, the rheological parameters that characterize a model are determined by using analytical equations based on a data set of measurements from the rotational viscometer. How- ever, to improve the accuracy of the calculation on the rheological parameters, statistical regression methods are used. They are applied to the complete set (τ,γ) of measurements performed on a sample of the fluid in the rota- tional viscometer. Outcomes are a higher accuracy in determining the rheological parameters that characterize the behavior of the tested fluid, and as a consequence a better evaluation of flow parameters such as velocity profile, flow regime, and pressure drop.

Merlo et al. presented a hydraulic computation model based on the Herschel-Bulkley rheological fluid model. They treated the annulus as a slot and considered also the temperature and pressure influence on drilling hydraulics. Field test circulation results were obtained for 17.5 in and 12.25 in. openhole sections at various depths while measuring the rheology of mud. Predictions were made by computing pressure drop using one of the three rheological

一般来说，流变模型的流变参数可利用旋转黏度计和流变方程来确定。然而，为了提高准确性，通常使用统计回归方法来计算流变参数，这些统计方法适用于处理旋转黏度计对流体样品进行的整套 (τ,γ) 测量结果。确定的流变参数结果在表征被测流体性能方面具有更高的准确性，因此可以更好地评估流速剖面、流态和压降等流动参数。

梅林等人提出了一个基于赫—巴模型的水力计算模型。他们将环空视为一个槽，并充分考虑了温度和压力对钻井液水力参数的影响。在 17.5in 和 12.25in 裸眼段的不同深度处测量钻井液流变性，并取得了现场循环测试结果。利用三种流变模型中的一种

models [i.e., Bingham plastic—American Petroleum Institute (API) 13 (600 and 300 r/min read- ings), power-law—API 13 (600 and 300 r/min readings), and Herschel-Bulkley based on regression of all six rheological data points]. The Herschel-Bulkley rheological model, with parameters derived through regression analysis, gave the best fit to their rheological data and to the pressure-drop data. The errors in predicting the pressure drop and comparing to measurements using the Herschel-Bulkley model were smaller than the errors when using the two-speed models for Bingham plastic and power-law fluids, although the differences, seen in Fig. 4.8, were not that great.

来计算压降（即宾汉 API 13 标准 600 和 300r/min 读数、幂律 API 13 号标准 600 和 300r/min 读数、基于所有六个流变数据点回归的赫—巴模型）。通过回归分析得出的赫—巴流变模型的参数给出了压降和流变数据之间最吻合的结果。用赫—巴流变模型预测的压降与测量值之间的误差，要比用宾汉模型和幂律模型所采用的双速模型误差小，如图 4.8 所示，尽管差异不是很大。

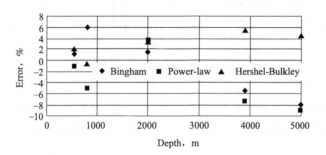

Fig.4.8 Comparison of Merlo et al. pressure-drop predictions with their field data

4.3.6 Other Rheological Models

The behavior of drilling fluids and cement slurries also can be simulated by other rheological models. The most used in the practice are the Casson model, for plastic fluids, and the Robertson-Stiff model, for yield-pseudoplastic fluids.

4.3.6.1 Casson Model

The Casson model (Casson 1959) is often used to simulate drilling fluids and cement slurries with plastic behavior, with a higher accuracy than the Bingham plastic model. The model is defined by

$$\sqrt{\tau} = \sqrt{\tau_y} + \sqrt{\mu_p}\sqrt{\dot{\gamma}}$$

(4.10)

Eq. 4.10 is valid only for laminar flow. Generally, the model is plotted with coordinates ($\sqrt{\tau}$ $\sqrt{\dot{\gamma}}$) instead of (τ, γ) to still maintain the linear trend. A graphical representation is shown in Fig. 4.9.

4.3.6 其他流变模型

钻井液和水泥浆的特性也可以通过其他流变模型来模拟。实践中最常用的是用于塑性流体的卡森模型，以及用于屈服假塑性流体的罗伯逊—斯蒂夫模型。

4.3.6.1 卡森模型

卡森模型通常用来模拟具有塑性特性的钻井液和水泥浆，其精度高于宾汉模型。该模型的定义如下：

(4.10)

式（4.10）仅适用于层流。一般用坐标（$\sqrt{\tau}$，$\sqrt{\dot{\gamma}}$）代替（τ，γ）绘制模型，仍然保持线性趋势，如图 4.9 所示。

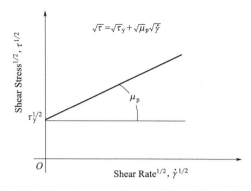

Fig.4.9 Shear stress vs.shear rate for a Casson fluid

Like the Bingham plastic model, the Casson model requires two parameters for fluid characterization. A fluid represented by this model requires a finite shear stress, τ_y below which it will not flow. Above this finite shear stress, referred to as the yield stress, changes in shear stress are proportional to changes in shear rate, and the constant of proportionality is called the plastic viscosity, μ_p. Note that the units of plastic viscosity are the same as the units of Newtonian or apparent viscosity. The yield stress has units of dyne/cm². In SI units, the yield stress is expressed in N/m² or Pa. However, yield stress usually is expressed in field units of lbf/100 ft²

It can be seen that the yield stress calculated by the Casson model is less than that observed with the Bingham plastic model. The plastic viscosity changes according to the model, and here it decreases substantially compared to that calculated by Bingham model.

4.3.6.2 Robertson-Stiff Model

The Robertson-Stiff model adequately describes the behavior of yield-pseudoplastic fluids, even though slightly less accurately than the Herschel-Bulkley model. The model is defined by

$$\tau = A(\dot{\gamma} + C)^B \tag{4.11}$$

Graphical representation of the model is similar to that for the Herschel-Bulkley model. Eq.4.11 can describe the behavior of yield-pseudoplastic fluids,

像宾汉模型一样,卡森模型也用两个参数来表征。卡森模型所代表的流体需要一个限定的切应力,低于这个切应力,它就不会流动;大于这个限定切应力(屈服应力),剪切应力与剪切速率的变化成正比,比例常数称为塑性黏度 μ_p。注意,塑性黏度的单位与牛顿流体的黏度或表观黏度单位相同。屈服应力以 dyn/cm² 为单位。在国际单位制中,屈服应力用 N/m² 或 Pa 表示。然而,现场通常用 lbf/100 ft² 来表示。

可以看出,用卡森模型计算的屈服应力小于用宾汉模型计算的屈服应力。塑性黏度随模型的变化而变化,与宾汉模型计算的塑性黏度相比,此处的塑性黏度明显下降。

4.3.6.2 罗伯逊—斯蒂夫模型

罗伯逊—斯蒂夫模型充分地描述了屈服假塑性流体的特性,其精度略低于赫—巴模型。该模型定义如下:

罗伯逊—斯蒂夫模型的图形表示类似于赫—巴模型。式(4.11)可以描述屈服假塑性流体、假塑性

pseudoplastic fluids ($C=0$), and Newtonian fluids ($B=1$, $C=0$). Eq. 4.11 is valid only for laminar flow. Fluids represented by the Robertson-Stiff model exhibit a yield stress once the flow is initiated. The yield stress is given by AC^B. A, B, and C are constants of the model. The units of the constant A depend on the value of the constant B. A has units of dyne $\cdot$ s^B/cm^2 org/cm $\cdot$ s^{2-B}. In SI units, the constant A is expressed in N $\cdot$ s^B/m^2 or Pa $\cdot$ s^B. The constant B is dimensionless, and the units of the constant C are s^{-1}.

The values of parameters A and B are comparable with the consistency index and the flow behavior index of the Herschel-Bulkley model, respectively. The yield stress τ_y determined with the Robertson-Stiff model is slightly higher (+32.0%), but still comparable

4.3.7 Advanced Models

In addition to the models previously reported, there are many other empirical mathematical descriptions that can describe with high accuracy the behavior of the viscous forces of drilling fluids and cement slurries. Very often the models have been developed to predict the properties of fluids not properly related to drilling operations, such as polymer solutions, suspensions, and blood. However, they also can be applied successfully to the flow of drilling fluids and cement slurries. They can be classified according to the number of constant parameters that the mathematical description contains. Generally, the higher the number of parameters that characterize the model, the better the approximation of the model to the fluid behavior, but the complexity of the flow equations becomes higher. Generally, the rheological parameters of these models are determined by non-linear regression techniques, and in very few cases analytical solutions can be available. The equations are valid only for laminar flow.

4.3.7.1 Three-Parameter Models

These models require three constant parameters for fluid characterization.

流体 ($C=0$) 和牛顿流体 ($B=1$, $C=0$)，式（4.11）仅对层流有效。当开始流动时，罗伯逊—斯蒂夫模型所代表的流体表现出屈服应力。屈服应力由 AC^B 给出。A、B、C 为模型的常数。常数 A 的单位取决于常数 B 的值。A 的单位是 dyn $\cdot$ s^B/cm^2 或 g/cm $\cdot$ s^{2-B}。在国际单位制中，常数 A 用 N $\cdot$ s^B/m^2 或 Pa $\cdot$ s^B 表示。常数 B 是无量纲的，常数 C 的单位是 s^{-1}。

参数 A、B 的值分别与赫—巴流变模型的稠度系数和流性指数相当。用罗伯逊—斯蒂尔模型确定的屈服应力 τ_y 略高 (+32.0%)，但仍具有可比性。

4.3.7 高级模型

除了前面所述的模型以外，还有许多其他经验数学描述可以准确地描述钻井液和水泥浆的流变特性。通常情况下，这些模型一般用来预测与钻井作业不相关的流体性质，如聚合物溶液、悬浮液和血液。但它们也可以用于钻井液和水泥浆。这些模型可以根据数学描述所包含的常数参数的个数来分类。通常，表征模型的参数数量越多，模型对流体性能描述越接近，但参数越多流动方程的复杂性越强。一般来说，这些模型的流变参数通过非线性回归技术来确定，在极少数情况下，可以获得解析解。这些方程也只对层流有效。

4.7.3.1 三参数模型

这些模型需要三个恒定参数来表征流体。

The Graves- Collins model is defined by

$$\tau = \left(1 - e^{-\beta \dot{\gamma}}\right)(\tau_0 + \mu \dot{\gamma}) \tag{4.12}$$

The constant parameters of the model are τ_0, μ and β. The model can approximate with good accuracy pseudoplastic fluids at low shear rates and plastic fluids at high shear rates.

The Gucuyener model (Gucuyener 1983) is defined by

$$\tau^{1/m} = \tau_y^{1/m} + \eta \dot{\gamma}^{1/2} \tag{4.13}$$

The constant parameters of the model are τ_y, η, and m. The model predicts the behavior of yield-pseudoplastic fluids. In addition, it can be used to represent pseudoplastic fluids ($\tau_y = 0$), plastic fluids ($m = 2$), and Newtonian fluids ($\tau_y = 0$, $m = 2$).

The Sisko model is defined by

$$\tau = a\dot{\gamma} + b\dot{\gamma}^c \tag{4.14}$$

The constant parameters of the model are a, b, and c. The model can describe the behavior of pseudoplastic fluids ($a = 0$), and Newtonian fluids ($b = 0$).

4.3.7.2 Four-Parameter Models

These models require four constant parameters for fluid characterization. The Shulman model is defined by

$$\tau^{1/n} = \tau_0^{1/n} + (\eta \dot{\gamma})^{1/m} \tag{4.15}$$

The constant parameters of the model are τ_0, η, n and m. The model approximates with high accuracy the proper-ties of yield-pseudoplastic fluids ($n = 1$), pseudoplastic fluids ($\tau_0 = 0$, $n = 1$), plastic fluids ($n = m = 1$ for Bingham plastic fluids, and $n = m = 2$ for Casson fluids), and Newtonian fluids ($\tau_0 = 0$, $n = m = 1$).

The Zhu model is defined by

$$\tau = \tau_0 \left(1 - e^{-m\dot{\gamma}}\right) + \eta_1 e^{-t_1 \dot{\gamma}} \dot{\gamma} \tag{4.16}$$

The constant parameters of the model are τ_0, η_1, m, and t_1. The model can approximate with high accuracy the behavior of yield-pseudoplastic fluids.

4.3.7.3 Five-Parameter Models

These models require five constant parameters for fluid characterization. The Maglione model is defined by

$$\tau^{1/n} = a^{1/n} + (b\dot{\gamma})^{1/m} + (c\dot{\gamma})^{\frac{1}{m}+1} \qquad (4.17)$$

The five constant parameters of the model are a, b, c, n, and m. The parameter a is the yield stress, parameters b and c are related to the fluid viscosity, and n and m are related to the with high accuracy the properties of yield-pseudoplastic flow behavior index of the fluid. The model approximates fluids ($c = 0$, $n = 1$), pseudoplastic fluids ($a = c = 0$, $n = 1$), plastic fluids ($c = 0$ and $n = m = 1$ for Bingham plastic fluids, $c = 0$ and $n = m = 2$ for Casson fluids), and Newtonian fluids ($a = c = 0$, $n = m = 1$).

4.4 Exercise

(1) An upper plate of 20cm² area is spaced 1 cm above a stationary plate. Compute the viscosity in cP of a Newtonian Fluid between the plates if a force of 100 dyn is required to move the upper plate at a constant velocity of 10 cm/s.

(2) An upper plate of 20cm² area is spaced 1 cm above a stationary plate. Compute the yield stress and plastic viscosity of a Bingham Plastic fluid between the plates if a force of 200 dyn is required to cause any movement of the upper plate and a force of 400 dyn is required to move the upper plate at a constant velocity of 10 cm/s.

(3) An upper plate of 20 cm² is spaced 1 cm above a stationary plate. Compute the Power-Law Fluids consistency index and flow behavior index if a force of 50 dyn is required to move the upper plate at a constant velocity of 4 cm/s and a force of 100 dyn is required to move the upper plate at a constant velocity of 10 cm/s.

(4) An upper plate of 20 cm² is spaced 1 cm above a stationary plate. Compute the three rheological parameters of the Herschel-Bulkley model if a force of

60 dyn is required to move the upper plate at a constant velocity of 5 cm/s, a force of 130 dyn is required to move the upper plate at a constant velocity of 12 cm/s, and a force of 250 dyn is required to move the upper plate at a constant velocity of 25 cm/s.

(5) An upper plate of 20cm² area is spaced 1 cm above a stationary plate. Compute the yield stress and plastic viscosity of a Casson fluid between the plates, if a force of 200 dyn is required to cause any movement of the upper plate and a force of 400 dyn is required to move the upper plate at a constant velocity of 10 cm/s.

(6) An upper plate of 20 cm² is spaced 1 cm above a stationary plate. Compute the three rheological parameters of the Robertson-Stiff model and the yield stress if a force of 60 dyn is required to move the upper plate at a constant velocity of 5 cm/s, a force of 130 dyn is required to move the upper plate at a constant velocity of 12 cm/s, and a force of 250 dyn is required to move the upper plate at a constant velocity of 25 cm/s.

速度移动上板，需要130dyn的力才能以12cm/s的恒定速度移动上板，需要250dyn的力才能以25cm/s的恒定速度移动上板，请计算赫—巴模型的三个流变参数。

（5）上平板面积为20cm²，和固定的下平板间隔为1cm。如果需要200dyn的力才能引起上板的移动，并且需要400dyn的力才能以10cm/s的恒定速度移动上板，请计算板之间卡森模型的屈服应力和塑性粘度。

（6）上平板面积为20cm²，和固定的下平板间隔为1cm。如果以5cm/s的恒定速度移动上板需要60dyn的力，以12cm/s的恒定速度移动上板需要130dyn的力，以25cm/s的恒定速度移动上板需要250dyn的力，则计算罗伯逊—斯蒂夫模型的三个流变参数和屈服应力。

Drilling Hydraulics 钻井水力学 5

5.1 Introduction to Drilling Hydraulics

5.1 钻井水力学简介

Audio 5.1

The science of fluid mechanics is very important to the drilling engineer. Extremely large fluid pressures are created in the long slender wellbore and tubular pipe strings by the presence of drilling mud or cement. The presence of these subsurface pressures must be considered in almost every well problem encountered.

In this chapter, the relations needed to determine the subsurface fluid pressures will be developed for three common well conditions. These well conditions are (1) a static condition in which both the well fluid and the central pipe string are at rest, (2) a circulating operation in which the fluids are being pumped down the central pipe string and up the annulus, and (3) operations in which a central pipe string is being moved up or down through the fluid. The second and third conditions listed are complicated by the non-Newtonian behavior of drilling muds and cement slurries. Also included in this chapter are the relations governing the transport of rock fragments and immiscible formation fluids to the surface by the drilling fluid.

Applications include (1) calculation of subsurface hydrostatic pressures tending to fracture exposed formations, (2) displacement of cement slurries, (3) bit nozzle size selection, (4) surge pressures due to a vertical pipe movement, and (5) cuttings-carrying capacity of drilling fluids.

流体力学对于钻井工程师来说非常重要。因为钻井液或固井水泥浆在细长的井眼和管柱中流动时会产生极大的流体压力。所以，在处理遇到的每个钻井复杂问题时，钻井工程师们都不能忽视这些井下压力的存在。

本章将针对三种常见的钻井工况，建立能够确定井下流体压力的关系式。这三种工况分别是：(1) 静止状态，即井内流体和管柱均处于静止状态；(2) 循环作业，即将钻井液向下泵入钻柱，并沿环空向上返出；(3) 起下钻，即将钻柱起出或下入井筒的作业过程。由于钻井液和水泥浆具有非牛顿流体的性质，这导致上述第二种和第三种工况变得复杂。本章还将讨论控制岩屑和非混溶地层流体通过钻井液运移至地面的关系式。

钻井水力学的用途包括：(1) 计算导致裸露地层破裂的井下静水压力；(2) 水泥浆的顶替；(3) 钻头喷嘴尺寸的选择；(4) 由管柱垂直运动产生的波动压力；(5) 钻井液的携岩能力。

The three primary functions of a drilling fluid (the transport of cuttings out of the wellbore, prevention of fluid influx, and the maintenance of wellbore stability) depend on the flow of drilling fluids and the pressures associated with that flow. For instance, if the wellbore pressure exceeds the fracture pressure, fluids will be lost to the formation. If the wellbore pressure falls below the pore pressure, fluids will flow into the wellbore, perhaps causing a blowout. It is clear that accurate wellbore pressure prediction is necessary. To properly engineer a drilling-fluid system, it is necessary to be able to predict pressures and flows of fluids in the wellbore. The purpose of this chapter is to describe in detail the calculations necessary to predict the flow performance of various drilling fluids for the variety of operations used in drilling and completing a well.

Fluid-mechanics problems range from the simplicity of a static fluid to the complexity of dynamic surge pressures associated with running pipe or casing into the hole.

This chapter will first present each specific wellbore flow problem in detail, starting from the simplest and progressing to the most complicated.

These problems will be considered in the following order: (1) Hydrostatic pressure calculations; (2) Steady flow of fluids.

Following these basic problems, we will present a series of special topics: (1) Fluid rheology; (2) Laminar flow; (3) Turbulent flow; (4) Eccentric annulus flow; (5) Flow with moving pipe; (6) Steady-State wellbore flow; (7) Dynamic wellbore pressure prediction; (8) Cuttings transport.

5.2 Hydrostatic Pressure Calculations

Subsurface well pressures are determined most easily for static well conditions. The variation of pressure with depth in a fluid column can be obtained

Audio 5.2

by considering the free-body diagram (Fig. 5.1) for the vertical forces acting on an element of fluid at a depth Z in a hole of cross-sectional area A. The downward force on the fluid element exerted by the fluid above is given by the pressure p times the cross-sectional area of the element, A:

度的变化，该微元位于一个横截面积为 A 的井眼中，深度为 Z。微元上方的流体对它施加向下的力，大小等于压力 p 乘以微元的横截面积 A：

$$F_1 = pA \tag{5.1}$$

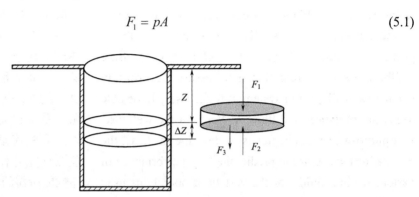

Fig.5.1 Forces acting on fluid element

Likewise, there is an upward force on the element exerted by the fluid below, given by

同样，微元下方的流体对它施加向上的力，即：

$$F_2 = \left(p + \frac{dp}{dZ} \Delta Z \right) A \tag{5.2}$$

In addition, the weight of the fluid element is exerting a downward force given by

此外，该微元的自重产生一个向下的力：

$$F_3 = \rho g A \Delta Z \tag{5.3}$$

where, ρ is the density of the fluid, and g is the acceleration due to gravity, which at sea level has a value of 9.81 m/s².

式中，ρ 是流体的密度，g 是重力加速度，取 9.81 m/s²。

Because the fluid is at rest, no shear forces exist and the three forces shown must be in equilibrium; hence,

由于流体处于静止状态，因此不存在剪切力，但上述三个力必须处于平衡状态。由此得到：

$$F_1 - F_2 + F_3 = 0 \tag{5.4}$$

$$pA - \left(p + \frac{dp}{dZ} \Delta Z \right) A + \rho g A \Delta Z = 0 \tag{5.5}$$

Expansion of the second term and division by the element volume $A\Delta Z$ gives

将第二项展开并除以微元的体积 $A \Delta Z$，得到：

$$\frac{dp}{dZ} = \rho g \tag{5.6a}$$

In SI units, Eq. 5.6a has density ρ in kg/m³, depth Z in m, and pressure p is given in Pascals (Pa), while in field units ρ is given in lbm/gal, Z in ft, g is 1 lbf/lbm, and p is lbf/in²(psi). With the conversion factors applied, Eq.5.6a takes the following form in field units:

用国际单位制表示，式（5.6a）中密度 ρ 的单位是 kg/m³，深度 Z 的单位是 m，压力 p 的单位是帕斯卡（Pa）。如果用现场单位制，则 ρ 的单位是 lbm/gal，Z 的单位是 ft，g 的单位是 1lbf/lbm，p 的单位是 lbf/in²（psi）。应用转换系数后，式（5.6a）用现场单位表示为：

$$\frac{\mathrm{d}p}{\mathrm{d}Z} = 0.05195\rho \quad (5.6b)$$

A note on units: The drilling industry uses an odd mixture of unconventional English units combined with assorted metric units, depending on where in the world you are drilling. As a result, almost any calculation requires unit conversions, or formulae will have strange coefficients, like 0.05195 in Eq. 5.6b. SI units have the advantage of not needing conversions because the units are consistent, meaning that use of SI units in the equation gives results in SI units. The downside is that no one uses SI units in the drilling industry. When in doubt, convert everything to SI units, calculate, and then convert the result back to field units.

注意：钻井行业常使用英制单位和公制单位，使用哪种单位制取决于钻井作业所处地区惯例。几乎所有的计算都需要进行单位转换，或者在公式中出现转换系数 [如式（5.6b）中的 0.05195]。国际单位制的优点是不需要单位转换，由于单位一致，计算过程和结果都是国际单位。但钻井行业的人不太习惯使用国际单位。因此，有时需要将所有数据转换为国际单位再进行计算，然后再将结果转换为油田单位。

5.2.1 Incompressible Fluids

If we are dealing with a liquid such as drilling mud or salt water, fluid compressibility is negligible for low temperatures, and specific weight can be considered constant with depth. Integration of Eq. 5.6a for an incompressible liquid gives

5.2.1 不可压缩流体

低温下，钻井液或盐水之类液体的可压缩性可忽略不计，并且其密度可视为随深度的变化保持恒定。对于不可压缩流体，由式（5.6a）积分得到：

$$p = \rho g Z + p_0 \quad (5.7a)$$

where, p_0, the constant of integration, is equal to the surface pressure at $Z = 0$. If we are interested in absolute pressure, p_0 will be, at least, atmospheric pressure. Absolute pressure in English units is designated psia. Often, we are interested only in incremental pressure relative to atmospheric, which is called gauge pressure.

式中，p_0 是积分常数，它等于深度为零时的地面压力。如果取绝对压力，则 p_0 等于大气压。绝对压力的英制单位是 psia。通常只关注相对于大气压的压力增量，即所谓表压，表压的英文单

Gauge pressure in English units is designated psig. In this case, p_0 can be negative. Eq. 5.7a becomes, in field units,

$$p = 0.05195\rho Z + \rho_0 \qquad (5.7b)$$

Normally the static surface pressure p_0 is zero (gauge pressure) unless the blowout preventer of the well is closed and the well is trying to flow. An important application of the hydrostatic pressure equation is the determination of the proper drilling-fluid density. The fluid column in the well must be of sufficient density to cause the pressure in the well opposite each permeable stratum to be greater than the pore pressure of the formation fluid in the permeable stratum. This problem is illustrated in the schematic drawing shown in Fig.5.2. However, the density of the fluid column must not be sufficient to cause any of the formations exposed to the drilling fluid to fracture. A fractured formation would allow some of the drilling fluid above the fracture depth to leak rapidly from the well into the fractured formation.

位为 psig。在这种情况下，p_0 可以为负。式（5.7a）用油田单位表示为：

通常，静态地表压力 p_0 为零（表压），除非发生溢流时关闭了地面的防喷器组合。静水压力方程的一个重要用途是确定适当的钻井液密度。因为井筒中的液柱必须具备足够的密度，才能形成能够支撑井壁的井筒压力，且该压力大于可渗透地层的孔隙压力，如图 5.2 所示。然而，该液柱的压力也必须小于裸露地层的破裂压力。一旦某处地层发生破裂，则该深度以上的钻井液将从井中迅速漏失到破裂的地层中。

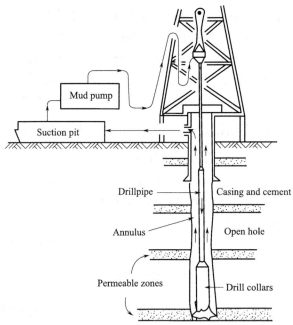

Fig. 5.2 The well-fluid system

5.2.2 Compressible Fluids

In many drilling and completion operations, a gas is present in at least a portion of the well. In some cases,

5.2.2 可压缩流体

在许多钻完井作业中，部分井筒中可能存在气体。在某些情

gas is injected in the well from the surface, while in other cases gas may enter the well from a subsurface formation.

The variation of pressure with depth in a static gas column is more complicated than in a static liquid column because the gas density changes with changing pressure. The gas behavior can be described using the real gas equation, defined by

$$p = \rho z \frac{RT}{M} \tag{5.8}$$

where, p=absolute pressure, r=gas density, z=gas compressibility factor, R= universal gas constant, T= absolute temperature, and M = gas molecular weight.

The gas compressibility factor z is a measure of how much the gas behavior deviates from that of an ideal gas, with z = 1 for ideal gases. Gas compressibility factors for natural gases have been determined experimentally as a function of temperature and pressure and are readily available in the petroleum and chemical engineering literature.

The gas density can be expressed as a function of pressure by rearranging Eq. 5.8. Solving this equation for the gas density ρ yields

$$\rho = \frac{pM}{zRT} \tag{5.9a}$$

Changing units from consistent units to common field units gives

$$\rho = \frac{pM}{80.3zT} \tag{5.9b}$$

where, ρ is expressed in lbm/gal, p is in lbf/in^2, M is in lbm/lbm-mol, and T is in degrees Rankine. The gas constant R is given in various unit systems as

$$R = 1\,545\,\frac{(\text{lbf}/\text{ft}^2)\cdot\text{ft}^3}{\text{lbm}\cdot\text{mol}\cdot\text{R}} = 80.3\,\frac{\text{psia}\cdot\text{gal}}{\text{lbm}\cdot\text{mol}\cdot\text{R}} = 8.3144\,\frac{\text{Pa}\cdot\text{m}^3}{\text{mol}\cdot\text{K}} \tag{5.10}$$

When the length of the gas column is not great and the gas pressure is above 6.895×10^6 Pa (1,000 psia), the hydrostatic equation for incompressible liquids given by Eq. 5.5 can be used together with Eq. 5.9a without much loss in accuracy. However, when the gas column is not short or highly pressured, the variation of gas density with depth within the gas column should be taken into account. Using Eqs. 5.6a and 9a, we obtain

$$\frac{dp}{dZ} = \frac{gpM}{zRT} \quad (5.11)$$

If the variation in z within the gas column is not too great, we can treat z as a constant, $\bar{z}$. Separating variables in the above equation and integrating gives

$$\ln\left(\frac{p}{p_0}\right) = \frac{gM}{\bar{z}RT}\int_{z_0}^{z}\frac{1}{T(\xi)}d\xi \quad (5.12a)$$

If we assume that T is relatively constant over the depth range, then Eq. 5.12a can be expressed

$$p = p_0 \exp\left(\frac{gM\Delta z}{\bar{z}RT}\right) \quad (5.12b)$$

or, in field units,

$$p = p_0 \exp\left(\frac{M\Delta z}{1.544\bar{z}T}\right) \quad (5.12c)$$

where, p is in psi, M in lbm/lbm-mol, Δz in ft, and T in degrees R.

5.2.3 Hydrostatic Pressure in Complex Fluid Columns

During many drilling operations, the well fluid column contains several sections of different fluid densities. The variation of pressure with depth in this type of complex fluid column must be determined by separating the effect of each fluid segment. For example, consider the complex liquid column shown in Fig. 5.3.

If the pressure at the top of Section 1 is known to be p_0, then the pressure at the bottom of Section 1 can be computed from Eq. 5.7a:

$$p_1 = \rho_1 g(Z_1 - Z_0) + p_0 \quad (5.13)$$

由式（5.5）表示的不可压缩流体静力学方程式可与式（5.9a）一起使用，该计算过程的精度损失不大。但是，当气柱较高或高度压缩时，应考虑气体密度在气柱内随深度的变化。通过式（5.6a）和式（5.9a），可以得到：

(5.11)

如果气柱内 z 的变化不太大，则可将 z 视为常数 $\bar{z}$。分离上述方程中的变量并积分，得到：

(5.12a)

假设 T 在整个深度范围内是相对恒定的，则式（5.12a）可表示为：

(5.12b)

或者，以油田单位表示为

(5.12c)

式中，p 的单位是 psi，M 的单位是 lbm/(lbm·mol)，Δz 的单位是 ft，T 的单位是 R。

5.2.3 复杂流体柱中的静水压力

在许多钻井作业期间，井筒液柱会包含多段不同密度的流体。在这种复杂的液柱中，确定压力随深度的变化时必须分别考虑每个流体段的作用。如图 5.3 中所示的复杂液柱。

如果已知第 1 段流体顶部的压力为 p_0，则可以根据式（5.7a）计算第 1 段流体底部的压力：

(5.13)

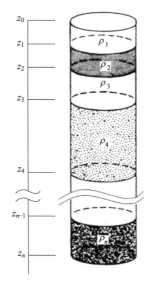

Fig. 5.3 A complex fluid column

The pressure at the bottom of Section 1 is equal to the pressure at the top of Section 2. Thus, the pressure at the bottom of Section 2 can be expressed in terms of the pressure at the top of Section 2:

第 1 段流体底部的压力又等于第 2 段流体顶部的压力。由此，第 2 段流体底部的压力可以用第 2 段流体顶部的压力表示为：

$$p_2 = \rho_2 g(Z_2 - Z_1) + \rho_1 g(Z_1 - Z_0) + p_0 \qquad (5.14)$$

In general, the pressure at any vertical depth Z can be expressed by

推广到一般，即任何垂直深度 Z 处的压力可表示为：

$$p = p_0 + g\sum_{i=1}^{i=n}\rho_i(Z_i - Z_{i-1}) + g\rho_n(Z - Z_{n-1}) \qquad (5.15)$$

where $Z_{n-1} < Z < Z_n$. It is frequently desirable to view the well fluid system shown in Fig. 5.3 as a manometer when solving for the pressure at a given point in the well. The drillstring interior usually is represented by the left side of the manometer, and the annulus usually is represented by the right side of the manometer. A hydrostatic pressure balance can then be written in terms of a known pressure and the unknown pressure using Eq. 5.15.

式中 $Z_{n-1} < Z < Z_n$。在求解井筒中给定深度的压力时，通常将图 5.3 所示的钻井液系统视为一个 U 形管。U 形管的左侧代表钻柱内的液柱压力，右侧代表环空压力。则根据式（5.15）可以用已知压力和未知压力写出静水压力平衡方程。

5.2.4 Equivalent Density Concept

5.2.4 当量密度

Field experience in a given area often allows guidelines to be developed for the maximum mud density that formations at a given depth will withstand without fracturing during normal drilling operations. It is sometimes helpful to compare a complex-well-fluid

在常规钻井作业期间，通常会根据所在区域的现场经验，确定给定深度处地层可承受的最大钻井液密度。有时，会将复杂钻井液柱等效为与大气连通的单一

column to an equivalent single-fluid column that is open to the atmosphere. This is accomplished by calculating the equivalent mud density ρ_e, which is defined by

$$\rho_e = \frac{\rho}{gZ} \qquad (5.16)$$

or, in field units,

$$\rho_e = \frac{p}{0.05195Z} \qquad (5.17)$$

The equivalent mud density always should be referenced to a specified depth.

Drilling activity has extended in recent years to high-pressure/high-temperature (HPHT) wells, where drilling fluids experience both hot and cold temperature extremes, thus undergoing changes in density. The term mud weight usually refers to measurements performed at the surface, and while it could be assumed constant at tem-peratures for standard drilling activity in the past, it certainly does not hold for these HP/HT and deepwater wells. Zamora and Roy have put forward this issue, and instead of the term mud weight they have proposed the use of the term equivalent static density (ESD) for static wells and equivalent circulating density (ECD) for cir- culating wells. Simulation results are shown in Fig. 5.4 for a water-based mud (WBM) and a synthetic-based mud (SBM) in 8,000 ft of water and onshore HP/HT environments.

液柱进行比较。这可以通过计算等效钻井液密度或当量密度 ρ_e 来完成，其定义为

(5.16)

或以油田单位表示为：

(5.17)

当量钻井液密度应始终对应具体的深度。

近年来，钻井作业已经扩展到高压高温 (HPHT) 井。在这些井中，钻井液会经历高温和低温两种极端温度，从而导致密度发生变化。钻井液密度通常是地面测量值，虽然过去可以假设钻井液密度在常规钻井作业的温度下保持恒定，但在高温高压和深水井中，这一假设就不适用了。Zamora 和 Roy 提出用等效静态密度 (ESD) 和当量循环密度 (ECD) 分别替代静态条件和循环条件下的钻井液密度。图 5.4 为水基钻井液 (WBM) 和合成基钻井液 (SBM) 在 8000ft 水下和陆地高温高压环境下的 ESD 模拟结果。

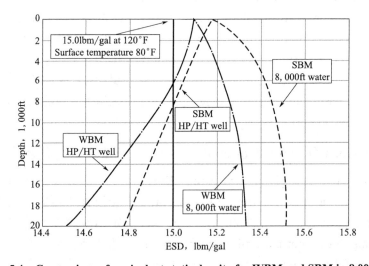

Fig. 5.4 Comparison of equivalent static density for WBM and SBM in 8,000 ft

5.2.5 Effect of Entrained Solids and Gases in Drilling Fluid

Drilling engineers seldom deal with pure liquids or gases. For example, both drilling fluids and cement slurries are primarily a mixture of water and finely divided solids. The drilling mud in the annulus also contains the drilled solids from the rock broken up by the bit and the formation fluids that were contained in the rock.

As long as the foreign materials are suspended by the fluid, or settling through the fluid at their terminal velocity, the effect of the foreign materials on hydrostatic pressure can be computed by replacing the fluid density in Eq. 5.7a with the density of the mixture. However, particles that have settled out of the fluid and are supported by grain-to-grain contact do not influence hydrostatic pressure. The average density of an ideal mixture of N components is given by

5.2.5 钻井液中夹带固体和气体的影响

钻井工程师应对的大都不是纯液体或纯气体，钻井液、水泥浆主要是水和细颗粒固体的混合物。环空中的钻井液还含有钻头破碎的岩屑及岩屑中包含的地层流体。

当杂质被流体悬浮或以自由沉降速度在流体中沉降时，可以用混合物密度替代式（5.7a）中的流体密度来计算杂质对静水压力的影响。然而，当颗粒沉降分离并由颗粒间的相互接触提供支撑时，颗粒将不再影响静水压力。由 N 种成分组成的理想混合物平均密度为：

$$\rho = \frac{\sum m_i}{\sum V_i} = \frac{\sum \rho_i V_i}{V} = \sum \rho_i \frac{V_i}{V} = \sum \rho_i f_i \tag{5.18}$$

$$V = \sum_{j=1}^{N} V_j, f_i = \frac{V_i}{V} \tag{5.19}$$

where, m_i, V_i, ρ_i and f_i are the mass, volume, density, and volume fraction of component i, respectively, and V is the total volume. As long as the components are liquids and solids, the component density is essentially constant through-out the entire length of the column. Thus, the average density of the mixture also will be essentially constant.

If one component is a finely divided gas, the density of the gas component does not remain constant but decreases with the decreasing pressure. A drilling fluid that is measured to have a low density due to the presence of gas bubbles is said to be gas-cut.

The determination of the density of a gas-cut mud can be made in the following way. If N_v moles of gas are dispersed in (or associated with) 1 m³ of drilling fluid,

式中，m_i、V_i、ρ_i 和 f_i 分别是成分 i 的质量、体积、密度和体积百分数，V 是总体积。如果成分是液体和固体，那么成分的密度在整个液柱长度上基本恒定。因此，混合物的平均密度也将保持恒定。

如果有一种成分是气体，那么气体的密度将随压力的减小而减小。由于气泡的存在，钻井液的测量密度将变小，此时称为气侵。

可通过如下方法进行气侵钻井液密度的测定。如果 N_v 摩尔气体分散在 1m³ 的钻井液中（或与

the volume fraction of gas at a given point in the column is given by

$1m^3$ 的钻井液相混合），液柱中给定点的气体体积百分数为：

$$f_g = \frac{V_g}{V_m + V_g} = \frac{\dfrac{zN_vRT}{p}}{1 + \dfrac{zN_vRT}{p}} = \frac{zN_vRT}{p + zN_vRT} \tag{5.20}$$

In addition, the gas density ρ_g at that point is defined by Eq.5.9a. Thus, the effective density of the mixture, $\bar{\rho}$ is given by

此外，该点的气体密度 ρ_g 由式（5.9a）定义。因此，混合物的有效密度 $\bar{\rho}$ 为：

$$\bar{\rho} = \rho_f(1-f_s) + \rho_g f_g = \rho_f \left[1 - \frac{zN_vRT}{p + zN_vRT}\right] + \frac{pM}{zRT}\left[\frac{zN_vRT}{p + zN_vRT}\right] = \frac{p(\rho_f + MN_v)}{p + zN_vRT} \tag{5.21}$$

where, f_g is volume fraction of gas; V_m is volume of mud; V_g isvolume of ga; p is pressure; N_v is moles of gas per volume; ρ_f is fluid density; f_s is the volume fraction of solid; ρ_g is gas density; M is the average molecular weight of the gas. Combination of this expression with Eq. 5.6a yields

式中，f_g 为气体体积分数；V_m 为钻井液体积；V_g 为气体体积；p 为压力；N_v 为每体积中气体的物质的量；ρ_f 为流体密度；f_s 为固体密度；ρ_g 为气体密度；M 是气体的平均分子量。代入式（5.6a）可得：

$$\Delta Z = \int_{p_1}^{p_2} \frac{p + zN_vRT}{g(\rho_f + MN_v)} \frac{dp}{p} \tag{5.22}$$

If the variation of z and T is not too great over the column length of interest, they can be treated as constants of mean values $\bar{z}$ and $\bar{T}$ Integration of Eq. 5.22 gives

如果 z 和 T 随液柱高度变化不大，则可以将它们视为常数 $\bar{z}$ 和 $\bar{T}$，对式（5.22）积分得到：

$$\Delta Z = \frac{\Delta p}{a} + \frac{b}{a}\ln\left(1 + \frac{\Delta p}{p_1}\right) \tag{5.23}$$

$$\Delta p = p_2 - p_1, \Delta Z = Z_2 - Z_1 \tag{5.24}$$

where

其中

$$a = g(\rho_f + MN_v) \tag{5.25}$$

$$b = \bar{z}N_vR\bar{T} \tag{5.26}$$

It is unfortunate that the pressure delta p appears within the logarithmic term in Eq. 5.24. This means that an iterative calculation procedure must be used for the determination of the change in pressure with elevation for a gas-cut fluid column. However, if the gas-liquid mixture is highly pressured and not very long,

压力 p 出现在式（5.24）的对数项中。这意味着必须采用迭代计算来确定气侵液柱的压力随液柱高度的变化。然而，如果气液混合物的压力很大，而液柱不是很高，则有：

$$\ln\left(1+\frac{\Delta p}{p_i}\right) \cong \frac{\Delta p}{p_i} - \frac{1}{2}\left(\frac{\Delta p}{p_i}\right)^2 + \frac{1}{3}\left(\frac{\Delta p}{p_i}\right)^3 \tag{5.27}$$

Which shows that if $\Delta p/p_1 \ll 1$, Eq.5.24 can be approximated by a linear or a quadratic equation.

5.2.6　Effect of Well Deviation

Today, drilling engineers seldom deal with vertical wells. For a variety of reasons, wells are deviated from the vertical, often by as much as 90. We have been considering Z as the true vertical depth (TVD) of the well. In a real well, we use measured depth, s, rather that Z to determine our location in the wellbore. What is even more important is that pressure changes due to friction from fluid flow vary with measured depth, not with TVD. The relationship between TVD and measured depth for a well with constant azimuth is given by:

$$\frac{dZ}{ds} = \cos\varphi$$

where φ is the angle of inclination of the wellbore with the vertical. For $\varphi=0$, a vertical well, the TVD varies as the measured depth; that is, $Z=s$.

5.3　Steady Flow of Drilling Fluids

The determination of pressure at various points in the well can be quite complex when either the drilling mud or the drillstring is moving. Frictional forces in the well system can be difficult to describe mathematically. However, in spite of the complexity of the system, the effect of these frictional forces must be determined for the calculation of (1) the flowing bottomhole pressure or ECD during drilling or cementing operations; (2) the bottomhole pressure or ECD during tripping operations; (3) the optimum pump pressure, flow rate, and bit nozzle sizes during drilling operations; (4) the cuttings-carrying capacity of the mud;

Audio 5.3

5.2.6　井斜的影响

现在，钻直井的情况很少。由于各种原因，需要钻斜井，且井斜角常高达90°。一直将 Z 看作是井的垂直深度（TVD）。在实际井中，常用测量井深 s 来确定井筒中的具体位置。更重要的是，流体流动摩擦引起的压力变化将随测量井深而变化，而不是随 TVD 变化。当井斜方位角不变时，TVD 与测量井深的关系为：

(5.28)

式中，φ 为井斜角。当 $\varphi=0$ 时为直井，TVD 等于测量井深，即 $Z=s$。

5.3　钻井液稳态流动

当钻井液或钻柱运动时，要确定井中各深度处的压力会非常复杂。井筒中的摩擦力很难用数学模型来描述。尽管井筒系统如此复杂，但仍必须确定这些摩擦力的影响，以便计算出：(1) 钻井或固井作业时的井底压力或 ECD；(2) 起下钻过程中的井底压力或 ECD；(3) 钻井作业期间的最佳泵压、排量和钻头喷嘴尺寸；(4) 钻井液的携岩能力；(5) 井控作业中不同钻井液排量下的立压和井底压力。

and (5) the surface and downhole pressures that will occur in the drillstring during well- controloperations for various mud flow rates.

The basic physical laws commonly applied to the movement of fluids are conservation of mass, conservation of momentum, and conservation of energy. All of the equations describing fluid flow are obtained by application of these physical laws using an assumed rheological model and an equation of state. Rheological models will be studied in a later section. Example equations of state are the incompressible fluid model, the slightly compressible fluid model, the ideal gas equation, and the real gas equation.

5.3.1 Mass Balance

The law of conservation of mass states that the net mass rate into any volume V is equal to the mass rate out of the volume. The balance of mass for single-phase flow is given by

$$\dot{m} = \rho v A = \text{constant} \tag{5.29}$$

where, $\dot{m}$ =massflowrate,kg/s; ρ=density,kg/m³; v= average velocity,m/s; A=area,m².

Where steady-state flow has been assumed. The drilling engineer normally considers only steady-state conditions. Note also that for constant area, which is usually the case, the product of the density and the average velocity is constant. As pressure decreases, so does density, which implies that the average velocity increases. In other words, pressure decreases will accelerate a gas in a constant-area pipe. With the exception of air, gas, or foam drilling, the drilling fluid usually can be considered incompressible. In the absence of any accumulation or leakage of well fluid in the surface equipment or underground formations, the flow rate of an incompressible well fluid must be the same at all points in the well. For an incompressible fluid, Eq. 5.29 takes an even simpler form

一般应用于流体运动的基本物理定律是质量守恒定律、动量守恒定律和能量守恒定律。所有描述流体流动的方程都可通过假定的流变模型和状态方程，应用这些物理定律得到。流变模型将在后面的章节中研究。状态方程有不可压缩流体模型、微可压缩流体模型、理想气体状态方程和实际气体状态方程。

5.3.1 质量守恒

质量守恒定律表明，进入任何体积V的净质量速率等于流出体积V的质量速率。单相流的质量平衡方程为：

式中，$\dot{m}$为质量流量，kg/s；ρ为密度，kg/m³；v为平均速度，m/s；A为面积，m²。

这里假设是稳态流动。钻井工程师通常只考虑稳态流。但也要注意，对于面积恒定的情况，通常密度和平均速度的乘积是恒定的。当压力减小时，密度也随之减小，这意味着平均速度将增大。换句话说，压力降低会使面积恒定的管道中气体加速。除空气、气体或泡沫钻井外，钻井液通常被看作是不可压缩流体。在地面设备或地层中没有钻井液积聚或漏失的情况下，不可压缩的钻井液在井筒中所有点的流量必须相等。因此，对于不可压缩流体，式（5.29）可采用更简单的形式：

$$q = vA \tag{5.30}$$

where, q=volume flow rate, m³/s.

Knowledge of the average velocity at a given point in the well is often desired. For example, the drilling engineer frequently will compute the average upward flow velocity in the annulus to ensure that it is adequate for cuttings removal.

5.3.2 Momentum Balance

The balance of momentum for single-phase flow has the form

$$\Delta p + \rho v \Delta v = \int_{\Delta Z} \rho g \mathrm{d}Z \pm \int_{\Delta s} \left(\frac{\mathrm{d}p_f}{\mathrm{d}s}\right) \mathrm{d}s \tag{5.31}$$

where, $\frac{\mathrm{d}p_f}{\mathrm{d}s}$ =pressure change due to friction, Pa/m; Δs=Length of flow increment, m.

where steady flow has been assumed again. The Δv term is called the fluid acceleration, and it is nonzero only for compressible fluids. Also note that the term ρv is constant, from Eq. 5.29. The ρg term is the fluid weight term, which has been discussed in detail in Section 5.2. The fluid friction term is often expressed using the friction factor concept:

$$\Delta p + \rho v \Delta v = \int_{\Delta s} \left[\rho g \cos\varphi \pm \frac{2f\rho v^2}{D_\mathrm{h}}\right] \mathrm{d}s \tag{5.32}$$

where, f= Fanning friction factor; D_h =hydraulic diameter, m; Δs =Length of flow increment, m.

We have also used Eq. 5.28 to make everything dependent on measured depth s, though it may be convenient to retain the TVD relationship in some calculations. The Fanning friction factor f depends on the fluid density, velocity, viscosity, fluid type, and pipe roughness.

Appropriate models for f, considering a variety of different fluid types, will be considered in detail in the section on rheology. The sign of the friction term is counter to the flow direction. The hydraulic diameter D_h

is defined as

$$D_h = \frac{4A_F}{W_p} \tag{5.33}$$

For a pipe cross-sectional area

$$D_h = \frac{\pi d^2}{\pi d} = d$$

where, A_F is flow area; W_p is wetted perimeter; d is the ID of the pipe. For the annulus formed by two pipes

$$D_h = \frac{\pi(d_o^2 - d_i^2)}{\pi(d_o + d_i)} = d_o - d_i$$

where, d_o is the inside diameter of the outer pipe and d_i is the outside diameter (OD) of the inner pipe. Notice that there is no effect of pipe eccentricity, so that also has to be accounted for in the friction factor. The friction factor we have defined is the Fanning friction factor. The student needs to be aware that there is an alternate definition called the Darcy friction factor that equals four times the Fanning friction factor. Be cautious when using friction factor graphs, tables, or formulas to be sure that you know which friction factor is being defined. For an incompressible fluid, Eq. 5.32 takes the following simple form:

$$\Delta p = \rho g \Delta Z \pm \frac{2 f \rho v^2}{D_h} \Delta s \tag{5.34}$$

where all the coefficients are constant. From his experience with static compressible fluids, the student should expect that solutions to Eq. 5.32 will be solutions to first-order nonlinear differential equations, and that analytic solutions will not, in general, be available.

5.3.3 Energy Balance

The law of conservation of energy states that the net energy rate out of a system is equal to the time rate of work done within the system. Consider the generalized flow system shown in Fig. 5.5. The work done by the fluid is equal to the energy per unit mass of fluid given by the fluid to a fluid engine (or equal to minus the work done by a pump on the fluid). Thus, the law of conservation of energy yields

等于负的泵对流体所做的功）。因此，由能量守恒定律可得：

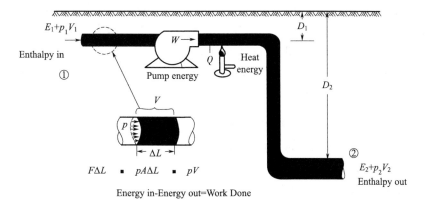

Fig.5.5 Generalized flow system

$$(E_2 - E_1) + (p_2V_2 - p_1V_1) - g(Z_2 - Z_1) + 1/2(v_2^2 - v_1^2) = W + Q \tag{5.35}$$

where E is the internal energy of the fluid, W is the work done by the fluid, and Q is the heat per unit mass added to the fluid, with subscript 1 indicating inlet properties and subscript 2 indicating outlet properties. Simplifying this expression using differential notations yields

式中 E 为流体的内能，W 为流体所做的功，Q 为单位质量流体吸收的热量，下标 1 表示入口，下标 2 表示出口。用微分符号简化这个表达式得到：

$$\Delta E - g\Delta Z + \Delta v^2/2 + \Delta(pv) = W + Q \tag{5.36}$$

Eq. 5.36 is the first law of thermodynamics applied to a steady flow process. This equation is best suited for flow systems that involve either heat transfer or adiabatic processes involving fluids. This form of the equation seldom has been applied by drilling engineers. The change in internal energy of the fluid and the heat gained by the fluid usually is considered using a friction loss term, which can be defined in terms of Eq. 5.36 using the following expression:

式（5.36）是应用于稳定流动过程的热力学第一定律。该方程最适用于传热或流体绝热过程的流动系统。然而，钻井工程师们几乎不用这种形式的方程。流体内能的变化和所获得的热量通常用一个摩擦损失项来考虑，该摩擦损失项可以用公式（5.36）来定义，其表达式如下：

$$F_{\text{fric}} = \Delta E + \int_1^2 p\,\mathrm{d}V - Q \tag{5.37}$$

The friction loss term can be used conveniently to account for the lost work or energy wasted by the viscous forces within the flowing fluid. Substitution of Eq. 5.37 into Eq. 5.36 yields

摩擦损失项可用于解释由流体黏性力所耗费的功或能量。将式（5.37）代入式（5.36）可以得到：

$$\int_1^2 V\,\mathrm{d}p - g\Delta Z + \frac{\Delta v^2}{2} = W - F_{\text{fric}} \tag{5.38}$$

Eq. 5.38 often is called the mechanical energy balance equation. This equation was in use even before heat flow was recognized as a form of energy transfer by Carnot and Joule and is a completely general expression containing no limiting assumptions other than the exclusion of phase boundaries and magnetic, electrical, and chemical effects. The effect of heat flow in the system is included in the friction loss term.

The first term in Eq. 5.38 $\int_1^2 V dp$ may be difficult to evaluate if the fluid is compressible unless the exact path of compression or expansion is known. Fortunately, drilling engineers deal primarily with essentially incompressible fluids having a constant specific volume V. For incompressible fluids, it holds that

$$\int_1^2 V dp = \frac{\Delta p}{\rho} \tag{5.39}$$

Eq. 5.38 also can be expressed by

$$\Delta p/\rho - g\Delta Z + \Delta v^2/2 = W - F_{fric} \tag{5.40}$$

or

$$p_1 + 9.81\rho(Z_2 - Z_1) - 0.5(v_2^2 - v_1^2) + \Delta p_p - \Delta p_f = p_2 \tag{5.41}$$

Expressing this equation in practical field units of lbf/in², lbm/gal, ft/sec, and ft gives

$$p_1 + 0.052\rho(Z_2 - Z_1) - 8.074 \times 10^{-4}(v_2^2 - v_1^2) + \Delta p_p - \Delta p_f = p_2 \tag{5.42}$$

5.3.4　Flow Through Jet Bits

A schematic of incompressible flow through a short constriction, such as a bit nozzle, is shown in Fig. 5.6. In practice, it generally is assumed that (1) the change in pressure due to a change in elevation is negligible; (2) the velocity v_0 upstream of the nozzle is negligible, compared with the nozzle velocity v_n; and (3) the frictional pressure loss across the nozzle is negligible. Thus, Eq. 5.42 reduces to

$$p_1 - 8.074 \times 10^{-4} \rho v_n^2 = p_2 \tag{5.43}$$

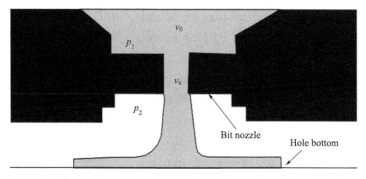

Fig.5.6 Flow through a bit nozzle

Substituting the symbol Δp_b for the pressure drop (p_1-p_2) and solving this equation for the nozzle velocity v_n yields

用 Δp_b 代替压降 (p_1-p_2)，求解喷嘴速度 v_n，得

$$v_n = \sqrt{\frac{\Delta p_b}{8.074\times 10^{-4}\rho}} \quad (5.44)$$

In SI units, Eq. 5.44 is given by

式（5.44）转换成国际单位制为：

$$v_n = \sqrt{2\frac{\Delta p_b}{\rho}} \quad (5.45)$$

where, v_n has units of m/s, Δp_b has units of Pa, and ρ has units of kg/m³. Unfortunately, the exit velocity predicted by Eq. 5.42 for a given pressure drop across the bit, Δp_b never is realized. The actual velocity is always smaller than the velocity computed using Eq. 5.42, primarily because the assumption of frictionless flow is not strictly true. To compensate for this difference, a correction factor or discharge coefficient C_d usually is introduced so that the modified equation,

式中，v_n 的单位是 m/s，Δp_b 的单位是 Pa，ρ 的单位是 kg/m³。然而，用公式（5.42）预测出口速度时钻头压降（Δp_b）总是未知。由于无摩擦流动的假设并不完全真实，实际速度总小于式（5.42）的计算值。为了补偿这种差异，通常引入一个修正系数或流量系数 C_d，这样修正后的方程为：

$$v_n = C_d\sqrt{\frac{\Delta p_b}{8.074\times 10^{-4}\rho}} \quad (5.46)$$

or, in SI units,

或者用国际单位制表示为：

$$v_n = C_d\sqrt{2\frac{\Delta p_b}{\rho}} \quad (5.47)$$

will result in the observed value for nozzle velocity. The discharge coefficient has been determined experimentally for bit nozzles by Eckel and Bielstein. These authors

将会得到喷嘴速度的观测值。埃克尔和比尔斯坦用实验方法测定了钻头喷嘴的流量系数。他们指

indicated that the discharge coefficient may be as high as 0.98 but recommended a value of 0.95 as a more practical limit.

A rock bit has more than one nozzle, usually the same number of nozzles and cones. When more than one nozzle is present, the pressure drop applied across all of the nozzles must be the same (Fig. 5.7).

According to Eq. 5.46, if the pressure drop is the same for each nozzle, the velocities through all nozzles are equal. Therefore, if the nozzles are of different areas, the flow rate q through each nozzle must adjust so that the ratio q/A is the same for each nozzle. If three nozzles are present

出，流量系数可以高达0.98，但建议用0.95更符合实际。

一个牙轮钻头有多个喷嘴，通常喷嘴的数量和牙轮数量相等。当存多个喷嘴时，施加在所有喷嘴上的压降必须相等（图5.7）。

根据式（5.46），如果每个喷嘴的压降相同，则通过所有喷嘴的流体速度相等。因此，如果喷嘴面积不同，则必须调整每个喷嘴的流量q，以使每个喷嘴的q/A比值相等。如果有三个喷嘴，则：

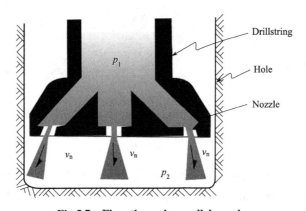

Fig.5.7 Flow through parallel nozzles

$$\overline{v}_n = \frac{q_1}{A_1} = \frac{q_2}{A_2} = \frac{q_3}{A_3} \tag{5.48}$$

Note also that the total flow rate of the pump, q, is given by

需要注意，泵的总流量q由下式得到：

$$q = q_1 + q_2 + q_3 = \overline{v}_n A_1 + \overline{v}_n A_2 + \overline{v}_n A_3 \tag{5.49}$$

Simplifying this expression yields

简化这个表达式可以得到：

$$q = \overline{v}_n (A_1 + A_2 + A_3) = \overline{v}_n A_t \tag{5.50}$$

where, A_t is total nozzle area.

式中，A_t为总喷嘴面积。

Thus, the velocity of flow through each nozzle is also equal to the total flow rate divided by the total nozzle area:

因此，通过每个喷嘴的流速也等于总流量除以总的喷嘴面积：

$$\frac{q}{A_t} = \frac{q_1}{A_1} = \frac{q_2}{A_2} = \cdots = \frac{q_i}{A_i} \tag{5.51}$$

In field units, the nozzle velocity v_n is given by

$$v_n = \frac{q}{3.117 A_t} \quad (5.52)$$

where, v_n has units of m/s, q has units of m³/s, and A_t has units of m². In SI units, v_n is given by

$$v_n = \frac{q}{A_t} \quad (5.53)$$

Combining Eqs. 5.46 and 5.52 and solving for the pressure drop across the bit, Δp_b yields

$$\Delta p_b = \frac{8.311 \times 10^{-5} \rho q^2}{C_d^2 A_t^2} \quad (5.54)$$

in SI units, Eq. 5.54 is given by

$$\Delta p_b = \frac{\rho q^2}{2 C_d^2 A_t^2} \quad (5.55)$$

Because the viscous frictional effects are usually negligible for flow through short nozzles, Eqs. 5.54 and 5.55 are valid for both Newtonian and non-Newtonian fluids, but should be used with caution. For critical applications, experimentally determined discharge coefficients should be determined for a specific mud.

Bit nozzle diameters often are expressed in 32nds of an inch. For example, if the bit nozzles are described as 12-13-13, this denotes that the bit contains one nozzle having a diameter of $^{12}/_{32}$ in and two nozzles having a diameter of $^{13}/_{32}$ in.

5.3.5 Bit Hydraulic Power

Because power is the rate of doing work, pump energy W can be converted to hydraulic power P_H by multiplying W by the mass flow rate ρq. Thus,

$$P_H = \rho W q = \Delta p_p q \quad (5.56)$$

In SI units P_H is expressed in watts (W), Δp_p in Pa, and q in m³/s. In field units, if the flow rate q is expressed in gal/min and the pump pressure Δp_p is expressed in lbf/in², then

$$P_H = \frac{\Delta p_p q}{1714} \tag{5.57}$$

Likewise, other terms in Eq. 5.42, the pressure balance equation can be expressed as hydraulic horsepower by multiplying the pressure term by $q/1,714$.

5.3.6 Bit Hydraulic Impact Force

The purpose of the jet nozzles is to improve the cleaning action of the drilling fluid at the bottom of the hole. Before jet bits were introduced, rock chips were not removed efficiently, and much of the bit life was consumed regrinding the rock fragments. Further improvements in the cleaning action have been obtained with the introduction of a central nozzle. This avoids the bit balling phenomenon in drilling soft formations.

The rheological properties of drilling fluids can affect the hole bottom cleaning, particularly in the area around the bit nozzles and cones. In addition, the apparent viscosity can affect the overall bit performance. An increase in the frictional pressure loss inside the drillstring, because of higher values of the viscosity, reflects in a decrease of the hydraulic power available at the bit. Several investigators have concluded that the cleaning action is maximized by maximizing the total hydraulic impact force of the jetted fluid against the hole bottom. If it is assumed that the jet stream impacts the bottom of the hole in the manner shown in Fig.5.7, all of the fluid momentum is transferred to the hole bottom.

Because the fluid is travelling at a vertical velocity v_n before striking the hole bottom and is travelling at zero vertical velocity after striking the hole bottom, the time rate of change of momentum (in field units) is given by

$$F_j = \frac{\Delta(m\bar{v})}{\Delta t} \approx \left(\frac{m}{\Delta t}\right)\Delta \bar{v} = \frac{(\rho q)\bar{v}_n}{32.17 \times (60)} = \frac{(\rho q)\bar{v}_n}{1930.2} \tag{5.58}$$

or, in SI units,

$$F_j = (\rho q)\bar{v}_n \tag{5.59}$$

where, ρq is the mass rate of the fluid. Combining Eqs. 5.46 and 5.58 yields

同理,式(5.42)中的其他项可以通过乘以压力项 $q/1714$,将压力平衡方程表示为水功率。

5.3.6 钻头水力冲击力

射流喷嘴的作用是提高钻井液在井底的清洁作用。在引入射流钻头之前,无法有效清除岩屑,导致钻头重复研磨岩屑,消耗大量的钻头寿命。中心喷嘴的引入,进一步改善了清洁作用,也避免了在钻软地层时出现钻头泥包。

钻井液的流变性能影响井底的清洁,特别是钻头喷嘴和牙轮周围的区域。此外,钻井液的表观黏度也会影响钻头的整体性能。由于钻井液黏度较高,钻柱内摩擦力损失大,导致钻头水功率降低。一些研究人员已得出结论,最大限度地提高喷射流体对井底的总水力冲击力,可以使清洁作用达到最佳。假设射流以图5.7所示的方式冲击井底,则所有的流体动量都将转移到井底。

由于流体在接触井底前以垂直速度 v_n 流动,在接触井底后以零垂直速度流动,故动量的变化率(以油田单位表示)为:

或者用国际单位制表示:

式中,ρq 是流体的质量流量。联立式(5.46)和式(5.58)可得:

$$F_j = 0.01823 C_d q \sqrt{\rho \Delta p_b} \qquad (5.60)$$

where, F_j is given in lbf. In SI units, Eq. 5.60 is given by

$$F_j = 1.4142 C_d q \sqrt{\rho \Delta p_b} \qquad (5.61)$$

5.3.7 Jet Bit Nozzle Size Selection

The determination of the proper jet bit nozzle sizes is one of the more frequent applications of the frictional pressure-loss equations by drilling personnel. Significant increases in penetration rate can be achieved through the proper choice of bit nozzles. In relatively competent formations, the penetration rate increase is believed to result from mainly improved cleaning action at the hole bottom. Wasteful regrinding of cuttings is prevented if the fluid circulated through the bit removes the cuttings as rapidly as they are made. In soft formations, the jetted fluid also may aid in the destruction of the hole bottom.

The true optimization of jet bit hydraulics cannot be achieved yet. Before this can be done, accurate mathematical relations must be developed that define the effect of the level of hydraulics on penetration rate, operational costs, bit wear, potential hole problems such as hole washout, and drilling-fluid carrying capacity. At present, there is still disagreement as to what hydraulic parameter should be used to indicate the level of the hydraulic cleaning action. The most commonly used hydraulic design parameters are bit nozzle velocity, bit hydraulic horsepower, and jet impact force. Current field practice involves the selection of the bit nozzle sizes that will cause one of these parameters to be a maximum.

5.3.7.1 Maximum Nozzle Velocity

Before jet bits were introduced, rig pumps usually were operated at the flow rate corresponding to the estimated minimum annular velocity that would lift the cuttings. To some extent, this practice continues even today. If the jet nozzles are sized so that the surface pressure at this flow rate is equal to the maximum

式中，F_j 以 lbf 为单位。用国际单位制，式（5.61）变为：

5.3.7 喷射钻头喷嘴尺寸的选择

钻井人员常使用摩擦压力损失方程确定合适的钻头喷嘴尺寸。通过选择适当的钻头喷嘴，可以显著提高钻速。人们认为，在相对较硬的地层中，机械钻速的提高主要是由于井底清洁作用的改善。如果钻井液循环流经钻头时能在岩屑形成的同时迅速将其清除，就可以避免钻头对岩屑的重复性研磨。在软地层中，喷射流体还能在井底辅助破岩。

目前还无法真正实现钻头的水力参数优化。为了实现这一目标，必须建立精确的数学关系，确定水力参数对机械钻速、作业成本、钻头磨损、井眼潜在问题（如井眼冲蚀）和钻井液携岩能力的影响。目前，关于用什么水力参数来表明水力清洁作用的程度仍存在分歧。最常用的水力设计参数是钻头喷嘴速度、钻头水功率和射流冲击力。现场实践中，可以通过选择适当的钻头喷嘴尺寸，实现某一个参数的最优。

5.3.7.1 最大喷射速度

在引入喷射式钻头之前，钻井泵通常采用举升岩屑所需的最小环空流速相对应的排量。这种做法一直沿用至今。如果确定了喷嘴的尺寸，并使该流速下的地面压力等于最大允许地面压力，

allowable surface pressure, then the fluid velocity in the bit nozzles will be the maximum that can be achieved and still lift the cuttings. This can be proved using Eq. 5.62, the nozzle velocity equation. As shown in this equation, nozzle velocity is directly proportional to the square root of the pressure drop across the bit.

$$\bar{v}_n \propto \sqrt{\Delta p_b} \qquad (5.62)$$

Thus, the nozzle velocity is a maximum when the pressure drop available at the bit is a maximum. The pressure drop available at the bit is a maximum when the pump pressure is a maximum and the frictional pressure loss in the drillstring and annulus is a minimum. The frictional pressure loss is a minimum when the flow rate is a minimum.

5.3.7.2 Maximum Bit Hydraulic Horsepower

Speer pointed out that the effectiveness of jet bits could be improved by increasing the hydraulic power of the pump. Speer reasoned that penetration rate would increase with hydraulic horsepower until the cuttings were removed as fast as they were generated. After this "perfect cleaning" level was achieved, there should be no further increase in penetration rate with hydraulic power.

Shortly after Speer published his paper, several authors pointed out that, because of the frictional pressure loss in the drillstring and annulus, the hydraulic power developed at the bottom of the hole was different from the hydraulic power developed by the pump. They concluded that bit horsepower rather than pump horsepower was the important parameter. Furthermore, it was concluded that bit horsepower was not necessarily maximized by operating the pump at the maximum possible horsepower.

The pump pressure is expended by frictional pressure losses in the surface equipment, Δp_s; frictional pressure losses in the drillpipe Δp_{dp} and drill collars Δp_{dc}; pressure losses caused by accelerating the drilling fluid through the nozzle; and frictional pressure losses in the drill collar annulus Δp_{dca} and drillpipe annulus Δp_{dpa}. Stated mathematically,

$$p_p = \Delta p_s + \Delta p_{dp} + \Delta p_{dc} + \Delta p_b + \Delta p_{dca} + \Delta p_{dpa} \tag{5.63}$$

If the total frictional pressure loss to and from the bit is called the parasitic pressure loss, Δp_d, then

如果将除钻头外的总摩擦压力损失称为附加压力损失 Δp_d，则：

$$\Delta p_d = \Delta p_s + \Delta p_{dp} + \Delta p_{dc} + \Delta p_{dca} + \Delta p_{dpa} \tag{5.64}$$

and

且

$$p_d = \Delta p_b + \Delta p_d \tag{5.65}$$

Because each term of the parasitic pressure loss can be computed for the usual case of turbulent flow,

因为在通常的紊流情况下附加压力损失的每一项都可以计算出来：

$$\Delta p_f \propto q^{1.75} \tag{5.66}$$

We can represent the total parasitic pressure loss using

可以用下列公式来表示总的附加压力损失

$$\Delta p_d \propto q^m = cq^m \tag{5.67}$$

where, m is a constant that theoretically has a value near 1.75, and c is a constant that depends on the mud properties and wellbore geometry. Substitution of this expression for into Eq. 5.65 and solving for Δp_d yields

式中，m 是常数，理论值接近 1.75；c 是一个常数，取决于钻井液性能和井眼几何形状。将这个表达式代入方程（5.65），并求解得到：

$$\Delta p_b = \Delta p_p - cq^m \tag{5.68}$$

Because the bit hydraulic horsepower P_{Hb} is given by Eq. 5.56,

由于钻头水功率 P_{Hb} 由式（5.69）给出：

$$P_{Hb} = \frac{\Delta p_b q}{1714} = \frac{\Delta p_p q - cq^{m+1}}{1714} \tag{5.69}$$

using calculus to determine the flow rate at which the bit horsepower is a maximum gives

用微积分来确定钻头水功率最大时的流量：

$$\frac{dP_{Hb}}{dq} = \frac{\Delta p_p - (m+1)cq^m}{1714} = 0 \tag{5.70}$$

Solving for the root of this equation yields

解这个方程，得到：

$$\Delta p_p = (m+1)cq^m = (m+1)\Delta p_d \tag{5.71}$$

or

或者：

$$\Delta p_d = \frac{\Delta p_p}{(m+1)} \tag{5.72}$$

Because $(d^2 P_{HB})/dq^2$ is less than zero for this root, the root corresponds to a maximum. Thus, bit hydraulic

由于 $(d^2 P_{HB})/dq^2$ 小于零，所以此根对应一个最大值。因此，当附加

horse power is a maximum when the parasitic pressure loss is [1/(m+1)] times the pump pressure.

From a practical standpoint, it is not always desirable to maintain the optimum $\Delta p_d/\Delta p_p$ ratio. It is usually convenient to select a pump liner size that will be suitable for the entire well rather than periodically reducing the liner size as the well depth increases to achieve the theoretical maximum. Thus, in the shallow part of the well, the flow rate usually is held constant at the maximum rate that can be achieved with the convenient liner size. For a given pump horsepower rating P_{HP} this maximum rate is given by

$$q_{max} = \frac{1714 P_{HP} E}{p_{max}}$$ (5.73)

where, E is the overall pump efficiency, and p_{max} is the maximum allowable pump pressure set by the contractor. This flow rate is used until a depth is reached at which $\Delta p_d/\Delta p_p$ is at the optimum value. The flow rate then is decreased with subsequent increases in depth to maintain $\Delta p_d/\Delta p_p$ at the optimum value. However, the flow rate never is reduced below the minimum flow rate to lift the cuttings.

5.3.7.3 Maximum Jet Impact Force

Some rig operators prefer to select bit nozzle sizes so that the jet impact force is a maximum rather than the bit hydraulic horsepower. McLean concluded from experimental work that the velocity of the flow across the bottom of the hole was a maximum for the maximum impact force. Eckel, working with small bits in the laboratory, found that the penetration rate could be correlated to a bit Reynolds number group so that

$$\frac{dD}{dt} \propto \left(\frac{\rho \bar{v}_n d_n}{\mu_a}\right)^{a8}$$ (5.74)

where, $\frac{dD}{dt}$ =penetration rete; ρ=fluid density; $\bar{V}_n$ =nozzle velocity; d_n =nozzle diameter; μ_a =apparent viscosity of the fluid at a shear rate of 10,000 seconds^{-1}; a= constant.

压力损失为 [1/(m+1)] 倍的泵压时,钻头水功率最大。

从实际应用的角度来看,始终保持最优的 $\Delta p_d/\Delta p_p$ 值并不可取。为了达到理论上的最大值,通常实用的做法是选择适合整口井的缸套尺寸,而不是随着井深的增加周期性地减小缸套尺寸。因此,在较浅井段时,流量通常保持在实用缸套尺寸所能达到的最大流量。对于给定的泵功率 P_{HP},这个最大流量为:

(5.73)

式中,E 为泵的总效率,p_{max} 为承包商设定的最大允许泵压。该流量一直使用至 $\Delta p_d/\Delta p_p$ 达到最优值的深度。之后,为了保持 $\Delta p_d/\Delta p_p$ 在最优值,流量将随着深度的增加而减小。然而,为了举升岩屑,流量不应低于携岩所需的最低流量。

5.3.7.3 最大射流冲击力

一些作业者采用优选钻头喷嘴尺寸(而不是钻头水功率)来获得最大射流冲击力。麦克莱恩从实验研究中得出结论,当流体流经井底的流速最大时获得最大冲击力。埃克尔在实验室对小钻头进行研究,发现钻头的机械钻速与钻头的雷诺数相关,从而得到:

(5.74)

式中,$\frac{dD}{dt}$ 为机械钻速;ρ 为钻井液密度;$\bar{V}_n$ 为喷射速度;d_n 为喷嘴直径;μ_a 为每秒10000次剪切速率下的表观黏度;a 为常数。

It can be shown that when nozzle sizes are selected so that jet impact force is a maximum, the Reynolds number group defined by Eckel is also a maximum. The derivation of the proper conditions for maximum jet impact was published first by Kendall and Goins.

The jet impact force is given by Eq. 5.75.

$$F_j = 0.01823 C_d q \sqrt{\rho \Delta p_b} = 0.01823 C_d q \sqrt{\rho (\Delta p_p - \Delta p_d)} \quad (5.75)$$

Because the parasitic pressure loss is given by Eq. 5.76

$$F_j = 0.01823 C_d \sqrt{\rho \Delta p_p q^2 - (m+2)\rho c q^{m+1}} \quad (5.76)$$

Using calculus to determine the flow rate at which the bit impact force is a maximum gives

$$\frac{dF_j}{dq} = \frac{0.009115 C_d \left[2\rho \Delta p_p q - (m+2)\rho c q^{m+1}\right]}{\sqrt{\rho \Delta p_p q^2 - (m+2)\rho c q^{m+1}}} \quad (5.77)$$

Solving for the root of this equation yields

$$\begin{aligned} 2\rho \Delta p_p - (m+2)\rho c q^{m+1} &= 0 \\ \rho q \left[2\Delta p_p - (m+2)\Delta p_d\right] &= 0 \end{aligned} \quad (5.78)$$

or

$$\Delta p_d = \frac{2\Delta p_p}{m+2} \quad (5.79)$$

because $(d^2 F_j)/dq^2$ is less than zero for this root, the root corresponds to a maximum. Thus, the jet impact force is a maximum when the parasitic pressure loss is $[2/(m+2)]$ times the pump pressure.

5.4 Exercise

(1) Calculate the static minimum mud density required to prevent flow from a permeable stratum at 12,200 ft if the pore pressure of the formation fluid is 8,500 psig.

(2) A well contains a tubing filled with methane gas (molecular weight = 16) to a vertical depth of 10,000 ft. The annular space is filled with a 9.0 lbm/gal brine. Assuming

ideal gas behavior (*z* = 1), compute the mount by which the exterior pressure on the tubing exceeds the interior tubing pressure at 10,000 ft if the surface tubing pressure is 1,000 psia and the mean gas temperature is 140°F. If the collapse resistance of the tubing is 8,330 psi, will the tubing collapse due to the high external pressure?

(3) In intermediate casing string is to be cemented in place at a depth of 10,000 ft. The well contains 10.5lbm/gal mud when the casing string is placed on bottom. The cementing operation is designed so that the 10.5 lbm/gal mud will be displaced from the annulus by ① 300 ft of 8.5 lbm/gal mud flush, ② 1,700 ft of 12.7 lbm/gal filler cement, and ③ 1,000 ft of 16.7 lbm/gal high-strength cement. The high-strength cement will be displaced from the casing with 9 lbm/gal brine. Calculate the pump pressure required to completely displace the cement from the casing.

(4) A 12-lbm/gal mud is being circulated at 400 gal/min. The 5.0in drillpipe has an inside diameter (ID) of 4.33in, and the drill collars have an ID of 2.5 in. The bit has a diameter of 9.875 in. Calculate the average velocity in the drillpipe, the drill collars, and the annulus opposite the drillpipe.

(5) Determine the pressure at the bottom of the drillstring if the frictional pressure loss in the drill string is 1,400 psi, the flow rate is 400 gal/min, the mud density is 12 lbm/gal, and the well depth is 10,000 ft. The ID of the drill collars at the bottom of the drillstring is 2.5 in, and the pressure increase developed by the pump is 3,000 psi.

(6) A 12.0lbm/gal drilling fluid is flowing through a bit containing three $^{13}/_{32}$ in nozzles at a rate of 400 gal/min. Calculate the pressure drop across the bit.

假设该处为理想气体性质 (z=1)，如果地面油管压力为 1000 psia，平均气体温度为 140°F，请计算在 10000ft 处油管的内外压差。若油管的抗挤强度为 8330psi，油管会因外部压力过高而挤毁吗？

（3）中间套管在 10,000ft 深处进行固井，当套管柱下到井底时，井内钻井液密度为 10.5lbm/gal。固井设计中用以下液柱顶替环空中密度为 10.5lbm/gal 的钻井液：① 300ft 密度为 8.5lbm/gal 的隔离液，② 1700ft 密度为 12.7lbm/gal 填充水泥，③ 1000ft 密度为 16.7lbm/gal 的高强度水泥。套管内用 9lbm/gal 的盐水将高强度水泥顶替到环空。请计算完全顶替套管内水泥所需的泵压。

（4）用 400 gal/min 的速度循环密度为 12lbm/gal 的钻井液。5in 钻杆的内径为 4.33in，钻铤的内径为 2.5in，钻头的直径是 9.875in。请计算钻杆内、钻铤内和钻杆段环空的钻井液平均速度。

（5）钻井泵的压力提升至 3000psi，钻井液流量为 400gal/min，密度为 12lbm/gal，井深为 10000ft，如果钻杆内压力损失为 1400psi，钻柱底部钻铤的内径为 2.5in，请计算确定钻柱底部的压力。

（6）密度为 12.0lbm/gal 钻井液以 400 gal/min 的排量流过含有三个 $^{13}/_{32}$in 喷嘴的钻头。请计算钻头上的压降。

Directional Drilling 6

定向钻井

6.1 Fundamentals of Trajectory Design

6.1 井眼轨道设计基础

6.1.1 Introduction

The term directional drilling is a broad term that refers to all activities that are required to design and drill a wellbore to reach a target, or a number of targets, located at some horizontal distance from the top of the hole. In other words, the purpose of directional drilling is to connect the surface location with oil/gas reservoirs that are not located right below it. Any well that is intentionally nonvertical is also called a directional well. Deviation control comprises all activities needed to drill a hole as required by the well plan and geological data.

Audio 6.1

Today, much oil and natural-gas production comes from directional wells drilled onshore and offshore. To enhance production, many wells are drilled with a high inclination angle or even horizontally. At first, horizontal wells were only a few hundred feet long. Thanks to continuous improvements in drilling technology, the horizontal departure was gradually increased to enable drilling of so-called extended-reach wells (ERWs).

Extended-reach drilling (ERD) is commonly used nowadays to reach onshore and offshore oil and gas deposits; the length of such a well can be 20,000 to 40,000 ft or even more. If the stepout is greater than 40,000 ft, the well is classified as ultra-extended-reach drilling (uERD). Drilling ERD and uERD wells creates

6.1.1 引言

定向钻井是一个广义的概念，是指沿着预先设计的井眼轨道和方向，钻达一个或多个目标的钻井工艺方法，这些目标距离井口有一定的水平位移。换言之，定向钻井的目的是将井口与不在同一条铅垂线上的油气储层连通起来。斜井通常也被称为定向井。钻井时需要按照钻井设计和地质数据进行井斜控制。

现在，许多石油、天然气都产自陆地和海上的定向井。为了提高产量，许多井被钻成了大斜度井，甚至是水平井。起初，水平井的长度只有几百英尺。随着钻井技术的不断发展，井的水平位移逐渐增加，直至大位移井（ERW）。

如今，大位移井已广泛应用于陆地和海上油气藏的开发；水平位移可达20000～40000ft，甚至更长。如果水平位移大于40000ft，则称为超大位移井（uERD）。钻大位移井和超大位移

a number of challenges for drilling personnel. To drill such wells effectively required significant improvements in drilling fluids, cuttings transport, and mechanical performance of the drillstring and other elements of a drilling system.

Development of logging-while-drilling (LWD) tools triggered development of so-called geosteering methods and maximum-reservoir-contact wells. Geosteering involves guiding the wellbore path based on real-time measurements of formation properties rather than following a predetermined trajectory.

Directional wells are also sometimes drilled to control a blowing well or to bypass (sidetrack) a portion of a vertical well that is impossible to drill [e.g., due to wellbore stability problems or loss of portions of a drillstring (fish) in a hole]. Some typical applications of directional drilling are shown schematically in Fig. 6.1.

井会给钻井作业人员带来许多挑战。要进行有效钻进，需要大大提高钻井液性能、岩屑运移效率及钻柱和钻井系统的机械性能。

随钻测井（LWD）工具的进步促进了地质导向钻进技术和最大储层接触面积井的发展。地质导向钻井技术根据随钻测量得到的地层特征，实现井眼轨道的实时调整和控制，而不依赖于预先设计的井眼轨道。

有时定向井也可用作井喷时的救援井，或者井下复杂情况时（如直井段出现井壁稳定问题或存在落鱼）的侧钻井。定向钻井的应用如图6.1所示。

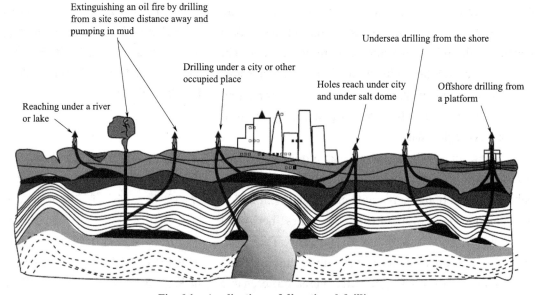

Fig. 6.1 Applications of directional drilling

Directional drilling is also widely used for geothermal and civil-engineering applications. Many geothermal projects involve drilling directional wells in hot hard rocks such as granite and other igneous and metamorphic rocks. Civil engineers frequently use directional-drilling techniques for drilling under rivers,

定向井也广泛地应于地热和土木工程领域。许多地热项目都需要在花岗岩或其他火成岩和变质岩中打定向井。为了绕开河流、高速公路和其他障碍物，土木工程师也常常需要应用定向钻

highways, and other obstacles.

In response to economic and environmental pressures, the use of directional wells is increasing in oil and other industries. Not only has the number of directionally drilled wells increased, but also the well trajectories have become increasingly complex, resulting in a need for more-sophisticated drilling tools and technologies. First, however, some basic concepts will be discussed, followed by some useful mathematical formulations that apply to all directionally drilled wells.

6.1.2 Basic Concepts

The conventional visual representation of a directional well consists of a horizontal and vertical cross-section, as shown in Fig. 6.2. For the sake of simplicity, a straight segment A-B is used here to represent the wellbore. The distance from the rotary table [the rotary kelly bushing (RKB)] to Point A or Point B as measured along the wellbore is called a measured depth (MD). The vertical distance from the rotary table to Point A or Point B is called true vertical depth (TVD) or simply vertical depth. The inclination angle φ is the angle between the vertical and the wellbore. The direction angle θ is specified as the azimuth between the geographic north and the projection of the wellbore onto a horizontal plane. A number of devices exist to measure the hole inclination and azimuth angles. Such devices are called surveying instruments.

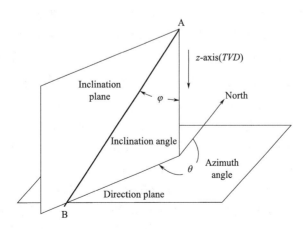

Fig. 6.2 Hole inclination and azimuth angles

The industry is currently using a number of surveying instruments, ranging from magnetic (single-shot and multishot) to more-sophisticated gyroscopic devices. Magnetic instruments use an inclinometer, a compass, a timer, and a camera, while gyroscopic instruments work on the principle of a spinning mass. The use of magnetic instruments requires installation in the bottomhole assembly (BHA) of special drill collars made of stainless steel (approximately 68% nickel and 28% copper, with small additions of iron and manganese-monel metals) with nonmagnetic properties. On the other hand, gyroscopic instruments do not require nonmagnetic drill collars because they are immune to magnetic influence.

The angle between geographic (true) north and magnetic north is called the declination angle. A location where measurements are taken is called a station. At each station, an MD, hole inclination angle, and azimuth are recorded.

Modern technology makes it possible to send directional-survey information to the surface using mud-pulse telemetry. Information on hole inclination and azimuth, as well as certain other information (e.g., downhole WOB, deflection of tool-face angle), is transmitted as pressure pulses and decoded at the surface while drilling. Typically, the measurements are taken at intervals of between 30 and 300 ft or even less, depending on the complexity of the well path and the purpose of the wellbore.

It is conventional practice to use four 90° quadrants, north, east, south, and west (N-E-S-W) to report well direction. For example, the azimuth angles $\theta_1 = 27°$ and $\theta_2 = 215°$ can also be reported as $\theta_1 = N27E$ and $\theta_2 = S35W$. In other words, if the hole direction is reported as E26S, the azimuth is 116° Figs. 6.3a and 6.3b are examples of a horizontal (plane) view and a vertical (section) view of a wellbore trajectory, represented for the sake of simplicity as straight segments. The x (north) and y (east) axes in Fig.6.3a intersect in the center of the rotary table (the RKB). The vertical cross section is

目前石油行业使用多种测斜工具，包括磁性测斜仪（单点和多点测斜仪）和更复杂的陀螺测斜仪。磁性测斜仪利用测斜仪、指南针、定时器和照相机进行测量，而陀螺测斜仪则根据旋转质量原理进行工作。磁性测斜仪需要安装到底部钻具组合（BHA）上，BHA由专门的不锈钢（由大约68％的镍和28％的铜组成，并添加少量的铁和锰）无磁钻铤组成的；而陀螺测斜仪不需要无磁钻铤，因为它们不受磁干扰。

正北方位和磁北方位之间的夹角称为磁偏角。测量的位置称为测点。在每个测点记录一次井深、井斜角和井斜方位角。

随着现代测井技术的发展，可以利用钻井液脉冲遥测技术向地面传送定向测井信息。钻井时，井斜角、方位角和其他井眼信息（如钻压、工具面角的偏差）将作为压力脉冲信号传输到地面并进行解码。通常会根据井眼轨迹的复杂程度和井的用途，按300ft或30ft甚至更短的间隔进行测量。

通常用东南西北四个90°象限（N-E-S-W）来表示井的方位。例如，方位角 $\theta_1 = 27°$ 和 $\theta_2 = 215°$ 也可分别记录为 $\theta_1 = N27E$ 和 $\theta_2 = S35W$。换句话说，如果将井眼方位记录为E26S，则方位角等于116°。图6.3（a）和图6.3（b）分别是一口井井眼轨迹的水平投影图和垂直投影视图。为简单起见，各井段用直线段表示。图6.3（a）中的 x（北）轴和 y（东）轴

drawn through the centers of the RKB and the target. Figs. 6.3a and 6.3b are qualitative in nature to help define some commonly used terms in directional drilling. The kickoff point (KOP) is the depth at which the well trajectory departs from the vertical in the direction of the target or is modified by the lead angle. The lead angle is usually to the left of the target horizontal departure line (the line from the initial point 0 to the target T). The departure is the horizontal distance between the surface location and the point on the traverse (trajectory). The closure is the horizontal distance between the rotary table and the center of the target. Fig. 6.4 shows a 3D view of a wellbore composed of three segments in x, y, z coordinates. The origin of the coordinate system is located at Point P_1; Points P_2, P_3, and P_4 lie on the trajectory at the coordinates shown in Fig. 6.4. For the sake of simplicity, the segments P_1P_2, P_2P_3, and P_3P_4 are assumed to be straight, and their inclination and direction angles are also shown in Fig. 6.4. In reality, a wellbore is composed of curved rather than straight segments, and it is useful to introduce the concepts of build and turn rates as discussed below.

相交于转盘（RKB）的中心。垂直投影面是通过 RKB 和目标点中心的铅垂面。图 6.3（a）和图 6.3（b）是定性的描述，可以帮助定义一些定向井中的常用术语。造斜点（KOP）是井眼轨迹沿目标点方向开始偏离垂直方向的深度或是被超前角修正的角度。超前角通常在目标点水平位移线（从初始点 0 到目标 T 的线）的左侧。水平位移是井眼轨迹上某点到井口所在铅垂线的距离。闭合距是完井时目标点至井口铅垂线的水平距离。图 6.4 展示了一个三段式轨迹在 x、y、z 坐标中三维视图。坐标原点位于点 P_1；井眼轨迹上点 P_2、P_3 和 P_4 的位置如图 6.4 所示。为简单起见，假设线段 P_1P_2、P_2P_3 和 P_3P_4 是直线段，它们的井斜角和方位角也如图 6.4 所示。实际上，井眼轨迹是由多个弯曲段而不是直线段组成的，因此有必要讨论造斜率和方位变化率的概念。

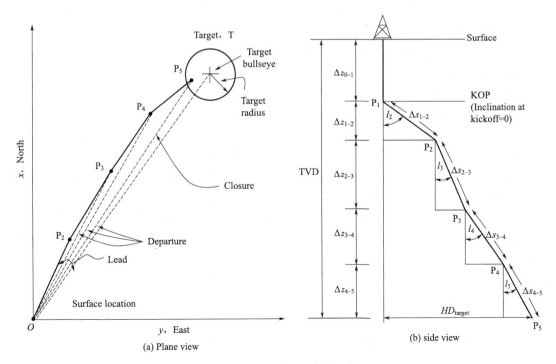

Fig. 6.3 Plane view and side view

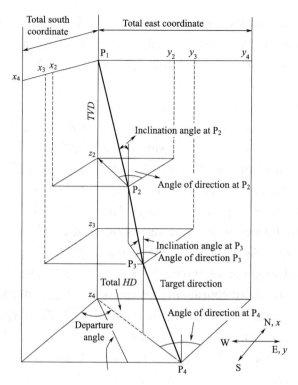

Fig. 6.4 Wellbore composed of three segments

6.1.3 Fundamental Mathematical Formulations

Let us consider a wellbore trajectory as shown schematically in Fig. 6.5 in a conventional rectangular x,y,z (right-hand) system of coordinates consistent with the north, east, and vertical directions, N, E, and V. The continuous curve Os in Fig. 6.5 represents the well trajectory (wellbore centerline). Consider a small element ds with components dx, dy, and dz. Because ds is small, it can be approximated as a straight segment with inclination angle φ, and azimuth θ. The projection of the small element ds onto a horizontal plane is denoted as dl. With motion along the well path, in general, both the inclination angle and azimuth will change. In other words, the hole inclination angle and azimuth are functions of the MD s. For purposes of directional-well trajectory planning, it is useful to define the following fundamental quantities:

（1）Rate of change of the hole inclination angle along the well path, or the so-called build rate:

6.1.3 基本数学公式

图 6.5 为常规直角三维坐标系中的井眼轨迹，该坐标系的 x、y、z 轴分别与北、东和垂直方向一致。图中的连续曲线 Os 表示井眼轨迹（井眼中心线）。在井眼轨迹上选取一个微元 ds，它的三维坐标分别是 dx、dy、dz。由于 ds 非常小，所以可将其近似为一个具有井斜角 φ 和方位角 θ 的直线段。微元段 ds 在水平面上的投影为 dl。通常，井斜角和方位角都将随着井眼轨迹的变化而变化。换言之，井斜角和方位角是测量井深 MD 的函数。为了进行定向井轨道设计，一些基本概念定义如下：

（1）井斜角随井眼轨迹的变化率，称为造斜率：

$$B(s) = \frac{d\varphi(s)}{ds} \tag{6.1}$$

(2) Rate of change of the azimuth angle along the projection of the well path onto a horizontal plane, or the so-called horizontal turn rate:

（2）方位角随水平投影长度的变化率，称为水平方位变化率：

$$H(s) = \frac{d\theta(s)}{dl} \tag{6.2}$$

(3) Rate of change of the azimuth along the well path, or the turn rate:

（3）方位角随井眼轨迹的变化率，称为方位变化率：

$$T(s) = \frac{d\theta(s)}{ds} \tag{6.3}$$

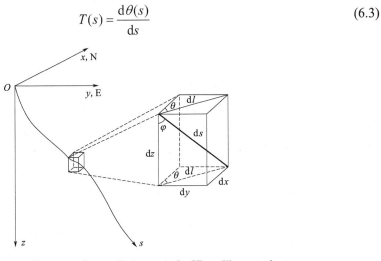

Fig. 6.5 Schematic diagram of a small element of a 3D wellbore trajectory

In other words, the build and turn rates are the first derivatives of the hole inclination and azimuth angle as functions of *MD*. In directional drilling, the build and turn rates are usually expressed in degrees/100 ft (or degrees/30m), and care should be exercised to carry out calculations in a consistent system of units (radians rather than degrees). The rate of change can be positive or negative, depending on whether the angles increase or decrease with *MD*. For example, a negative build rate indicates that the inclination angle decreases with depth, in which case it is usually called a drop rate. Note that, in general, the hole inclination and azimuth angles as well as the build and turn rates are functions of the measured hole depth s. Examination of the right triangles in Fig. 6.5 immediately reveals the following useful relationships:

换句话说，造斜率、方位变化率是井斜角和方位角相对于测量井深的一阶导函数。在定向钻井中，造斜率和方位变化率通常以（°）/100ft 或（°）/30m 表示，计算时注意使用统一的单位制（弧度而非度）。变化率可以为正也可以为负，这取决于角度随测量井深增加还是减小。例如，造斜率为负意味着井斜角随深度的增加而减小，这通常称为降斜率。注意，井斜角和方位角、造斜率和方位变化率都是相对于井深s的函数。分析图6.5中的直角三角形后，可以得到以下关系：

$$\frac{dx}{dl} = \cos\theta(s) \tag{6.4}$$

$$\frac{dy}{dl} = \sin\theta(s) \tag{6.5}$$

$$\frac{dl}{ds} = \sin\varphi(s) \tag{6.6}$$

$$\frac{dz}{ds} = \cos\varphi(s) \tag{6.7}$$

$$\frac{dx}{ds} = \frac{dx}{dl}\frac{dl}{ds} = \sin\varphi(s)\cos\theta(s) \tag{6.8}$$

$$\frac{dy}{ds} = \frac{dy}{dl}\frac{dl}{ds} = \sin\varphi(s)\sin\theta(s) \tag{6.9}$$

In calculus, the quantities defined by Eqs. 6.7, 6.8, and 6.9 are called directional cosines and are frequently denoted by the letters l, m, and n. It can be shown that the sum of the squares of the directional cosines is equal to one.

Differentiating Eqs. 6.7, 6.8, and 6.9 yields the second derivatives:

由式（6.7）、式（6.8）、式（6.9）定义的值称为方向余弦，通常用字母 l、m 和 n 表示。可以证明，方向余弦的平方和等于1。

对式（6.7）、式（6.8）、式（6.9）求微分得到二阶导数：

$$\frac{d^2x}{ds^2} = \cos\varphi\cos\theta\frac{d\varphi}{ds} - \sin\varphi\sin\theta\frac{d\theta}{ds} = B\cos\varphi\cos\theta - T\sin\varphi\sin\theta \tag{6.10}$$

$$\frac{d^2y}{ds^2} = \cos\varphi\sin\theta\frac{d\varphi}{ds} + \sin\varphi\cos\theta\frac{d\theta}{ds} = B\cos\varphi\sin\theta + T\sin\varphi\cos\theta \tag{6.11}$$

$$\frac{d^2z}{ds^2} = -\sin\varphi\frac{d\varphi}{ds} = -B\sin\varphi \tag{6.12}$$

A good understanding of the definitions formulated above and of their derivatives is essential for solving a number of practical directional-drilling problems.

In well-trajectory design, the coordinates of the initial point are usually known [i.e., the top of the hole (the RKB) or the KOP]. The target-point coordinates are generally also known. In some instances, such as when the well must pass through multiple targets, the directions in terms of inclination and azimuth angles are also specified. The task of the designer is to calculate the coordinates of all other points on the well path. This usually is accomplished in a stepwise manner by

充分理解上述定义式及其导数，对于解决许多实际的定向钻井问题至关重要。

在井眼轨道设计中，通常已知起始点坐标[即井口转盘（RKB）或KOP]和目标点坐标。在某些情况下，例如当一口井必须通过多个目标时，也会以井斜角和方位角的形式指定各目标方向。设计人员的任务是计算井眼轨道上所有其他点的坐标。通常会从初始点（例如KOP）开始，依

selecting a subsequent point on the trajectory a distance Δs (measured along the trajectory) from the initial point (e.g., the KOP). To carry out the required calculations, assumptions frequently must be made about well path build and turn rates as well as hole inclination and direction angles. The calculations are repeated until a smooth well path is obtained that will reach the target or targets. In principle, to obtain the coordinates (x, y, z) of an arbitrary point on a well path, Eqs. 6.7 through 6.9 must be integrated.

次选择和前一点距离为 Δs（沿井眼轨迹测量）的后续点逐步完成。为了进行所需的计算，必须经常假设井眼轨迹的井斜角、方位角、造斜率、方位变化率。重复计算上述内容直到获得一条直达目标点的光滑井眼轨道为止。原则上，要获得井眼轨道上任意点的坐标（x，y，z)，必须对式（6.7）、式（6.8）、式（6.9）积分。

$$\Delta x = x_2 - x_1 = \int_{x_1}^{x_2} \sin\varphi(s)\cos\theta(s)\mathrm{d}s \tag{6.13}$$

$$\Delta y = y_2 - y_1 = \int_{y_1}^{y_2} \sin\varphi(s)\sin\theta(s)\mathrm{d}s \tag{6.14}$$

$$\Delta z = z_2 - z_1 = \int_{z_1}^{z_2} \cos\varphi(s)\mathrm{d}s \tag{6.15}$$

Clearly, if the coordinates at Point 1 are known, the coordinates at Point 2 can be calculated using Eq. 6.13, Eq.6.14, and Eq.6.15. At times, the integrals are difficult to calculate, and for practical applications, designers use various assumptions to obtain closed-form solutions. If a closed-form solution is not available, the integrals are evaluated numerically.

显然，如果已知点 1 的坐标，用式（6.13）、式（6.14）和式（6.15）可以计算点 2 的坐标。有时积分很难，所以在实际应用中，设计人员常用各种假设来获得解析解。如果闭合解不存在，则对积分进行数值计算。

6.1.4 Bends in the Vertical and Horizontal Planes

6.1.4 垂直面和水平面中的弯曲段

Consider two important special cases of curved wellbore sections located in the vertical and horizontal planes. If a well path is confined to a vertical plane, its azimuth θ is constant along the trajectory. Then Eq. 6.13, Eq. 6.14, and Eq. 6.15 take the form

Audio 6.2

接下来将讨论位于垂直面和水平面中的弯曲井段。如果井眼轨迹局限在一个垂直面内，其方位角 θ 沿井眼轨迹保持不变，则式（6.13）、式（6.14）和式（6.15）可采用以下形式：

$$\Delta x = \cos\theta\int_{s_1}^{s_2}\sin\varphi(s)\mathrm{d}s = \cos\theta\int_{\varphi_1}^{\varphi_2}\frac{1}{B}\sin\varphi\mathrm{d}\varphi \tag{6.16a}$$

$$\Delta y = \sin\theta\int_{s_1}^{s_2}\sin\varphi(s)\mathrm{d}s = \sin\theta\int_{\varphi_1}^{\varphi_2}\frac{1}{B}\sin\varphi\mathrm{d}\varphi \tag{6.16b}$$

$$\Delta z = \int_{\varphi_1}^{\varphi_2}\frac{1}{B}\cos\varphi\mathrm{d}\varphi \tag{6.16c}$$

Furthermore, for the case where a wellbore segment is a circular arc with radius *R*, the build rate is constant and equal to the reciprocal of the radius (1/*R*). Then, Eq. 6.16 can be integrated to obtain:

$$\Delta x = \frac{\cos\theta}{B}(\cos\varphi_1 - \cos\varphi_2) = \xi R\cos\theta(\cos\varphi_1 - \cos\varphi_2) \tag{6.17a}$$

$$\Delta y = \frac{\sin\theta}{B}(\cos\varphi_1 - \cos\varphi_2) = \xi R\sin\theta(\cos\varphi_1 - \cos\varphi_2) \tag{6.17b}$$

$$\Delta z = \frac{1}{B}(\sin\varphi_2 - \sin\varphi_1) = \xi R(\sin\varphi_2 - \sin\varphi_1) \tag{6.17c}$$

The parameter ξ is chosen to be positive (+1) for a positive turn rate and negative (−1) for a negative rate. Consequently, the horizontal departure, *HD*, between Points 1 and 2 is

$$HD_{1-2} = \sqrt{\Delta x^2 + \Delta y^2} = \xi R(\cos\varphi_1 - \cos\varphi_2) \tag{6.18}$$

Clearly, in this case, the departure is independent of the hole azimuth.

6.1.5 Wellbore Curvature and Dogleg Severity

For several practical reasons, in addition to build and turn rates, it is also useful to determine wellbore curvature and torsion along the well trajectory. In directional-drilling, wellbore curvature is frequently called dogleg severity (*DLS*) and expressed in degrees/100 ft, as mentioned earlier.

From calculus, the curvature of a 3D curve can be calculated as

$$\kappa(s) = \left[\left(\frac{d^2x}{ds^2}\right)^2 + \left(\frac{d^2y}{ds^2}\right)^2 + \left(\frac{d^2z}{ds^2}\right)^2\right]^{\frac{1}{2}} \tag{6.19}$$

Eq. 6.19 gives curvature in any consistent system of units (e.g., 1°/ft, 1°/m). In everyday directional-drilling terminology, the term DLS, expressed in degrees per unit length, is often used rather than curvature. If the *DLS* is expressed in degrees/100 ft, then

$$DLS = \frac{18000\kappa(s)}{\pi} \tag{6.20}$$

In other words, Eq. 6.20 gives wellbore *DLS* in degrees/100 ft if the curvature is expressed as 1/ft. Sometimes the *DLS* is called the dogleg rate. The radius of curvature *R* is defined as the inverse of curvature: $R(s)=\kappa(s)^{-1}$. By substituting Eqs. 6.10, 6.11, and 6.12 into Eq. 6.19 and performing some rearrangements, it is possible to obtain the wellbore curvature in terms of build rate, turn rate, and hole inclination angle, as follows:

$$\kappa(s) = \sqrt{B^2 + T^2 \sin^2\varphi(s)} \tag{6.21}$$

Sometimes the build rate *B* is called a vertical build rate (vertical curvature) and denoted as B_V and the product ($T\sin\varphi$) is called a lateral curvature and denoted as B_L. The curvature expressed by Eq. 6.21 is called the total curvature. Because, $T = \frac{d\theta}{dl}\frac{dl}{ds} = H\sin\varphi(s)$ it is also possible to write the curvature equation in terms of build rate and horizontal turn rate:

$$\kappa(s) = \sqrt{B^2 + H^2 \sin^4\varphi(s)} \tag{6.22}$$

Wellbore curvature provides information about the rate of overall change in angle due to simultaneous changes in hole inclination and azimuth along the well path. The overall angle change between two points on a well path is defined as the angle between the tangent lines at the two points under consideration. The curvature is the rate of change of the overall angle along the trajectory, and therefore the overall angle change β between two neighboring points on the trajectory located Δs apart can be obtained by integrating the curvature along the trajectory as follows:

$$\beta = \int_0^{\Delta s} \kappa(s)\,ds \tag{6.23}$$

The overall angle change is frequently called a dogleg. For example, if the turn rate is nil (no change in azimuth along the well path), the dogleg (*DL*) will be:

换句话说，如果井眼曲率用1/ft表示，则式（6.20）给出狗腿严重度以（°）/100ft表示。有时 *DLS* 称为狗腿率。将曲率的倒数定义为曲率半径 R：$R(s) = \kappa(s)^{-1}$。把式（6.10）、式（6.11）和式（6.12）代入式（6.19），并整理后就能得到用造斜率、方位变化率和井斜角表示的井眼曲率，即：

有时，造斜率 *B* 被称为垂直造斜率（垂直曲率），用 B_V 表示；乘积（$T\sin\varphi$）称为横向曲率，用 B_L 表示。由式（6.21）表示的曲率称为总曲率。由于 $T = \frac{d\theta}{dl}\frac{dl}{ds} = H\sin\varphi(s)$，所以还可以用造斜率和水平方位变化率来表示曲率方程：

井眼曲率提供的是全角变化率的信息，全角变化既反映了井斜角的变化，又反映了井斜方位角的变化。将井眼轨迹上两点处的井眼方向线（切线）夹角定义为这两点间的全角变化。而井眼曲率是全角变化沿井眼轨迹的变化率，因此，可以按如下方式沿轨迹对井眼曲率进行积分，便得到相距 Δs 的两个相邻点之间的全角变化：

全角变化常称为狗腿角。例如：若方位变化率为零（即井眼的方位角不变），则狗腿角（*DL*）将为：

$$\beta = DL = \int_0^{\Delta s} B\,\mathrm{d}s = \int_{\varphi_1}^{\varphi_2} \mathrm{d}\varphi = \varphi_2 - \varphi_1 \qquad (6.24)$$

Clearly, if the well path is in the vertical plane, the *DL* is simply equal to the difference between the inclination angles at the two adjacent points.

Lubinski et al. was the first to derive an equation for DL of the form

$$\beta = 2\arcsin\sqrt{\sin^2\left(\frac{\varphi_2-\varphi_1}{2}\right)+\sin\varphi_1\sin\varphi_2\sin^2\left(\frac{\theta_2-\theta_1}{2}\right)} \qquad (6.25)$$

Analysis of E.q.6.25 shows that for $\theta_1=\theta_2$, E.q.6.25 reduces to E.q.6.24.

6.1.6 Directional-Well Profiles

Typically, the design of a directional-well profile consists of two phases. First, a well path is constructed to connect the target with the surface location, and then adjustments are made to account for factors that will eventually influence the final trajectory. In other words, the location of the target and of the drilling rig must be decided before the trajectory shape is designed. The location of the target is the first and most important step. In principle, the well should be placed in the reservoir to optimize production if the purpose of drilling is to recover oil and gas. The optimal well bore trajectory (traverse) should result in minimum drilling and completion cost or time. Some oil and gas wells are designed to be confined to a vertical plane and are referred to as 2D wells. The well path shape should be considered simultaneously with casing (casing sizes and setting depths, cementing). completion program(perforating, fracturing, gravel packs), well bore stability, cuttings transport, and any anticipated hole problems. Frequently, to optimize the well path, the geoscientists and engineers must work together from the outset of the project.

After the base well trajectory has been calculated. The designer needs to make corrections to compensate for anticipated effects related to drill pipe rotation (bit walk), formation hardness and dip angle, type of drill bit, and other factors.

For example, drill pipe rotation typically results in right-hand bit walk, and therefore a left lead angle is used to compensate for this tendency, as schematically shown earlier in Fig. 6.3a. The optimum lead angle results in the closest approach to the base trajectory. The required information on the directional tendencies of various drilling systems can be obtained by analyzing drilling data from similar wells drilled under similar geological conditions. Lack of such data can lead to considerable discrepancies between calculated well trajectories and those actually observed while drilling the well. This situation places the designer of an exploratory well in a difficult situation. In such cases, as well as during more typical drilling jobs, it is essential to have a contingency plan. Under more complex geological conditions, drilling a pilot hole should be considered to obtain at least some preliminary information about geological conditions, including types of rock, formation dip and strike angles, and possible hole problems. There are three basic 2D directional well trajectories, as shown in Fig. 6.6.

Type 1 consists of a vertical part, a build section, and a tangent that is also called a hold part or slant section. This well profile is also called a slant well. Type 2, also called an S-shaped pattern, consists of live segments: vertical, buildup, tangent, drop-off, and another vertical at the bottom. A modified S-shaped trajectory has a tangent segment (not vertical) at the bottom of the drop-off part. The S-shaped pattern penetrates the target vertically, and the modified S-shaped pattern penetrates the formation at some desired inclination angle. Type 3 is called a continuous-build trajectory and consists of a vertical part and a buildup section. Horizontal wells and ERWs are additional types.

眼方位漂移)、地层硬度和倾角、钻头类型及其他因素造成的影响，设计人员还需要对设计进行校正。

例如，钻杆旋转通常会导致钻头向右漂移，因此要使用左超前角来补偿这种趋势，如图6.3（a）所示。最佳超前角可使实际井眼轨迹最接近基本井眼轨道。通过分析在类似地质条件下所钻邻井的钻井数据，可以获得各种钻井系统的定向趋势信息。如果缺乏这些数据，则可能导致计算出的井眼轨道与钻井时实际观测到的井眼轨迹之间出现很大差异。这就是探井设计较为困难的原因所在。因此，不论是在设计情况下，还是在平常的钻井作业中，都必须准备一个应急计划。在复杂的地质条件下，应考虑先钻一个导眼，获得一些有关地质条件的初步信息，包括岩石类型、地层倾角和走向角及可能出现的井眼复杂情况。如图6.6所示，二维定向井有三种基本轨道类型。

类型1由直井段、增斜段和稳斜段组成，也称为三段式轨道。类型2，也称为S形轨道，由多井段组成，包括直井段、增斜段、稳斜段、降斜段和另一个位于底部的直井段。改进后的S形轨道在降斜段底部是一个稳斜段（非直井段）。S形轨道垂直穿透目标层，而改进后的S形轨道则以一定井斜角穿过目的层。类型3称为双增式轨道，由一个直井段和两个增斜段组成。水平井和大位移井则是另外的轨道类型。

6.1.6.1 The Ideal Slant-Type Well Profile

The ideal slant-type well trajectory is confined to a vertical plane, resulting in a 2D well profile. Consider the slant well shown in Fig.6.7, using the following notations: KOP depth; the vertical depth of target (VDT); the horizontal departure of target (HDT); and β [inclination angle of slant part(tangential part)]. For the slant-type well, the tangent angle is equal to the DL of the curved section. Examination of Fig. 6.7 reveals the following geometric relationships:

① $VDT-KOP=L_{ab}+L_{bd}$, ② $HDT=L_{de}+L_{ef}$

The segments ab, bc, and bd can be calculated as

③ $L_{ab}=R\sin\beta$, ④ $L_{bc}=R(1-\cos\beta)$, ⑤ $L_{bd}=L_{ce}=\dfrac{L_{ef}}{\tan\beta}$

Substituting Lines ③, ④, and ⑤ as calculated above for Lines ① and ② gives

⑥ $VDT-KOP=R\sin\beta+\dfrac{L_{ef}}{\tan\beta}$ ⑦ $HDT=R(1-\cos\beta)+L_{ef}$ (6.26)

Solving the equation for Line ⑥ for ef and substituting the result into the equation for Line ⑦, after some rearrangements, gives

$$(VDT-KOP)\sin\beta+(R-HDT)\cos\beta=R \tag{6.27}$$

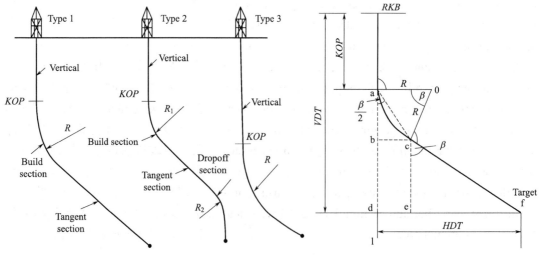

Fig.6.6 Three major types of 2D well bore trajectories   Fig.6.7 Slant-type well profile

Eq.6.27 describes the desired relationship between the departure of the tangent, the *VDT*, the kickoff point(*KOP*) depth, the radius of curvature, and the inclination angle of the tangent section.

If the target *VDT*, the *KOP* depth, the *HDT*, and the radius of the build section are given, then Eq 6.27 can be solved for the DL angle *β*:

$$\beta = \arcsin\frac{R}{\sqrt{(R-HDT)^2 + (VDT-KOP)^2}} - \arctan\frac{R-HDT}{VDT-KOP} \quad (6.28)$$

Once the DL angle has been calculated, it is possible to determine the length of the curved section and the build rate.

On the other hand, if the quantities *HDT*, *VDT*, *KOP*, and *β* are given, then Eq.6.27 can be solved for the radius of curvature of the build section and subsequently for the required build rate.

6.1.6.2　Horizontal Well Profiles

In practical applications, horizontal wells are high-angle wells with inclination angles of approximately 80° to 100°. In an ideal horizontal well, as the name indicates, the inclination angle is equal to 90°. Wells with inclination angles greater than 90° are sometimes drilled to recover oil and gas located in the upper part of a formation as well as to enhance production rates (gravity helps to counteract the frictional pressure losses). Most horizontal wells are drilled in a reservoir partly to maximize wellbore contact with the formation in anticipation of higher-production wells. Horizontal wells are also drilled for enhanced oil recovery purposes (waterflooding) and for water and gas control. Fig. 6.8 shows a schematic diagram of a horizontal well that consists of a vertical segment, a first buildup segment, a tangent part, and a second buildup segment, followed by a horizontal section. Here, the departure is defined as the displacement from vertical until the well reaches the beginning of the horizontal section. Horizontal displacement is the sum of the length of the horizontal section and the departure. Some horizontal wells consist of one build

式（6.27）描述了目标点水平位移、目标点垂深、造斜点深度、曲率半径和稳斜段井斜角之间的关系。

如果给出了目标点垂深、水平位移，造斜点深度和造斜段的曲率半径，则可以利用式（6.27）解出狗腿角 *β*。

一旦计算出狗腿角，就可以确定弯曲井段的长度和造斜率。

另外，如果给出了目标点的水平位移、垂直井深、造斜点深度和狗腿角，也可以利用式（6.27）求解出造斜段的曲率半径和随后所需的造斜率。

6.1.6.2　水平井剖面

在实际应用中，水平井是一种井斜角为 80°～100° 的大斜度井。一口理想的水平井，它的井斜角应等于 90°。但有时为了开采产层上部的油气和提高采收率（重力有利于抵消摩擦压力损失），也会钻井斜角大于 90° 的水平井。由于水平井的长水平段穿入储层，能极大地增加井筒和储层的接触面积，从而提高单井产量，同时还能提高采收率（水驱），并能进行气和水的控制。图 6.8 所示为水平井的示意图，示例中的水平井由垂直段、第一增斜段、稳斜段、第二增斜段及水平段组成。如果将从井口铅垂线到水平段起始点的位移定义为靶前位移，那么水平井的水平位移则是水平段长度与靶前位移之和。也有一些水平井由垂直段、增斜段和水平段

section connecting the vertical part with a horizontal section.

组成。

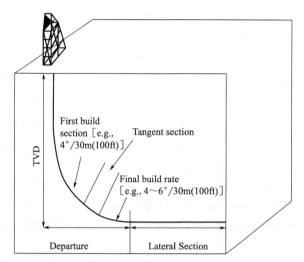

Fig. 6.8 Horizontal well profile consisting of two build sections

Typically, horizontal wells are classified by their radius of curvature as:

(1) Long-radius, with a radius of approximately 1000 ~ 3000 ft;

(2) Medium-radius, with a radius of 200 ~ 1000 ft;

(3) Short-radius, with a radius of 30 ~ 200 ft.

There are also ultrashort-radius systems that use high-pressure jetting techniques to turn the well from a vertical to a horizontal orientation.

The distinction between the three horizontal well categories is arbitrary, and in engineering practice, the build rates overlap. Some wells can be a combination of long and medium build rates or of medium and short. For example, a 3°/100 ft build rate may be used in the upper section of a well, followed by a tangent section, with a 10°/100 ft buildup rate below the tangent section to reach a horizontal section.

The build curve can be classified as ideal, simple-tangent, or complex-tangent. The ideal build curve connects the *KOP* with the beginning of the horizontal section using one or two circular arcs. One circular arc

通常，水平井按曲率半径可分为三类：

（1）长半径：曲率半径为1000 ~ 3000ft；

（2）中半径：曲率半径为200 ~ 1000ft；

（3）短半径：曲率半径为30 ~ 200ft。

还有一些超短半径水平井，通常会利用高压喷射技术将油井从垂直方向转到水平方向。

这三类水平井的区别并非十分清晰，在工程实践中，造斜率常互搭重叠。有些水平井可以是长半径和中半径造斜率的组合，也可以是中半径和短半径造斜率的组合。例如，一口井的上部井段使用3°/100ft的造斜率，然后是稳斜段，稳斜段结束时紧接着一个10°/100ft的造斜段直达水平段。

造斜曲线可分为理想型、简单切线型或复杂切线型。理想的造斜曲线使用一个或两个圆弧将造斜点与水平段起点连接起来。

located in a vertical plane is possible if *TVD−KOP=HD*. Then the radius of the circular arc is simply equal to *TVD−KOP*.

The simple-tangent build curve consists of an upper circular arc, a tangent section, and a lower circular arc.

The designer must decide on the buildup rates, the angle of the tangent section, and the length of the tangent section. The length of the tangent part should not be less than approximately 120 ～ 150 ft so that adjustments can be made if the performance of downhole tools (e.g., mud motors with bent housing) differs from that assumed by the designer.

A complex-tangent build curve uses the first build interval in a manner similar to the simple-tangent method. However, the second build curve is designed with a lower build rate and also involves turning the curve to the left or right reach the desired target. Eventually, a 3D curve is drilled to connect the tangent part with the horizontal segment. This task is accomplished by proper orientation of the tool-face deflection angle. The well path can be designed to have the entire turn in one direction (right or left) or to turn in one direction for some distance and then in the opposite direction to complete the curve. Of course, turning the well path involves a change in azimuth, which may change the direction (azimuth) of the end of curve. In general, the major factors affecting a horizontal-well profile are as follows:

(1) Anticipated reservoir production performance, existence of fractures and their orientation, depth of gas-oil and water-oil contact (WOC);

(2) Completion type (open-hole, slotted liner, etc.) and anticipated workover requirements;

(3) Casing and cementing program;

(4) Drilling fluids and wellbore stability;

(5) Drillstring design (torque and drag);

(6) Anticipated hole problems (cuttings transport, washouts, others).

6.1.7 Three-Dimensional Well Profiles

In engineering practice, any well trajectory that is not located in a vertical plane is considered to be a 3D well. Under favorable drilling conditions, trajectories can be restricted to a vertical plane; however, in many instances the well path must move in 3D space to meet the well objectives. A 3D well trajectory is designed for a variety of geological and engineering reasons—for example, to avoid some difficult-to-drill subsurface formations (e.g., drilling around salt domes) or to avoid faults. The well path must nearly always be in 3D if the well needs to intersect multiple targets, as is frequently required in horizontal drilling and geosteering applications. Most wells drilled offshore are three-dimensional to avoid intersecting with other wells. An example of the situation that often exists is the 3D view of a multiwell offshore platform as shown schematically in Fig. 6.9. Three-dimensional wells are also drilled onshore for environmental reasons (to minimize the footprint of drilling rigs) or when drilling under buildings and other constructions. A group of wells is called a cluster. Frequently, drilling a cluster of wells is not only environmentally friendly, but also more economical because of its higher efficiency and reduced footprint. New generations of onshore drilling rigs that can slide on rails make it easier to implement a multiwell (15～20 wells or more) directional-drilling program.

Another example is the so-called designer wells, as shown in Fig. 6.10. Such wells originally were drilled in the geologically complex Gullfaks field in the Norwegian sector of the North Sea. The field has a very complex oil reservoir, with many normal and reverse faults. Typically, a designer-type well path involves a strong change in the hole azimuth combined with some change in hole inclination angle. To be classified as a "designer well", a well must turn in a horizontal plane through not less than 30°, consist of both right turns (positive turn rates) and left turns (negative turn rates), and not have

6.1.7 三维井眼剖面

在工程实践中，任何超出垂直剖面的井眼轨迹都被看作是三维井眼轨迹。在钻井条件允许的情况下，井眼轨迹可以只分布在一个垂直剖面内。但多数情况下，井眼轨迹必须采用三维形式才能钻达目标。此外，当考虑到各种地质和工程因素时，例如为了避开一些难以钻进的地层（如盐丘）或断层时，也需要设计三维的井眼轨道。如果需要钻多目标井，就像钻水平井和地质导向那样频繁调整目标，井眼轨迹也必须是三维的。再有就是为了防碰，海上钻的大多都是三维井眼，如图6.9给出的就是海洋平台上常见的多分支井三维剖面。此外，由于环境因素（为了最小钻机占地面积），或者需要在建筑物或其他障碍物下方钻井时，也会在陆地上钻三维井。通常将同一个井场钻出的井组称为丛式井。丛式井不仅有利于环境保护，而且效率更高、占地面积更小，也就更经济。现在，能在轨道上滑动的新一代陆地钻机使丛式井组（15～20口井）定向钻井变得更加容易。

另一个实例是所谓的三维多目标井，如图6.10所示。最初，这类井是在北海挪威海域地质条件复杂的哥尔法克斯油田钻成的。该油田的储层情况非常复杂，有许多正断层和逆断层。三维多目标井的特征是井眼轨道的方位和井斜角有一些复杂的变化。而要成为一口"三维多目标井"必须满足以下条件：在水平面内的方位变化不少于30°，且同时包括右旋（方位变化率为

the turns restricted by inclination. The need to optimize production or injection efficiency or to place a horizontal well underneath a platform frequently results in a designer well profile.

正）和左旋（方位变化率为负），此外扭方位不受井斜的限制。在需要优化生产和注入效率时，或者在平台下方钻水平井时，就需要这种三维多目标井剖面。

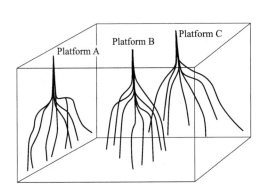

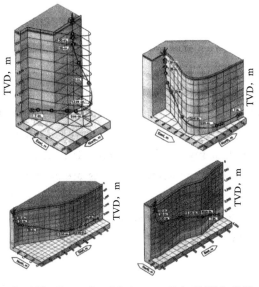

Fig. 6.9 Group of wells drilled for offshore applications   **Fig. 6.10** Examples of designer wells in Gullfaks fteld

In general, the design task is to construct a smooth 3D path that connects a surface or subsurface location to a known target or targets. In addition to 3D geometric requirements, the designer must also consider other factors related to the drilling process, such as drillstring mechanical integrity, wellbore stability, cuttings transport, running of casing, cementing and perforating operations, and other factors. For the purpose of well-path optimization, minimum drilling cost or minimum drilling time is usually used as the optimization criterion. The design process frequently requires a few iterations before the desired solution is found. Drilling data from offset wells (e.g., types of formations, drilling fluids, drilling problems, performance of drill bits, BHAs, downhole motors, and expected drilling rates) are very valuable for the designers. If the well is to produce both oil and gas, certain production aspects, such as two-phase flow hydraulics, a possible artificial-lift system, and formation-stimulation requirements must be considered. This paper will focus on only the 3D geometric

通常，轨道设计工作就是要构造一条平滑的三维轨道，将地面或地下某处与目标点连接起来。除了三维的几何要求以外，设计人员还必须考虑钻井过程中的钻柱机械完整性、井壁稳定性、岩屑运移、套管下入、固井和射孔作业等影响因素。为了优化井眼轨道，常常将最小钻进成本或最少钻进时间作为优化目标。设计过程通常需要进行几次迭代，才能找到解决方案。邻井的钻井数据（例如地层类型、钻井液、钻井复杂问题、钻头性能、下部钻具组合、井下马达及预期的钻井速度）对设计人员是非常有价值的。如果该井同时生产油和气，还必须考虑生产方面的因素，比如两相流动液压系统、合理的人工举升系统及油层增产的需求。

considerations and a discussion of five methods that are available from SPE literature. The five methods considered are average-angle method (AAM), radius-of-curvature (RCM), minimum-curvature (MCM), constant build and turn rate (CBTM), constant curvature and build rate (CCBM).

6.1.7.1 AAM

Several methods can be used to design a 3D well path. One of the first developed is the so-called AAM. In this approach, the well is modeled as a series of straight segments in a vertical and a horizontal plane, as shown earlier in Figs. 6.3a and 6.3b. It is assumed that the inclination and azimuth angles are constant and equal to the average value for two subsequent points on the trajectory. With this assumption, Eqs.6.13, 6.14, and 6.15 can be rewritten as:

$$x_2 = x_1 + \left(\sin\overline{\varphi}\cos\overline{\theta}\right)\Delta s \tag{6.29a}$$

$$y_2 = y_1 + \left(\sin\overline{\varphi}\cos\overline{\theta}\right)\Delta s \tag{6.29b}$$

$$z_2 = z_1 + \left(\cos\overline{\varphi}\right)\Delta s \tag{6.29c}$$

where the average values of the hole inclination and azimuth angles ($\overline{\varphi}$ and $\overline{\theta}$) are defined as $\overline{\varphi} = \dfrac{\varphi_2 + \varphi_1}{2}$ and $\overline{\theta} = \dfrac{\theta_2 + \theta_1}{2}$.

In other words, for a given increment in MD Δs between two subsequent points with inclination and azimuth angles φ_{i-1}, θ_{i-1} and φ_i, θ_i the arithmetic average values and the coordinates x_i, y_i, and z_i can be calculated if the coordinates, x_{i-1}, y_{i-1}, and z_{i-1} are known.

6.1.7.2 The Radius-of-Curvature Method (RCM)

The RCM was originally proposed by Wilson to replace earlier methods that used a series of straightline segments to represent the wellbore between survey

本文仅对三维几何因素进行分析，并讨论文献中提供的五种方法。这五种方法分别是平均角法（AAM）、曲率半径法（RCM）、最小曲率法（MCM）、恒定造斜率和方位变化率法（CBTM）及恒定曲率和造斜率法（CCBM）。

6.1.7.1 平均角法

设计三维井眼轨道的方法有多种，最早的一种就是平均角法（AAM）。该方法是在垂直剖面和水平平面上用一系列直线段对井眼轨道进行建模，如图6.3（a）和图6.3（b）所示。假设每个井段的井斜角和方位角为常量，且等于该井段上下两个点的角度平均值。以此假设为基础，式（6.13）、式（6.14）及式（6.15）可以重新整理得到：

式中，井斜角和方位角的平均值（$\overline{\varphi}$ 和 $\overline{\theta}$）定义为：$\overline{\varphi} = \dfrac{\varphi_2 + \varphi_1}{2}$ 和 $\overline{\theta} = \dfrac{\theta_2 + \theta_1}{2}$。

换句话说，如果已知上测点的坐标 x_{i-1}，y_{i-1} 和 z_{i-1}，测量井深的增量 Δs，以及上下两个测点的井斜角和方位角 φ_{i-1}，θ_{i-1} 和 φ_i，θ_i，则可以计算出下测点的坐标 x_i，y_i 和 z_i。

6.1.7.2 曲率半径法

曲率半径法最初由威尔逊提出，用于取代之前用一系列直线段来表示测点之间井眼轨迹的平

stations. In this method, it is assumed that the build rate B and the horizontal turn rate H are constant over the trajectory.

Typically, a few build and horizontal turn rates must be tried by the designer before finding a well path that will meet the desired objectives. Fig.6.11 shows a segment of wellbore between two points on the 3D trajectory. The MD between Points 1 and 2 is Δs.

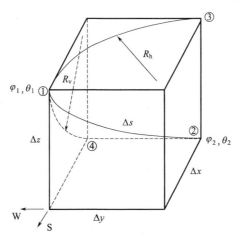

Fig.6.11 Schematic diagram of a wellbore segment for the radius-of-curvature method

It should be pointed out that, even if the build and turn rates are constant, the wellbore curvature is not constant between the two points on the well path because the hole inclination angle varies between Points 1 and 2. The assumptions of a constant build rate B and a constant horizontal turn rate H imply that the projections on the vertical plane (the segment between Points 1 and 4) and the horizontal plane (the segment bet-ween Points 1 and 3) have constant curvature. The radius of curvature in the vertical plane (R_v) is the reciprocal of the build rate, while the radius of curvature in the horizontal plane (R_h) is the reciprocal of the horizontal turn rate H.

Using the assumption that both build rate and turn rate are constant, by integration of Eqs. 6.13, 6.14, and 6.15, the following equations are obtained for calculating the desired rectangular trajectory coordinates:

$$\Delta x = \frac{1}{H}\int_{\theta_1}^{\theta_2}\cos\theta\,\mathrm{d}\theta = \frac{\sin\theta_2 - \sin\theta_1}{H} \tag{6.30a}$$

$$\Delta y = \frac{1}{H}\int_{\theta_1}^{\theta_2}\sin\theta\,\mathrm{d}\theta = \frac{\cos\theta_1 - \cos\theta_2}{H} \tag{6.30b}$$

$$\Delta z = \frac{1}{B}\int_{\varphi_1}^{\varphi_2}\cos\varphi\,\mathrm{d}\varphi = \frac{\sin\varphi_2 - \sin\varphi_1}{B} \tag{6.30c}$$

Clearly, four basic cases can be distinguished here (Case 1: B amd H are constant along the trajectory; Case 2: $H = 0$ and $B = $ constant; Case 3: $B = 0$ and $H = $ constant; and Case 4: both $B=0$ and $H=0$), resulting in four different types of well-trajectory segments. In practical designs, any combination of these can be used to connect the initial point smoothly with the target. A smooth trajectory is achieved if at the connection points (common points), the ends of the two segments have the same inclination and direction angles.

McMillian provided several examples of use of the RCM to design the well paths of slant and S-shaped wells as well as for 3D well configurations. He proposed a method by which a 3D problem can be transformed into 2D space. Once the 2D well path is determined in 2D space, it can be transferred back into 3D space.

After the base well trajectory is calculated, the designer needs to make corrections to compensate for anticipated effects related to drillpipe rotation (bit walk), formation hardness and dip angle, type of drill bit, and other factors. For example, as mentioned earlier, drillpipe rotation typically results in right-hand bit walk, and therefore a left lead angle is used to compensate for this tendency. The optimum lead angle results in the closest approach to the base trajectory. The required information on the directional tendencies of various drilling systems can be obtained by analyzing drilling data from similar wells drilled under similar geological conditions. Lack of such data can lead to considerable discrepancies between

显然，这里可以划分为四种基本情况（情况1：B和H沿轨迹保持不变；情况2：$H=0$和$B=$常数；情况3：$B=0$和$H=$常数；情况4：$B=0$且$H=0$），相应得到四种不同类型的井眼轨迹段。在实际设计中，起始点与目标点之间可以用任意组合的线段平滑连接。如果在连接点（公共点）处，两个井段的端点具有相同的井斜角和方位角，则得到的就是一条平滑轨迹。

麦克米兰提供了一些用曲率半径法设计斜井和S形井井眼轨道及三维井眼轨道的示例。他提出了一种可以将三维问题转化到二维空间的方法。一旦在二维空间中确定了二维井眼轨道，就可以再将其转换回三维空间。

基本的井眼轨道计算完成后，设计人员还需要对其进行修正，以消除钻杆旋转（方位漂移）、地层硬度和倾角、钻头类型及其他因素的影响。例如前面所说的，钻杆旋转通常会导致钻头向右漂移，需要使用左超前角来修正钻头向右漂移的趋势。最佳的超前角可以使实钻的井眼轨迹最接近设计轨道。通过分析相似地质条件下钻井的相关数据，可以获得各种钻井系统定向所需的信息。缺乏此类数据则可能导致计算出

calculated well trajectories and those actually achieved while drilling the well. This situation places the designer of an exploratory well in a difficult situation.

Under complex geological conditions, drilling a pilot hole should be considered to obtain at least some preliminary information about geological conditions, including types of rock, formation dip and strike angles, and possible hole problems.

6.1.7.3 Minimum-Curvature Method

An analytical formulation of the minimum-curvature method was originally proposed by Taylor and Mason and by Zaremba as a way to improve directional-survey analysis. Zaremba used the term circular-arc method and carried out the development using the method of vectors. More recently, Sawaryn and Thorogood published a compendium of algorithms useful for directional-well planning and deflection-tool orientation. Currently this method is widely used by the petroleum industry for both well-trajectory planning and directional-survey evaluation.

In this method, two successive points on the trajectory are assumed to lie on a circular are located in a plane, as shown schematically in Fig.6.12. In other words, Points 1, 2, and O in Fig.6.12 lie on the same plane, and the curvature of the segment between Points 1 and 2 is constant. The *MD* between Points 1 and 2 is Δs, and the radius of the circular are connecting the two points is R. The angle β is called the *DL*.

The reader is encouraged to analyze Fig.6.12 carefully and to prove that the equations for calculating changes in the rectangular coordinates on the trajectory are as given below, with the ratio factor quantity is represented by *RF*:

$$\Delta x = (\sin \varphi_1 \cos \theta_1 + \sin \varphi_2 \cos \theta_2) RF \qquad (6.31\text{a})$$

$$\Delta y = (\sin \varphi_1 \sin \theta_1 + \sin \varphi_2 \sin \theta_2) RF \qquad (6.31\text{b})$$

$$\Delta z = (\cos \varphi_1 + \cos \varphi_2) RF \qquad (6.31\text{c})$$

where

$$RF = \frac{\Delta s}{\beta} \tan \frac{\beta}{2} \qquad (6.32)$$

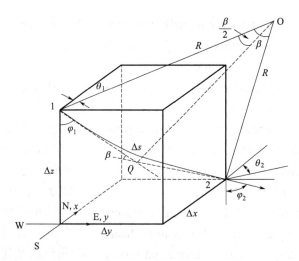

Fig.6.12 Schematic diagram of a wellbore segment for the minimum-curvature method

6.1.7.4 Constant Build and Turn Rate Method

In this method, proposed by Planeix and Fox, it is assumed that the bulid rate B and turn rate T are constant along the well trajectory. With this assumption, by integrating Eqs.6.13 and 6.14, the trajectory coordinates can be obtained as:

6.1.7.4 恒定造斜率和方位变化率法

普兰纳库斯和福克斯提出了假设造斜率 B 和方位变化率 T 沿井眼轨道保持恒定的计算方法。在此假设条件下，对式（6.13）和式（6.14）进行积分，可以得到以下的轨道坐标公式：

$$\Delta x = \frac{1}{T^2 - B^2}[T(\sin\varphi_2 \sin\theta_2 - \sin\varphi_1 \sin\theta_1) + B(\cos\varphi_2 \cos\theta_2 - \cos\varphi_1 \cos\theta_1)] \quad (6.33)$$

$$\Delta y = \frac{1}{T^2 - B^2}[B(\sin\theta_2 \cos\varphi_1 - \sin\theta_1 \cos\varphi_1) - T(\cos\theta_2 \sin\varphi_2 - \cos\theta_1 \sin\varphi_1)] \quad (6.34)$$

The equation for calculating the change in a vertical coordinate, Δz, remains the same as for the RCM (Eq.6.30c) and for this reason is not repeated here. It should be remembered that Eqs.6.33 and 6.34 are valid only for points on the bulid and turn curves. As for the RCM, one can consider several special cases such as $T=0$ and B=constant, $B=0$ and T=constant or perhaps $B=0$ and $T=0$.

Because the turn rate and the horizontal turn rate are functionally related ($T=H\sin\varphi$), the solutions for Δx, Δy, Δz are all the same as in the RCM. Here a special case of a wellbore trajectory composed of a segment of a circular helix will be discussed. Because a circular helix

垂直坐标变化 Δz 的计算公式与 RCM［式（6.30c）］相同，所以在此不再重复。注意，式（6.33）和式（6.34）仅对恒定造斜率和方位变化率的曲线有效。对于 RCM，可以考虑几种特殊情况，例如 $T=0$ 且 $B=$ 常数、$B=0$ 且 $T=$ 常数，或者 $B=0$ 且 $T=0$。

由于方位变化率和水平方位变化率存在关系（$T=H\sin\varphi$），所以 Δx、Δy、Δz 的解都与 RCM 中的解相同。下面讨论的特殊情况是由一段圆柱螺线组成的井眼

has constant curvature and constant inclination angle, the bulid rate is nil ($B = 0$) and the turn rate T is constant.

6.1.7.5 Constant Curvature and Build Rate Method

The constant-curvature method was proposed by Guo et al.(1992)to produce a well path that can be drilled with a constant tool face and to provide more flexibility in 3D well-path trajectory designs. In this method, it is assumed that the wellbore curvature(κ, DLS)and build rate B are constant with the MD. This method is also known as the constant tool-face method and was proposed by Schuh (1992). Again, to calculate the coordinates, Eqs.6.13.6.14.and 6.15 are needed. To perform the required integrations, the inclination and azimuth angles must be determined along the trajectory. Using Eqs.6.1 and 6.3.

$$\varphi(s) = \varphi_1 + B(s - s_1) \tag{6.35}$$

$$\theta(s) = \theta_1 + \int_{s_1}^{s_2} T(s) \mathrm{d}s \tag{6.36}$$

The turn rate in Eqs.6.36 can be expressed in terms of DLS and build rate as

$$T(s) = \frac{\sqrt{DLS^2 - B^2}}{\sin \varphi(s)} \tag{6.37}$$

With the assumption that the DLS and build rate are constant, integration of Eq.6.36 yields

$$\theta(s) = \theta_1 + \frac{\sqrt{DLS^2 - B^2}}{B} \ln \frac{\tan \frac{\varphi(s)}{2}}{\tan \frac{\varphi_1}{2}} \tag{6.38}$$

It is clear that because of the nonlinear form of Eq.6.38, the integrals of the trajectory equations (Eqs.6.13 and 6.14) need to be evaluated numerically. Closed-form solutions can be obtained for a case where the well path is part of a circular helix. For a circular helix, the build rate is nil, resulting in a constant hole-inclination angle and constant turn rate, as discussed earlier.Much simpler solutions are possible if the average values of turn rates are used piecewise for calculations. The average values are given by the following equations:

$$\overline{H} = \frac{\sqrt{DLS^2 - B^2}}{B\Delta s} \int_{\varphi_1}^{\varphi_2} \frac{d\varphi}{\sin^2 \varphi} = \frac{\sqrt{DLS^2 - B^2}}{B\Delta s}(\cot\varphi_1 - \cot\varphi_2) \qquad (6.39)$$

$$\overline{T} = \frac{\sqrt{DLS^2 - B^2}}{B\Delta s} \int_{\varphi_1}^{\varphi_2} \frac{d\varphi}{\sin\varphi} = \frac{\sqrt{DLS^2 - B^2}}{B\Delta s} \ln\frac{\tan\frac{\varphi_2}{2}}{\tan\frac{\varphi_1}{2}} \qquad (6.40)$$

Once the average values have been determined, the RCM or constant build-and-turn method can be used to calculate the desired rectangular coordinates x, y, and z along the well trajectory.

6.2 Deviation Control

Audio 6.3

A number of different methods have been invented to initiate new hole-inclination and azimuth angles (e.g., at KOPs) and to maintain control of a well path while drilling. In some applications, a whipstock or hydraulic jetting can be a cost-effective method to initiate a hole departure from the vertical and to make other required adjustments in hole direction. In buildup or drop-off wellbore segments, as well as in straight-hole drilling (e.g., drilling a tangent section), the inclination angle can be controlled to some extent using a conventional BHA by careful selection of BHA components. Considerable control of a well path, in terms of hole inclination and azimuth angles, can be achieved using downhole motors with a bent sub or bent housing and rotary-steerable tools.

These tools and techniques can be used to change the inclination or the azimuthal direction of the wellbore or both. All of these tools and techniques work on one of two basic principles. The first principle is to introduce a bit tilt angle into the axis of the BHA just above the bit and the second is to introduce a sideforce to the bit (See Fig. 6.13). The introduction of a tilt angle or sideforce to the bit will result in the bit drilling off at an angle to the current trajectory.

头施加侧向力都会导致钻头以一定的角度偏离当前的井眼轨迹轴线继续钻进。

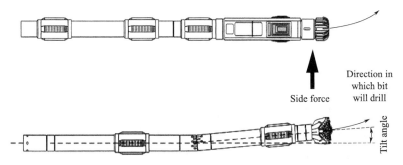

Fig. 6.13　Bit tilt angle and Sideforce

The major tools currently used for this purpose are:
（1）Bent Sub and Positive Displacement Motor;
（2）Non-Rotating Steerable Drilling Systems;
（3）Rotary Steering System;
（4）BHA;
（5）Whipstocks.

控制井眼轨迹的工具主要有：
（1）弯接头＋螺杆钻具；
（2）非旋转导向钻井系统；
（3）旋转导向系统；
（4）BHA；
（5）造斜器。

6.2.1　Bent Sub and Mud Motor

The most commonly used technique for changing the trajectory of the wellbore uses a piece of equipment known as a "bent sub" (Fig. 6.14) and a Positive Displacement (mud) motor. A bent sub is a short length of pipe with a diameter which is approximately the same as the drillcollars and with threaded connections on either end. It is manufactured in such a way that the axis of the lower connection is slightly offset (less than 3 degrees) from the axis of the upper connection. When made up into the BHA it introduces a "tilt angle" to the elements of the BHA below it and therefore to the axis of the drillbit. However, the introduction of a bent sub into the BHA means that the centre of the bit is also offset from the centre line of the drillstring above the bent sub and it is not possible therefore to rotate the drillbit by rotating thedrillstring from surface. Even if this were possible, the effect of the tilt angle would of course be eliminated since there would be no preferential direction for the bit to drill in.

6.2.1　弯接头和井下动力钻具

控制井眼轨迹最常用的技术是利用"弯接头"（图6.14）结合螺杆钻具（井下动力钻具）。弯接头是一段直径与钻铤大致相同的短管，两端都有螺纹连接。其结构特点是上下接头的轴线不在一条直线上（两条轴线的夹角小于3°）。当它与BHA组合时，它会使其下方的BHA和钻头轴线产生一个"倾斜角"。然而，将弯接头加入BHA就意味着钻头的中心线偏离了弯接头上方钻柱的中心线，因此定向控制时就不能通过转盘旋转钻柱来驱动钻头。如果使用转盘驱动，由于钻头没有了优先钻进的方向，"倾斜角"的作用也就不那么明显了。

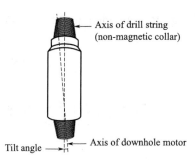

Fig. 6.14 Bent Sub

The bent sub must therefore be used in conjunction with a Positive Displacement Motor, PDM or a Drilling Turbine. The PDM is often called a mud motor and is used in far more wells than the turbine. The mud motor is made up into the BHA of the drillstring below the bent sub, between the bent sub and drillbit (See Fig. 6.15). When drilling fluid is circulated through the drillstring the inner shaft of the mudmotor, which is connected to the bit, rotates and therefore the bit rotates. It is therefore not necessary to rotate the entire drillstring from surface if a mud motor is included in the BHA. Mud motors and turbines are rarely used when not drilling directionally because they are expensive pieces of equipment and do wear out.

弯接头必须和螺杆钻具（PDM）或涡轮钻具配合使用。PDM常被称为井下动力钻具，它在油井中的应用比涡轮钻具多得多，是底部钻具组合的一部分，在弯接头下方，位于弯接头和钻头之间（图6.15）。当钻井液在钻柱中循环流动时，它将推动与钻头相连的PDM内轴旋转，从而带动钻头旋转。所以，如果BHA中包含了PDM，则无需从地面驱动整个钻柱旋转。在不进行定向钻井时，PDM和涡轮钻具尽量少用，因为它们比较昂贵，且容易磨损。

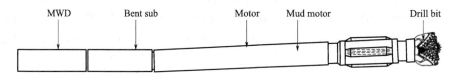

Fig.6.15 BHA with bent sub and mudmotor

A scribe line is marked on the inside of the bend of the bent sub, and this indicates the direction in which the bit will drill (this direction is known as the "tool face"). A directional surveying tool (quite often an MWD tool) is generally run as part of the BHA, just above the bent sub so that the trajectory of the well can be checked periodically as the well is deviating.The bent sub and PDM can of course only be used in the build up or drop off portion of the well since the bit will continue to drill in the direction of the tilt angle as long as the bent sub is in the assembly and the mud motor is being used to rotate the bit. This leads to the major disadvantage

在弯接头弯曲部分的内侧标记一条划线，这表示钻头将钻进的方向（该方向称为"工具面"）。定向测斜工具（通常是MWD工具）通常作为BHA的一部分安装在弯接头上方，以便在定向钻井时定期监测井眼轨迹。弯接头和PDM只能用于定向井的增斜或降斜段，因为只要弯接头在底部钻具组合中，同时用PDM驱动钻头，那么钻头将继续沿倾斜角的方向钻进。这是弯接头和PDM改

of using a bent sub and PDM to change the trajectory of the well. When drilling a well, the "conventional" assembly (without bent sub and mud motor) used to drill the straight portion of the well must be pulled from the hole and the bent sub and PDM assembly run in hole before the well trajectory can be changed. The bent sub and motor will then be used to drill off in a particular direction. When the well is drilling in the required direction (inclination and azimuth), the bent sub and PDM must then be pulled and the conventional assembly re-run. Otherwise the drillbit would continue to change direction. This is a very time consuming operation (taking approximately 8 hrs at 1,000 ft depth for each trip out of, and into, the hole). Remember however that the build up section of a well can be 1,000 ~ 2,000 ft long depending on the build up rate (typically 1 ~ 3 degrees/100ft) and the required inclination and therefore the bent sub and mud motor will be, depending on the rate of penetration, in the well for quite a long time time.

6.2.2 Steerable Drilling Systems

A steerable drilling system allows directional changes (azimuth and/or inclination) of the well to be performed without tripping to change the BHA, hence its name.It consists of: a drill bit; a stabilized positive displacement steerable mud motor; a stabilizer; and a directional surveying system which monitors and transmits to surface the hole azimuth, inclination and toolface on a real time basis (See Fig.6.16).

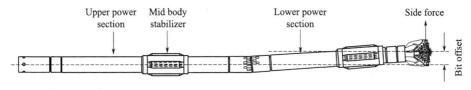

Fig. 6.16 Steerablre drilling system

The capability to change direction at will is made possible by placing the tilt angle very close to the bit, using a navigation sub on a standard PDM. This tilt angle can be used to drill in a specific direction, in the same way as the tilt angle generated by a bent sub with the the drillbit being rotated by the mud motor when circulating. However, since the tilt angle is much closer to the bit than a conventional bent sub assembly, it produces a much lower bit offset and this means that the drill bit can also be rotated by rotating the entire string at surface (in the same way as when using a conventional assembly). Hence the steerable assembly can be used to drill in a specific direction by orienting the bent sub in the required direction and simply circulating the fluid to rotate the bit (as in the bent sub assembly) or to drill in a straight line by both rotating and circulating fluid through the drillstring. When rotating from surface we will of course be circulating fluid also and therefore the rotation of the bit generated by the mud motor will be super-imposed on the rotation from surface. This does not alter the fact that the effect of the bit tilt angle will be eliminated by the rotation of the entire assembly.

When using the navigation sub and mud motor to drill a deviated section of hole (such as build up or drop off section of hole) the term "oriented or sliding" drilling is used to describe the drilling operation. When drilling in a straight line, by rotation of the assembly, the term "rotary" drilling is used to describe the drilling operation. The directional tendencies of the system are principally affected by the navigation sub tilt angle and the size and distance between the PDM stabilizer and the first stabilizer above the motor. The steerable drilling systems are particularly valuable where changes in the direction of the borehole are difficult to achieve; where directional control is difficult to maintain in the tangent sections of the well or where frequent changes may be required. The steerable systems are used in conjunction with MWD tools which contain petrophysical and directional sensors. These types of MWD tools are often called Logging Whilst Drilling, LWD tools. The petrophysical sensors are used to detect changes in the properties of the formations (lithology, resistivity or porosity) whilst drilling and therefore determine if a

一个特定的方向钻进，通过循环钻井液驱动井下动力钻具迫使钻头旋转。但是，由于弯曲点比传统的弯接头钻具组合更接近钻头，因此钻头偏离井眼中心的距离会大大减小，这意味着可以通过地面旋转整个钻柱，从而带动钻头旋转（与使用传统钻具组合时的方式相同）。这样，既可以利用导向钻具组合将弯接头定向到所需方向，通过循环钻井液驱动钻头旋转（就像使用弯接头钻具组合那样），进行定向钻进；也可以通过旋转整个钻柱并循环钻井液来钻直井段。当从地面旋转钻柱时，同时也会循环钻井液，地面旋转和井下动力钻具共同驱动钻头旋转。这时导向接头产生的倾斜角随着整个钻具组合的旋转几乎没有什么作用了。

当使用导向接头和井下动力钻具钻斜井段（如增斜段或降斜段）时，钻井作业用"导向或滑动"方式进行钻进。当利用旋转钻具组合沿直线钻进时，钻井作业用"旋转"方式进行钻进。系统的定向性能主要受导向接头弯曲角、PDM稳定器和井下动力钻具上方第一个稳定器之间的距离及尺寸的影响。在难以改变井眼方向、难以保持稳斜段定向控制或者井眼轨迹需要频繁变化时，使用导向钻井系统特别合适。导向钻井系统可以和随钻测量工具一起使用，当随钻测量工具包含岩石物性和方向传感器时，通常被称为随钻测井工具（LWD）。岩石物性传感器用来监测钻井过程中地层性质（岩性、电阻率或孔隙度）的变化，从而确定是否需要改变井眼方向。从油藏工程的

change in direction is required.Effectively the assembly is being used to track desirable formation properties and place the wellbore in the most desirable location from a reservoir engineering perspective. The term "Geosteering" is often used when the steerable system is used to drill a directional well in this way.

6.2.2.1 Components

There are five major components in a Steerable Drilling System. These components are: (1) Drill Bit; (2) Mud Motor; (3) Navigation Sub; (4)Navigation Stabilizers; (5)Survey System.

1. Drill Bit

Steerable systems are compatible with either tricone or PDC type bits. In most cases, a PDC bit will be used since this eliminates frequent trips to change the bit.

2. Mud Motor

The motor section of the system causes the bit to rotate when mud is circulated through the string. This makes oriented drilling possible. The motors may also have the navigation sub and a bearing housing stabilizer attached to complete the navigation motor configuration.

3. Navigation Sub

The navigation sub converts a standard Mud motor into a steerable motor by tilting the bit at a predetermined angle. The bit tilt angle and the location of the sub at a minimal distance from the bit allows both oriented and rotary drilling without excessive loads and wear on the bit and motor. The design of the navigation sub ensures that the deflecting forces are primarily applied to the bit face thereby maximizing cutting efficiency.

Two types of subs are presently available for steerable Systems: (1)The double tilted universal joint housing or DTU; (2)The tilted kick-off sub or TKO.

The DTU and TKO both utilize double tilts to produce the bit tilt required for hole deflection. The DTU's two opposing tilts reduce bit offset and sideload forces, and thereby maintaining an efficient cutting action. The TKO has two tilts in the same direction that are close to the bit.

角度来看，使用这种导向钻具组合可以有效地跟踪地层特性，并使井眼在地层中处于最理想的位置。当导向钻井系统按这种方式进行定向钻进时，通常称其为"地质导向"。

6.2.2.1 导向钻井系统组成

导向钻井系统有五个主要的部件。这些部件包括：（1）钻头；（2）井下动力钻具;(3)导向接头；（4）导向稳定器；（5）测量系统。

1. 钻头

导向钻井系统可以用三牙轮和PDC钻头。但大多用PDC钻头，这样可以避免频繁起下钻更换钻头。

2. 井下动力钻具

当钻井液在管柱中循环时，井下动力钻具驱动钻头旋转，进行定向作业。导向动力钻具由井下动力钻具、导向接头和轴承外壳稳定器组成。

3. 导向接头

导向接头通过将标准井下动力钻具换为导向动力钻具使钻头按预定角度倾斜，导向接头使钻头产生倾角，且弯曲点尽可能靠近钻头，这样定向和旋转钻进时就不会对钻头和动力钻具造成过度负载及磨损。导向接头的设计确保了侧向力主要作用于钻头刃面，从而最大限度地提高破岩效率。

目前常用的导向接头有两种：（1）反向双弯外壳，DTU；（2）同向双弯外壳，TKO。

DTU 和 TKO 都是利用两个弯曲角来产生钻头与井眼轴线间的倾斜角。DTU 两个方向相反的弯曲角降低了钻头的偏移和侧向力，从而保证了钻头有效的破岩作用。TKO 在靠近钻头的同一方向上有

4. Navigation Stabilisers

Two specially designed stabilizers are required for the operation of the system and influence the directional performance of a steerable assembly. The motor stabilizer or Upper Bearing Housing Stabilizer, UBHS is an integral part of the navigation motor. The upper stabilizer defines the third tangency point and is similar to a string stabilizer.

The size and spacing of the stabilizers also can be varied to fine-tune assembly reactions in both the oriented and rotary modes.

5. Survey System

A real time downhole survey system is required to provide continuous directional information. A measurement while drilling, MWD system is typically used for this purpose. An MWD tool will produce fast, accurate data of the hole inclination, azimuth, and the navigation sub toolface orientation. In some cases, a wireline steering tool may be used for this purpose.

6.2.2.2 Dogleg Produced by a Steerable System

When oriented drilling, the theoretical geometric dogleg severity or *TGD* produced by the system is defined by three points on a drilled arc (Fig. 6.17). The three points required to establish the arc are: (1) The Bit; (2) The PDM stabilizer or Upper Bearing Housing Stabilizer; (3) The first stabilizer above the mud motor (upper stabilizer).

两个弯曲角。

4. 导向稳定器

导向系统的操作需要两个专门设计的稳定器，它们会影响导向钻具组合的方向性能。马达稳定器或上轴承稳定器（UBHS）是导向动力钻具的一个基本组成部分。上稳定器限定了第三个接触点，它类似于钻柱稳定器。

通过改变稳定器的尺寸和间距，可以微调定向和旋转模式下钻具组合的反作用力。

5. 测量系统

导向钻井时，需要一个实时井下测量系统提供连续的井眼方向信息。随钻测量系统（MWD）的作用就在于此，它及时、准确地将产生的井斜、方位角和导向接头工具面方向等数据传递到地面。在某些情况下，也可以使用电缆导向工具来实现这一目的。

6.2.2.2 导向系统产生的狗腿

定向钻井时，导向系统产生的狗腿严重度（*TGD*）可以在理论上通过钻进井眼曲线空间几何的三个点来定义（图6.17）。这三个点是：（1）钻头位置；（2）PDM稳定器或上轴承稳定器所在位置；（3）井下动力钻具上方的第一个稳定器（上稳定器）位置。

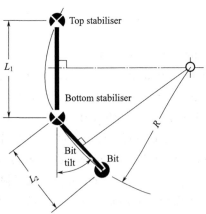

Fig. 6.17 Dogleg Srverity

The radius of the arc is further determined by the tilt of the navigation sub, as seen in the Fig. 6.17. The following basic relationship is produced by mathematical derivation.

$$TGD = \frac{200\varphi}{L_1 + L_2} \tag{6.41}$$

where, φ=Bit tilt in degrees; L=length between bit and upper stabilizer ($L_1 + L_2$); L_1=length between UBHS and the upper Stabilizer; L_2=length between UBHS and the bit.

This dogleg rate or TGDS is that created when the steerable system drills in the oriented mode.

6.2.2.3 Operation of a Steerable System

As described above, the steerable system can drill directionally or straight ahead, as required. This enables the driller to control the well's trajectory without making time-consuming trips to change bottomhole assemblies. To steer the hole during kickoffs or course corrections the system is oriented using MWD readings so the bit will drill in the direction of the navigation sub's offset angle. When drilling in this way the system is said to be drilling in the oriented or sliding (since the drillstring is not rotating) mode. The bit is driven by the downhole motor, and the rotary table is locked in place, as it is when conventional motor drilling. As mentioned previously, the system's two stabilizers and bit serve as the tangency points that define the curve to be drilled by the oriented assembly. The dogleg rate produced can be controlled by varying the placement and size of the stabilizers, by using a DTU with a different offset angle, or by alternating drilling with oriented and rotary intervals.

The system can also be used to drill straight ahead by simple string rotation. The rotary table is typically turned at 50～80 r/min while the motor continues to run. When drilling in this way the system is said to be drilling in the rotating mode. Through careful well planning and bottomhole assembly design, oriented sections are minimized and the assembly is rotated as

井眼曲线的半径可由导向接头的倾斜角进一步确定，并有以下基本关系式：

式中，φ 为钻头倾斜角度；L 为钻头和上稳定器间长度；L_1 为 UBHS 和上稳定器间长度；L_2 为 UBHS 和钻头间长度。

当导向系统按定向模式进行钻进时，会产生狗腿严重度。

6.2.2.3 导向系统的操作

如上所述，导向系统可根据需要进行定向钻井或直井段钻井。这样司钻不需要花费大量时间更换钻具组合就可以控制井眼轨迹。为了在造斜或修正过程中控制井眼，系统使用 MWD 进行随钻测量，保证钻头沿着导向接头偏移角的方向钻进。这时，系统的钻井方式被称为定向或滑动钻井（因为此时钻柱不旋转），此时钻头由井下动力钻具驱动，转盘与常规井下动力钻具钻井时一样锁定到位。如前所述，系统的两个稳定器和钻头作为切点，决定了定向钻具组合钻进的井眼曲线。通过改变稳定器的位置和尺寸、采用不同偏移角的 DTU，或转换定向和旋转钻进模式可以控制钻进过程中产生的狗腿度。

导向系统也可以通过旋转钻柱直接进行钻进。当动力钻具继续工作时，转盘通常以 50～80r/min 的转速旋转。这种钻进方式称为旋转模式下的钻进。钻井工程师通过精心的井眼轨道设计和底部钻具组合设计，可以使定向段最

much as possible. This maximizes penetration rates while keeping the well on course. Survey readings from an MWD tool enable efficient monitoring of directional data so the driller can maintain the well path close to the desired path. Slight deviations can be detected and corrected with minor oriented drilling intervals before they become major problems.

小，钻具组合尽可能多地处于旋转状态。这在控制井眼轨迹的同时大大提高了机械钻速。司钻依据MWD工具提供的有效定向监测数据，可以使井眼轨迹尽可能地接近设计轨道，避免出现严重的轨迹控制问题。轻微的轨迹偏离可以通过小段的定向钻井进行检测和纠偏，从而避免出现严重的轨迹控制问题。

6.2.3 Rotary Steering System

The rotary steering system (Fig.6.18) described here operates on the principle of the application of a sideforce in a similar way to the non-rotating systems described above. However, in these systems it is also possible to rotate the drillstring even when drilling directionally or as described above when in the "oriented mode" of drilling. It is therefore possible to rotate the string at all times during the drilling operation. This is desirable for many reasons but mostly because it has been found that it is much easier to transport drilled cuttings from the wellbore when the drillstring is rotating. When the drillstring is not rotating there is a tendency for the cuttings to settle around the drillstring and it may become stuck.

6.2.3 旋转导向系统

旋转导向系统（图6.18）与前面所述的非旋转导向钻井系统的工作原理相似，都是向钻头施加侧向力。然而，旋转导向系统即使在定向钻井或是按"定向模式"钻井时，也可以保证钻柱旋转。因此，在钻井作业期间，可以随时旋转钻柱。使用旋转导向系统的主要原因是当钻柱旋转时，可以更容易地将岩屑运移出井筒，避免由于钻柱不旋转，使岩屑容易在钻柱周围沉降，进而可能导致卡钻。

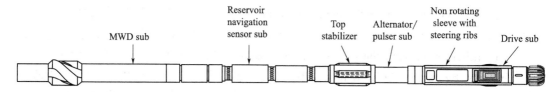

Fig. 6.18 Rotary steering system (Courtesy of Baker Hughes Inteq)

6.2.4 Directional Bottom Hole Assemblies

A conventional rotary drilling assembly is normally used when drilling a vertical well, or the vertical or tangent sections of a deviated well. When using a steerable assembly in a deviated well it is of course possible to drill the tangent sections of the well with the steerable assembly.

6.2.4 定向底部钻具组合

钻直井或斜井的直井段及稳斜段时，通常使用常规的旋转钻具组合。也可以使用导向钻具组合钻斜井的稳斜段。

The BHA of the conventional assembly can also be designed in such a way as to result in an increase or decrease in the inclination of the wellbore but it is very difficult to predict the rate at which the angle will increase or decrease with a conventional BHA and therefore this technique is not widely used today.

The tendency of a conventional BHA to result in an increase or decrease in hole angle is a function of the flexibility of the BHA. Since all parts of the drillstring are flexible to some degree (even large, heavy drill collars) the BHA will bend when weight is applied to the bit. This will introduce a tilt angle at the bit. The magnitude and orientation of the tilt angle will depend on the stiffness of the drillcollars, the WOB and the number and position of the stabilizers in the BHA. A great deal of research was conducted in the 1960s and 70s, in an attempt to predict the directional tendencies of BHAs with but it is very difficult to predict the impact of the above variables on the rate at which the angle will increase or decrease and therefore this technique is not widely used today.

Three types of drilling assemblies have been used in the past to control the hole deviation.

6.2.4.1 Packed Hole Assembly

This type of configuration is a very stiff assembly, consisting of drill collars and stabilizers positioned to reduce bending and keep the bit on course. This type of assembly is often used in the tangential section of a directional hole. In practice it is very difficult to find a tangent assembly which will maintain tangent angle and direction. Short drill collars are sometimes used, and also reamers or stabilizers run in tandem. A typical packed hole assembly is given in Fig 6.19a.

虽然常规的 BHA 也能使井斜角增加或减少，但井眼角度增加或减少的速率很难预测，因此这项技术在进行定向钻井时，目前还未得到广泛应用。

常规 BHA 导致井斜角增加或减少的程度和 BHA 弹性有关。由于钻柱都具有一定程度的柔韧性（即使是大直径厚壁的钻铤），所以当向井底施加钻压时，BHA 会发生弯曲。这样就会使钻头产生一个倾角。倾角的大小和方向取决于钻铤的刚度、钻压大小及 BHA 中稳定器的数量和位置。20 世纪 60 年代至 70 年代，有许多专家学者进行了大量的研究，试图预测 BHA 对井眼方向改变的作用，但事实上很难预测上述变量对增斜或降斜的影响，因此到现在也没有在定向钻井中得到广泛应用。

过去曾用于井斜控制的钻具组合有以下三种：

6.2.4.1 满眼钻具组合

这种钻具组合由几个外径与钻头直径相近的稳定器和一些外径较大的钻铤组成，它的刚度大，且能填满井眼，因而在大钻压下不易弯曲，并能保持钻具在井内居中。它通常用于钻定向井的稳斜段。但实际上，很难找到一个能保持井斜角和方位的稳斜组合。有时会将短钻铤、扩眼器或稳定器串联使用。典型的满眼钻具组合如图 6.19（a）所示。

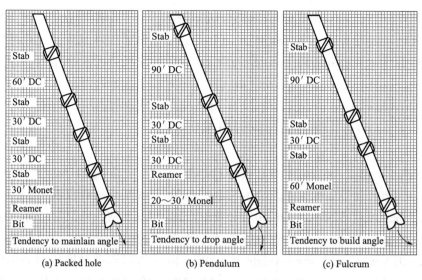

Fig. 6.19 Directional BHA's

6.2.4.2 Pendulum Assembly

The principles behind a Pendulum Assembly is that the unsupported weight of drill collars will force the bit against the low side of the hole. The resulting decrease or drop off in angle depends on WOB, RPM, stabilization and the distance between the bit and the first reamer (Fig. 6.19b). The basic drop off assembly is: Bit - DC- reamer - DC - stab - DC-stab - DC–stab

To increase the tendency to drop angle: (1) Apply less WOB (lower penetration rate); (2) Apply more RPM and pump pressure in soft formations.

6.2.4.3 Fulcrum Assembly

The principles behind a Fulcrum Assembly is to place a reamer near the bit (Fig. 6.19c) and apply a high WOB. When WOB is applied, the DCs above the reamer will tend to bend against the low side of hole, making the reamer act as a fulcrum forcing the bit upwards. The rate of build up depends on WOB, size of collars, position of reamer and stabilization above the reamer. The basic build-up assembly is: Bit - sub- reamer - DC - DC - stab - DC - stab - DC – stab.

6.2.4.2 钟摆钻具组合

钟摆钻具组合的工作原理是，利用斜井内切点以下钻铤重量的横向分力把钻头推向井壁下方（井眼低边）。降斜效果取决于钻压、转速、稳定性及钻头与第一个扩眼器之间的距离［图6.19（b）］。基本的降斜钻具组合是：钻头—钻铤—扩眼器—钻铤—稳定器—钻铤—稳定器—钻铤—稳定器。

为了增加降斜能力，可以采取以下措施：（1）采用低钻压（低钻速）；（2）在软地层中，提高转速和泵压。

6.2.4.3 支点钻具组合

支点钻具组合的工作原理是在近钻头处安装一个扩眼器，并提高钻压。当提高钻压时，扩眼器上方的钻铤将向井眼低边弯曲，使扩眼器起到支撑作用，迫使钻头向上产生倾斜的趋势。增斜率取决于钻压大小、钻铤尺寸、扩眼器及其上方稳定器的位置。基本的增斜钻具组合是：钻头—接头—扩眼器—钻铤—钻铤—稳定器—钻铤—稳定

To increase the build: (1) Add more WOB; (2) Use smaller size monel (increase buckling effect); (3) Reduce RPM and pump rates in soft formations.

6.2.5 Whipstocks

The whipstock is a steel wedge, which is run in the hole and set at the KOP. This equipment is generally used in cased hole when performing a sidetracking operation for recompletion of an existing well. The purpose of the wedge is to apply a sideforce and deflect the bit in the required direction. The whipstock is run in hole to the point at which the sidetrack is to be initiated and then a series of mills (used to cut through the casing) are used to make a hole in the casing and initiate the sidetrack. When the hole in the casing has been created a drilling string is run in hole and the deviated portion of the well is commenced.

6.3 Exercise

(1) Consider two points on a curved part of a trajectory located in a vertical plane with azimuth $\theta=60°$. The hole inclination angle at Point 1 is $\varphi_1=60°$ and at Point 2 is $\varphi_2=32°$. The drop-off rate is 6.5°/100 ft ($B=-6.5°/100$ ft). The rectangular coordinates of Point 1 are $x_1=1,650$ ft, $y_1=2,858$ ft, and $z_1=4,250$ ft. Calculate:

① the x, y, z coordinates at Point 2;
② the radius of curvature R;
③ the HD between Points 1 and 2;
④ the differences in MD between Points 1 and 2.

(2) Design the trajectory of a slant-type offshore well for the conditions stated below:

① Elevation of the rotary table = 180 ft;
② Target depth =5,374 ft;
③ Target south coordinate = 2,147 ft;
④ Target east coordinate = 3,226 ft;
⑤ Declination = 6° E;

⑥ *KOP* depth = 1,510 ft;
⑦ Buildup rate = 2°/100 ft.

A vertical section of this well is shown in Fig. 6.20(a) and a horizontal view in Fig. 6.20(b). Find the following:

① Slant angle;
② Vertical depth at the beginning of the tangent part;
③ Departure at the beginning of the tangent part;
④ *MD* to the target.

(3) Design a simple-tangent horizontal-well profilile, given the following:

① *KOP* = 8,206 ft;
② *VDT* = 9,000 ft;
③ Tangent length = 120 ft;
④ Tangent angle = 50°;
⑤ Target angle = 90° at VDT;
⑥ Expected build rate = 8°/100 ft.

(4) Calculate the rectangular coordinates of a well for the depth range from 8,000 to 8,400 ft. The KOP is at 8,000 ft, and the build rate is 1°/100 ft, using a lead of 10° and a right-hand walk rate of 1°/100 ft (the turn rate in a horizontal plane). The direction of the target is N30E. Assume that the first 200 ft is to set the lead, where the direction is held constant to 8,200 ft and then turns right at a rate of 1°/100 ft.

⑥ 造斜点深度 =1510ft;
⑦ 造斜率 =2°/100ft。

垂直剖面如图 6.20（a）所示，水平投影图如图 6.20（b）所示，请计算：

① 稳斜段井斜角；
② 稳斜段开始处的垂直深度；
③ 稳斜段开始处的水平位移；
④ 目标点的井深测深。

（3）设计一个带有单稳斜段的水平井剖面，已知数据如下：

① 造斜点深度 8206ft；
② 靶点垂深 9000ft；
③ 稳斜段长度 120ft；
④ 稳斜角 50°；
⑤ 靶点垂深处的井斜角 90°；
⑥ 预计造斜率 8°/100ft。

（4）造斜点位于 8000ft，造斜率为 1°/100ft，用 10° 的导眼和 1°/100ft 变化率向右扭方位（水平面上的方位角变化率）。目标的方位是 N30E。假设前 200ft 用来设置导眼，方位保持恒定至 8200ft，然后以 1°/100ft 的速率向右扭方位。请计算 8000～8400ft 深度范围内油井的直角坐标。

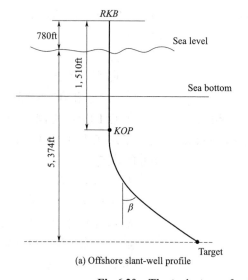
(a) Offshore slant-well profile

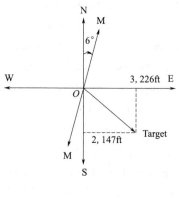
(b) Horizontal view

Fig.6.20 The trajectory of a slant-type offshore well

(5) At a certain point P_1 on a well path, the inclination angle and azimuth are φ_1=10.8° and θ_1=36.5°. Assuming an increment in the MD Δs = 200 ft (the distance between Points 1 and 2), to perform the calculations, assume a build rate B=5.14°/100 ft and a horizontal turn rate H=17°/100 ft. At Point 2 it is necessary to calculate the

① Hole inclination angle φ_2 and azimuth θ_2;
② Increments in x, y, z coordinates, Δx, Δy and Δz;
③ Wellbore curvature (i.e., DLS).

(6) List major applications of directional wells.

(7) Define hole inclination and azimuth angles.

(8) Define HD and closure.

(9) Define build and turn rates.

Well Control 井控 7

7.1 Introduction

7.1 引言

Audio 7.1

This chapter will introduce the procedures and equipment used to ensure that fluid (oil, gas or water) does not flow in an uncontrolled way from the formations being drilled, into the borehole and eventually to surface. This flow will occur if the pressure in the pore space of the formations being drilled (the formation pressure) is greater than the hydrostatic pressure exerted by the column of mud in the wellbore (the borehole pressure). It is essential that the borehole pressure, due to the column of fluid, exceeds the formation pressure at all times during drilling. If, for some reason, the formation pressure is greater than the borehole pressure an influx of fluid into the borehole (known as a *kick*) will occur. If no action is taken to stop the influx of fluid once it begins, then all of the drilling mud will be pushed out of the borehole and the formation fluids will be flowing in an uncontrolled manner at surface. This would be known as a Blowout. This flow of the formation fluid to surface is prevented by the secondary control system. Secondary control is achieved by closing off the well at surface with valves, known as Blowout Preventers - BOPs.

The control of the formation pressure, either by ensuring that the borehole pressure is greater than the formation pressure (known as Primary Control) or by closing off the BOP valves at surface (known as Secondary Control) is generally referred to as keeping the pressures in the well under control or simply well control.

When pressure control over the well is lost, swift

本章介绍的工艺和设备用于确保地层流体（油、气或水）不会以不受控制的方式从所钻地层流入井筒，并最终上返至地面。如果所钻地层孔隙压力（地层压力）大于井筒内液柱所产生的静水压力（井眼压力），就会发生地层流体侵入井眼的现象。在钻井过程中，要确保井内钻井液柱产生的压力始终大于地层压力。如果由于某些原因，使得地层压力大于井筒压力，就会发生地层流体侵入井筒的现象（称为井涌）。当井涌发生时，如果没有采取任何措施来阻止地层流体涌入，那么所有钻井液将会被顶替出井筒，地层流体会在地面以不受控制的方式喷出，称为井喷。二级井控系统可以利用防喷器组合的阀门从地面关闭井筒，阻止地层流体向地面流动。

地层压力控制通常是指保持井筒压力可控，简称为井控。井控既可以通过井筒压力大于地层压力（称为一级井控）来实现，也可以通过关闭地面防喷器组合的阀门（称为二级井控）来实现。

当井筒压力失去控制时，必

action must be taken to avert the severe consequences of a blow-out. These consequences may include: (1) Loss of human life; (2) Loss of rig and equipment; (3) Loss of reservoir fluids; (4) Damage to the environment; (5) Huge cost of bringing the well under control again.

For these reasons it is important to understand the principles of well control and the procedures and equipment used to prevent blowouts. Every operating company will have a policy to deal with pressure control problems. This policy will include training for rig crews, regular testing of BOP equipment, BOP test drills and standard procedures to deal with a kick and a blow-out.

One of the basic skills in well control is to recognize when a kick has occurred. Since the kick occurs at the bottom of the borehole its occurrence can only be inferred from signs at the surface. The rig crew must be alert at all times to recognize the signs of a kick and take immediate action to bring the well back under control. The severity of a kick (amount of fluid which enters the wellbore) depends on several factors including the: type of formation; pressure; and the nature of the influx. The higher the permeability and porosity of the formation, the greater the potential for a severe kick (e.g. sand is considered to be more dangerous than a shale). The greater the negative pressure differential (formation pressure to wellbore pressure) the easier it is for formation fluids to enter the wellbore, especially if this is coupled with high permeability and porosity. Finally, gas will flow into the wellbore much faster than oil or water.

7.2　Well Control Principles

There are basically two ways in which fluids can be prevented from flowing, from the formation, into the borehole:

(1)Primary Control. Primary control over the well is maintained by ensuring that the pressure

Audio 7.2

须迅速采取措施，避免井喷产生严重后果。这些后果可能包括：（1）造成人员生命损失；（2）损失钻机和钻井设备；（3）造成储层流体流失；（4）对环境产生破坏；（5）花费较大的成本。

基于上述原因，掌握井控工艺原理并了解防止井喷的设备至关重要。每个作业公司都会制定井筒压力控制策略。这项策略包括对钻井人员的培训、定期测试防喷器组合、防喷器组合演习及制定处理井涌和井喷的标准程序。

井控的基本内容之一就是井涌识别。由于井涌常发生在井底，只能通过地面的一些征兆推断是否发生。钻井作业人员必须时刻保持警惕，及时发现井涌征兆，并立即采取措施对油气井压力进行控制。井涌的严重程度（侵入井筒的流体量）取决于以下几个因素：地层类型、地层压力及侵入流体的性质。地层的渗透率和孔隙度越高，发生严重井涌的可能性就越大（如砂岩比页岩更危险）。负压差（地层压力和井筒压力之差）越大，地层流体就越容易进入井筒，尤其是在高孔、高渗情况下，侵入量会更大。当然，气体侵入井筒的速度要比油水侵入的速度快得多。

7.2　井控原则

阻止流体从地层侵入井筒的基本方法有两种：

（1）一级井控。一级井控是通过确保正钻井井筒内的钻井液

due to the column of mud in the borehole is greater than the pressure in the formations being drilled i.e. maintaining a positive differential pressure or overbalance on the formation pressures. (Fig. 7.1)

(2) Secondary Control. Secondary control is required when primary control has failed (e.g. an unexpectedly high pressure formation has been entered) and formation fluids are flowing into the wellbore. The aim of secondary control is to stop the flow of fluids into the wellbore and eventually allow the influx to be circulated to surface and safely discharged, while preventing further influx downhole.The first step in this process is to close the annulus space off at surface, with the BOP valves, to prevent further influx of formation fluids (Fig. 7.2). The next step is to circulate heavy mud down the drillstring and up the annulus, to displace the influx and replace the original mud. The second step will require flow the annulus but this is done in a controlled way so that no further influx occurs at the bottom of the borehole. The heavier mud should prevent a further influx of formation fluid when drilling ahead. The well will now be back under primary control.

Primary control of the well may be lost (i.e. the borehole pressure becomes less than the formation pressure) in two ways. The first is if the formation pressure in a zone which is penetrated is higher than that predicted by the reservoir engineers or geologist. In this case the drilling engineer would have programmed a mud weight that was too low and therefore the bottomhole pressure would be less than the formation pressure (Fig. 7.1). The second is if the pressure due to the column of mud decreases for some reason, and the bottomhole pressures drops below the formation pressure. Since the bottomhole pressure is a product of the mud density and the height of the column of mud. The pressure at the bottom of the borehole can therefore only decrease if either the mud density or the height of the column of mud decreases (Fig. 7.3 and Fig. 7.4).

液柱压力大于地层压力，即保持正压差或过平衡钻井，来维持井筒的压力控制（图 7.1）。

（2）二级井控。当一级井控失败（如钻遇意外的高压地层）导致地层流体侵入井筒时，就需要进行二级井控。二级井控的目的是阻止地层流体进一步侵入井筒，并最终安全地将侵入井筒的流体循环至地面排出。二级井控流程是首先用防喷器组合的阀门关闭地面的环空，阻止地层流体进一步侵入（图 7.2）。然后将加重钻井液从钻柱内向下循环，再从环空向上返出，循环排出受井侵污染的钻井液。这一步是在环空受控的方式下进行的，因此井底不会进一步发生井侵。继续钻进时，加重钻井液密度足够阻止地层流体的进一步侵入。这时油气井就恢复到一级井控状态。

失去一级井控（即井筒压力小于地层压力）可能有以下两方面的原因。第一种情况是实际钻遇的地层压力高于油藏工程师或地质工程师预测的地层压力。在这种情况下，钻井工程师设计的钻井液密度可能较低，导致井底压力小于地层压力（图 7.1）。第二种情况是由于一些原因导致钻井液液柱压力降低，从而井底压力小于地层压力。由于井底压力是钻井液密度和液柱高度的乘积，所以无论是钻井液密度还是液柱高度的减小，都会导致井底压力随之下降（图 7.3 和图 7.4）。

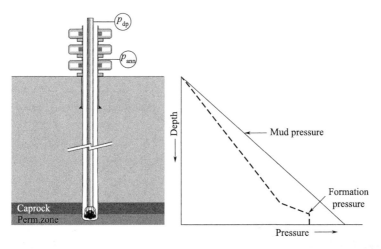

Fig. 7.1 Primary Control - Pressure due to mud column exceeds Pore Pressure

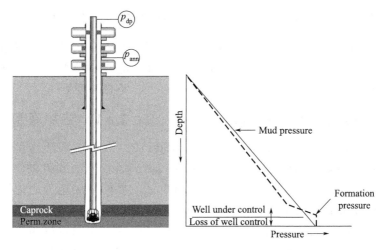

Fig.7.2 Secondary Control -Influx Controlled by Closing BOP's

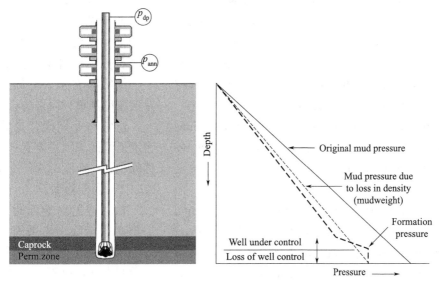

Fig. 7.3 Loss of primary control - due to reduction in mud weight

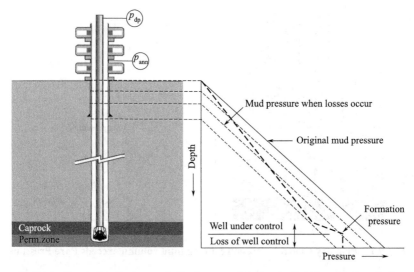

Fig. 7.4 Loss of primary control - due to reduction in fluid level in borehole

There are a number of ways in which the density of the mud (mud weight) and/or the height of the column of mud can fall during normal drilling operations.

7.2.1 Reduction in Mud Weight

The mud weight is generally designed such that the borehole pressure opposite permeable (and in particular hydrocarbon bearing sands) is around 200～300psi greater than the formation pore pressure. This pressure differential is known as the overbalance. If the mud weight is reduced the overbalance becomes less and the risk of taking a kick becomes greater. It is therefore essential that the mud weight is continuously monitored to ensure that the mud that is being pumped into the well is the correct density. If the mud weight does fall for some reason then it must be increased to the programmed value before it is pumped downhole.

The mud weight will fall during normal operations because of the following: (1) Solids removal; (2) Excessive dilution of the mud; (3) Gas cutting of the mud.

7.2.1.1 Solids Removal

The drilled cuttings must be removed from the mud when the mud returns to surface. If the solids removal equipment is not designed properly a large amount of the weighting solids (Barite) may also be removed. The solids removal equipment must be designed such that it

在正常的钻井作业过程中，有多种情况会导致钻井液密度和液柱高度下降。

7.2.1 钻井液密度降低

对于渗透性地层（特别是含烃砂岩），设计的钻井液密度通常要使井底压力比地层孔隙压力大200～300psi（1.38～2.07MPa），这种压力差称为正压差。如果钻井液密度降低，正压差减小，就会增大井涌风险。这时，就需要持续监测钻井液密度，确保泵入井中的钻井液密度合理。如果由于某种原因钻井液密度确实降低了，就需要在泵入井筒之前将钻井液密度加重到原来的设计值。

正常作业期间，导致钻井液密度下降的原因主要有：(1) 固相移除；(2) 稀释；(3) 气侵。

7.2.1.1 固相移除

钻井液返出地面时，其中的岩屑必须清除。如果固控设备设计不当，则可能同时将钻井液中加重材料（重晶石）移除。因此，固控设备尽可能是仅仅去除岩屑，

removes only the drilled cuttings. If Barite is removed by the solids removal equipment then it must be replaced before the mud is circulated downhole again.

7.2.1.2　Dilution

When the mud is being treated to improve some property (e.g. viscosity) the first stage is to dilute the mud with water in order to lower the percentage of solids. Water may also be added when drilling deep wells, where evaporation may be significant. During these operations mud weight must be monitored and adjusted carefully.

7.2.1.3　Gas Cutting

If gas seeps from the formation into the circulating mud (known as gas-cutting) it will reduce the density of the drilling fluid. When this is occurs, the mud weight measured at surface can be quite alarming.

It should be appreciated however that the gas will expand as it rises up the annulus and that the reduction in borehole pressure and therefore the reduction in overbalance is not as great as indicted by the mud weight measured at surface. Although the mud weight may be drastically reduced at surface, the effect on the bottom hole pressure is not so great. This is due to the fact that most of the gas expansion occurs near the surface and the product of the mud weight measured at surface and the depth of the borehole will not give the true pressure at the bottom of the hole. For example, if a mud with a density of 0.53 psi/ft. were to be contaminated with gas, such that the density of the mud at surface is 50% of the original mud weight (i.e. measured as 0.265 psi/ft.) then the borehole pressure at 10,000ft would normally be calculated to be only 2,650 psi. However, it can be seen from Fig. 7.5 that the decrease in bottom hole pressure at 10,000 ft. is only 40～45 psi.

It should be noted however that the presence of gas in the annulus still poses a problem, which will get worse if the gas is not removed. The amount of gas in the mud should be monitored continuously by the mudloggers, and any significant increase reported immediately.

如果同时清除了重晶石，那么钻井液就需要在再次循环入井之前进行更换。

7.2.1.2　稀释

有时为了改善钻井液性能（如黏度）需要对其进行处理，首选的方法是用水稀释，从而降低固相含量。深井钻井时由于蒸发显著，也可能会向钻井液中加水稀释。所有的稀释都必须谨慎监测和调整钻井液密度。

7.2.1.3　气侵

如果气体从地层渗入循环的钻井液中（称为气侵），钻井液的密度就会降低。这时，在地面测得的钻井液密度可能会大大降低。

然而应当认识到，由于气体沿环空上升时，其体积膨胀会引起井底压力的降低，降低的程度并不像地面测得的那样显著。尽管在地面测得的钻井液密度显著降低，但由于大部分的气体膨胀发生在地面附近，对井底压力的影响不是很大。因此地面测量的钻井液密度与井眼深度的乘积并不是真实的井底压力。例如，密度为 0.53psi/ft（1.223g/cm³）的钻井液被气体污染后，在地面测得的密度可能是初始值的 50%（即测量值为 0.265psi/ft），通过计算得出 10000ft（3048m）处的井筒压力仅为 2650psi 然而，从图 7.5 可以看出，在 10000ft 深度处的井底压力仅降低了 40～45psi。

注意，环空中存在的气体仍然会为钻井作业带来麻烦，如果不去除，情况会变得更糟。录井人员应持续监测钻井液中的气体含量，如有任何明显的气体含量增加应立即报告。

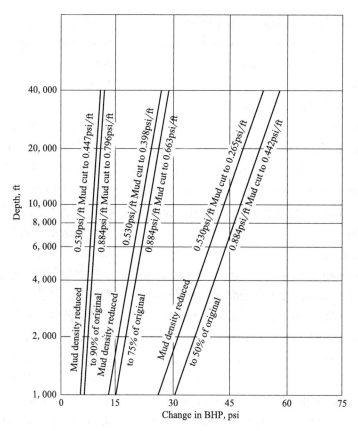

Fig. 7.5 Reduction in bottom hole pressure due to observed surface reduction caused by gas influx

7.2.2 Reduced Height of Mud Column

During normal drilling operations the volume of fluid pumped into the borehole should be equal to the volume of mud returned and when the pumps are stopped the fluid should neither continue to flow from the well (this would indicate that a kick was taking place) nor should the level of the mud fall below the mud flowline. The latter can be observed by looking down the hole through the rotary table.

If the top of the mud drops down the hole then the height of the column of mud above any particular formation is decreased and the borehole pressure at that point is decreased. It is therefore essential that the height of the column of mud is continuously monitored and that if the columm of mud does not extend to surface then some action must be taken before continuing operations.

The mud column height may be reduced by:

7.2.2 钻井液液柱降低

在正常的钻井作业中，钻井液返出体积应等于泵入体积。停泵时既不应出现钻井液继续从井中流出（可能表明正在发生井涌）的现象，也不应出现钻井液液面低于钻井液出口管线的情况。钻井液液面高度可以通过转盘向下观察井内情况进行观测。

如果井筒中的钻井液液面高度下降，那么井内任何地层上方的钻井液液柱高度就会相应减小，井筒压力随之降低。因此，对钻井液液柱高度进行连续监测至关重要，如果钻井液液柱未返至地面，则在继续作业之前必须采取措施进行补救。

钻井液柱高度降低的原因可

(1)Tripping; (2)Swabbing; (3)Lost circulation.

7.2.2.1 Tripping

The top of the column of mud will fall as the drillpipe is pulled from the borehole when tripping. This will result in a reduction in the height of the column of mud above any point in the wellbore and will result in a reduction in bottom hole pressure. The hole must therefore be filled up when pulling out of the hole. The volume of pipe removed from the borehole must be replaced by an equivalent volume of drilling fluid.

7.2.2.2 Swabbing

Swabbing is the process by which fluids are sucked into the borehole, from the formation, when the drillstring is being pulled out of hole. This happens when the bit has become covered in drilled material and the drillstring acts like a giant piston when moving upwards. This creates a region of low pressure below the bit and formation fluids are sucked into the borehole. (The opposite effect is known as Surging, when the pipe is run into the hole).

The amount of swabbing will increase with:

(1) The adhesion of mud to the drillpipe;

(2) The speed at which the pipe is pulled;

(3) Use of muds with high gel strength and viscosity;

(4) Having small clearances between drillstring and wellbore;

(5) A thick mud cake;

(6) Inefficient cleaning of the bit to remove cuttings.

7.2.2.3 Lost Circulation

Lost circulation occurs when a fractured, or very high permeability, formation is being drilled. Whole mud is lost to the formation and this reduces the height of the mud column in the borehole. Lost circulation can also occur if too high a mud weight is used and the formation fracture gradient is exceeded. Whatever the cause of lost circulation it does reduce the height of the column

能有：（1）起下钻作业；（2）抽汲作用；（3）井漏。

7.2.2.1 起下钻作业

起钻时，随着钻柱从井眼中起出，钻井液液柱高度会下降。这会导致井筒内钻井液液柱高度降低，并引起井底压力减小。因此，起下钻时应及时进行灌浆，用与井筒中起出管柱相同体积的钻井液来弥补由于钻柱引起钻井液液面降低。

7.2.2.2 抽汲作用

抽汲是指将钻柱从井眼中起出的同时会将流体从地层中吸入井眼的现象。当钻头泥包时，钻柱向上运动就像在拔一个巨大的活塞，这时就会产生抽汲作用。会在钻头下方形成一个低压区，将地层流体吸入井筒（下钻时，易发生与之相反的激动作用）。

下列因素会导致抽汲作用增大：

（1）钻井液对钻杆的黏附效应较大；

（2）起出钻柱的速度较快；

（3）使用高静切力和高黏度钻井液；

（4）钻柱和井筒之间的环空间隙很小；

（5）滤饼较厚；

（6）钻头清洁和携岩效率低。

7.2.2.3 井漏

当钻遇裂缝性或高渗透性地层时可能会发生井漏。钻井液漏失到地层，井筒内的钻井液液柱高度就会降低。如果使用的钻井液密度过高，超过了地层破裂压力梯度，也会发生井漏。无论何种原因造成的井漏，都会降低井筒中的钻

of mud in the wellbore and therefore the pressure at the bottom of the borehole. When the borehole pressure has been reduced by losses an influx, from an exposed, higher pressure, formation can occur.

Losses of fluid to the formation can be minimised by:

(1) Using the lowest practicable mud weight;

(2) Reducing the pressure drops in the circulating system therefore reducing the ECD of the mud;

(3) Avoid pressure surges when running pipe in the hole;

(4) Avoid small annular clearances between drillstring and the hole.

It is most difficult to detect when losses occur during tripping pipe into or out of the hole since the drillpipe is being pulled or run into the hole and therefore the level of the top of the mud column will move up and down. A Possum Belly Tank (or trip tank) with a small diameter to height ratio is therefore used to measure the amount of mud that is used to fill, or is returned from, the hole when the pipe is pulled from, or run into, the hole respectively. As the pipe is pulled from the hole, mud from the trip tank is allowed to fill the hole as needed. Likewise when tripping in, the displaced mud can be measured in the trip tank (Fig. 7.6). The advantage of using a tank with a small diameter to height ratio is that it allows accurate measurements of relatively small volumes of mud.

When the drillpipe is pulled out the hole the volume of mud that must be pumped into the hole can be calculated from the following:

The volume of mud that must be pumped into the hole=Length of Pipe×Displacement of Pipe

10 stands of 5 in, 19.5 lb/ft drillpipe would have a displacement of:

$10 \times 93 \times 0.00734$ (bbl/ft) = 6.8(bbl)

Therefore, the mud level in the hole should fall by an amount equivalent to 6.8bbl of mud. If this volume of mud is not required to fill up the hole when 10 stands have been pulled from the hole then some other

井液液柱高度，从而降低井底压力。当井筒内压力由于漏失而降低时，就可能会导致在较高地层压力的裸眼段发生井侵。

降低井漏风险的方法有：

（1）尽可能使用低密度钻井液；

（2）减少循环压降，从而降低钻井液的当量循环密度；

（3）下钻时，避免压力波动；

（4）避免在钻柱和井筒之间形成较小的环空间隙。

起下钻时井筒内的液柱高度会上下移动，因此起下钻过程中的井漏很难监测。为此，井队用一个直径与高度之比较小的计量罐来测量起钻时灌入或下钻时返出的钻井液量。当钻柱从井筒中起出时，可根据需要将计量罐中的钻井液灌入井筒内。同理，下钻时，可以通过计量罐测量井内排出的钻井液量（图 7.6）。之所以使用直径与高度之比较小的计量罐，原因在于这样可以精确地测量出体积变化较小的钻井液量。

起钻时，可以用以下公式计算出必须灌入井筒的钻井液量：

灌入井筒的钻井液量 = 钻柱长度 × 单位长度排开体积

例如 10 根外径为 5in、密度为 19.5lb/ft 的立柱体积应为：

$10 \times 93 \times 0.00734$(bbl/ft)= 6.8(bbl)

正常情况下，井筒钻井液液面下降量应为 6.8bbl（1.08m³）。若起出 10 根立柱后，不需要用 6.8bbl 的钻井液就可以填满井筒，

fluid must have entered the wellbore. This is a primary indicator of a kick.

则表明一定有其他流体进入井筒，这是井涌的重要征兆。

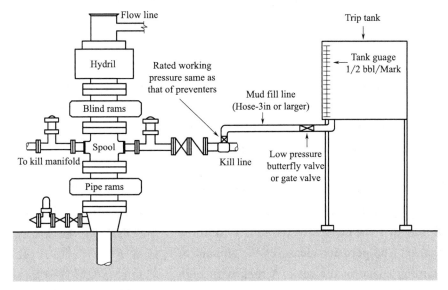

Fig. 7.6 Trip tank connected to BOP stack to closely monitor volume of mud required for fill-up

7.3 Warning Indicators of Kicks　7.3　井涌预警

7.3.1 Introduction to Kicks　7.3.1　井涌简介

Different drilling problems confront the operator on a day-to-day basis, including lost circulation, stuck pipe, deviation control, and well control. This discussion focuses on well control; other drilling problems will be considered here only in relation to some aspect of well control. A kick can be defined as a well-control problem in which the pressure encountered within the rock being drilled is greater than the mud hydrostatic pressure acting on the borehole or rock face. When this occurs, the greater formation pressure tends to force formation fluids into the wellbore. The result is an uncontrolled flow into a wellbore. This fluid flow is called a kick. If the flow is successfully controlled, the kick has been killed. A blowout is an uncontrolled flow out of a wellbore and often occurs because a kick was not properly controlled. For a kick to occur, three things must happen simultaneously. First, a mobile fluid must

Audio 7.3

钻井作业者每天都会面对井漏、卡钻、井斜控制和井控等复杂情况。本章讨论的重点是井控，在此仅考虑和井控有关的钻井复杂情况。井涌是由于所钻遇的地层压力大于井筒内钻井液液柱压力而产生的井控问题。发生井涌时，较大的地层压力会迫使地层流体侵入井筒，结果可能会引起地层流体不受控制地流入井眼，这种流动称为井涌。如果能成功控制侵入的流体，那么井涌就不会进一步发展。井喷是井筒中流体不受控制的流动现象，井涌发生必须同时满足三个条件：首先，井筒附近的多孔岩石中必须存在可流动的流体；其次，岩石的渗透率也要足够大，才具备流体流

be present in the porous rock adjacent to the borehole. Second, there must be enough permeability to sustain a flow into the wellbore. Third and most importantly, the pressure exerted in the wellbore, from a combination of hydrostatic, dynamic, and surface pressures, must be less than the pore pressure in the formation. If any one of these factors is missing, the well cannot kick. In one particular situation, gas can be entrained into a wellbore from inside the volume of rock being drilled, and this can lead to a kick even if permeability is not an important issue.

The severity of a kick depends on several factors. One is the ability of the rock to allow fluid flow. The perme- ability of rock describes its ability to allow fluid movement. The porosity measures the amount of fluid-containing space in the rock. A rock with high permeability and high porosity has a greater potential to generate a severe kick than a rock with low permeability and low porosity. For example, sandstone is considered to have a greater kick potential than shale because, in general, sand has a greater permeability and porosity than shale.

Another controlling variable for kick severity is the pressure differential involved. The pressure differential is the difference between the formation-fluid pressure and the mud hydrostatic pressure. If the formation pressure is much greater than the hydrostatic pressure, a large negative differential pressure exists. If this negative differential pressure is coupled with high permeability and porosity, a severe kick can occur.

A kick can be characterized in several ways. One way is by the type of formation fluid that has entered the borehole. Known kick fluids include gas, oil, salt water, magnesium chloride, water, hydrogen sulfide (sour) gas, and carbon dioxide. If gas has entered the borehole, the kick is called a gas kick. Another method of characterizing kicks is by the required mud-weight increase necessary to control the well and to kill a potential blowout.

An additional important well-control consideration

入井筒的条件；最后，也是最重要的一点，就是由静液压力、动态压力和井口回压共同作用形成的井筒压力必须小于地层孔隙压力。如果缺少这些因素中的任何一个，就不会形成井涌。需要说明的是渗透率并不是十分重要的因素，因为有些气体可能会从所钻岩石中侵入井筒，也会导致井涌。

井涌的严重程度取决于多个因素，其中之一就是渗透率和孔隙度的大小，渗透率表征岩石允许流体流动的能力，孔隙度衡量的是岩石能够容纳流体的空间总量。高孔、高渗地层比低孔、低渗地层发生严重井涌的可能性会更大。例如，人们通常认为砂岩比页岩更容易发生井涌，那是因为砂岩的渗透率和孔隙度通常比页岩要高。

影响井涌严重程度的另一个因素是压差。压差是地层流体压力和钻井液静液压力之间的差值。如果地层压力远大于钻井液静液压力，就会产生一个较大的负压差。如果这个负压差再加上高渗透率和高孔隙度，就会发生严重井涌。

井涌可以用几种方式来表述：一种方式是通过侵入井筒的地层流体类型来表述。已知的井涌流体包括天然气、原油、盐水、氯化镁、水、硫化氢（酸性）气体和二氧化碳。还有一种方式是通过实施井控进行压井作业的钻井液密度来表述。

井控还需要考虑的一个重要

is the pressure that the formation rock can withstand without generating an induced fracture. This rock strength is often called the fracture mud weight or gradient and is usually expressed in lbm/gal equivalent mud weight.The equivalent mud weight is the sum of the pressures exerted on the borehole wall and includes mud hydrostatic pressure, pressure surges resulting from pipe movement, frictional pressures applied against the formation as a result of pumping the drilling fluid, and any casing pressure cause by a kick. For example, if the fracture mud weight of a formation has been determined as 16.0 lbm/gal, the well can withstand any combination of these pressures that yields the same total pressure as a column of 16.0lbm/gal (1,920kg/m^3) mud extending to the depth in question. This combination could be (1) a 16.0lbm/gal (1,920kg/m^3) mud, (2) a 15.0lbm/gal (1,800kg/m^3) mud and some amount of casing pressure, (3) a 15.5lbm/gal (1,860kg/m^3) mud and a smaller amount of casing pressure, or (4) other combinations.

7.3.2 Causes of Kicks

Kicks occur because formation pressure is greater than mud hydrostatic pressure, which causes fluids to flow from the formation into the wellbore. In almost all drilling operations, the operator attempts to maintain a hydrostatic pressure greater than the formation pressure (a relationship that is called an overbalanced condition) and, thus, to prevent kicks. However, on occasion (and for various reasons), the formation will exceed the wellbore pressure, and a kick will occur. Following are the key causes of kicks.

7.3.2.1 Insufficient Mud Weight

Insufficient mud weight is the predominant cause of kicks. A permeable zone is drilled using a mud weight that exerts less pressure than the formation pressure within the zone. Fluids begin to flow into the wellbore, and a kick occurs.Abnormal formation pressures are often associated with kicks. They are defined as pressures that have an equivalent mud weight greater than normal conditions. Normal conditions are defined as equivalent mud weights ranging from a freshwater density of 8.34 lbm/gal (1,000 kg/m³) to a saturated NaCl-water density of 10 lbm/gal (1,200 kg/m³). In well-control situations, greater than normal formation pressures are the greatest concern. Because normal formation pressure is equal to a full column of native water, an abnormally pressured formation will exert more pressure than that column of water. If an abnormally pressured formation is encountered while drilling a with mud weight that is insufficient to control the zone being drilled, a potential kick situation is present. Whether a kick occurs depends on the permeability and porosity of the rock. A number of methods can be used to estimate formation pressures in an effort to prevent this type of kick. Some are listed below:

(1) Qualitative Methods: ①Paleontology; ②Offset well-log analysis; ③Temperature-anomaly analysis; ④Gas measurement; ⑤Mud- or cuttings-resistivity analysis; ⑥Cutting-characteristics analysis; ⑦Hole-condition analysis.

(2) Quantitative Methods: ①Shale density profile; ②d-exponent analysis; ③Normalized penetration-rate analysis; ④Other drilling equations.

Kicks caused by insufficient mud weight seem to require the obvious solution of drilling with high mud weight. However, this is not always a viable solution. First, a high mud weight may exceed the fracture mud weight of the formation and induce lost circulation. Second, a mud weight in excess of the formation pressure may significantly reduce penetration rates. In addition, pipe sticking becomes a serious concern when excessive mud weights are used. The best solution is to maintain a mud weight slightly greater than the formation pressure until the mud weight begins to approach the fracture mud weight.

7.3.2.2 Improper Hole Fill up on Trips

Improperly filling the hole during trips is another predominant cause of kicks. As the drill-pipe is pulled out of the hole, the mud level falls because the pipe steel had displaced some mud. With the pipe no longer in the hole, the overall mud level decreases. It is necessary to fill the hole with mud periodically to avoid reducing

（1000kg/m³）到饱和盐水密度10lbm/gal（1200kg/m³）。井控时，要特别关注异常高压。正常地层压力等于地层水的静液柱压力，而异常压力地层受到的压力会比水的静液柱压力更大。如果钻遇异常压力地层，而使用的钻井液不足以控制该层时，就存在潜在的井涌风险。是否发生井涌还取决于岩石的渗透率和孔隙度。为了防止井涌发生，可以使用多种方法来估算地层压力。具体如下：

（1）定性方法：①古生物学；②邻井测井分析；③异常温度分析；④气体测量；⑤钻井液或岩屑电阻率分析；⑥岩屑特性分析；⑦井况分析。

（1）定量方法：①泥页岩密度剖面；②d指数分析；③标准化机械钻速分析；④其他钻井方程。

由于钻井液密度不足引起的井涌问题似乎只需提高钻井液密度就可解决，但这样并非一直可行。首先，钻井液密度过大可能会超过地层破裂压力从而诱发井漏。其次，钻井液密度超过地层压力会使机械钻速显著降低。此外，当钻井液密度过大时，还会造成卡钻。最好的解决方案是将钻井液密度维持在安全窗口范围内，即略大于地层压力，略小于破裂压力。

7.3.2.2 起下钻过程中灌浆不当

起下钻期间灌浆不足是引起井涌的另一个主要原因。由于钻柱占了部分钻井液体积，所以当钻柱从井筒中起出时，钻井液液面会下降。随着钻柱起出井筒，总的钻井液液面高度会降低。必

the hydrostatic pressure and thereby allowing a kick to occur.

Several methods can be used to fill the hole, but all must be able to measure accurately the amount of mud required. It is not satisfactory under any conditions to allow a centrifugal pump to fill the hole continuously from the suction pit, because with this approach, accurate mud-volume measurement is not possible. The two methods most commonly used to monitor hole fill up are a trip tank and pump-stroke measurement. A trip tank includes a calibration device to monitor the volume of mud entering the hole. The tank can be placed above the preventer to allow gravity feed into the annulus, or a centrifugal pump can pump mud into the annulus, with the overflow returning to the trip tank. The advantages of a trip tank include ensuring that the hole remains full at all times and providing an accurate measurement of the amount of mud entering the hole.

Another method of keeping a full hole is to fill the hole periodically using a positive-displacement pump. A flowline device can be installed to measure the number of pump strokes required to fill the hole and to shut off the pump automatically when the hole is full.

7.3.2.3 Swabbing

Pulling the drillstring from the borehole creates swab pressures. Swab pressure is negative and reduces the effective hydrostatic pressure throughout the hole below the bit. If this pressure reduction lowers the effective hydrostatic pressure below the formation pressure, a potential kick situation has developed. The variables controlling swab pressures include pipe-pulling speed, mud properties and hole configuration. Some of these effects can be seen in Table 7.1.

须定期向井内灌注泥浆，避免由于静液压力降低而引发井涌。

灌浆有多种方法，但所有方法都必须能够准确地测量出所需的钻井液量。通常用的离心泵直接从钻井液池中抽吸进行连续灌浆的方法并不理想，这是因为它无法准确地测量钻井液体积。最常用的灌浆监测方法是利用计量罐和泵冲数进行计量。计量罐上装有一个校准装置，可以监测进入井筒的钻井液体积。计量罐通常位于防喷器组合上方，这样既可以通过重力作用也可以用离心泵将钻井液注入环空，多余的流体会返回计量罐。计量罐的优点是既能确保井筒充满钻井液，又可以提供灌浆的准确测量值。

另一种灌浆方法是用螺杆泵进行定期灌注。通过在钻井液出口管线上安装计量设备来监测灌满井筒所需的泵冲数，井筒充满钻井液时自动关闭泵。

7.3.2.3 抽汲

起钻会产生抽汲压力，抽汲压力会降低整个井筒的有效静液柱压力。如果静液柱压力降低到小于地层压力，就会形成潜在的井涌风险。控制抽汲压力的可变因素包括管柱的起出速度、钻井液性质和井身结构等。不同井眼尺寸和起钻速度对抽汲压力的影响见表7.1。

Table 7.1 Example swab pressures in various hole sizes at various pulling speeds for a 14.0-PPG mud and $4^1/_2$ in, drillpipe

Hole size, in	Pulling Speeds, s/stand					
	15	22	30	45	68	75
$8^1/_2$	267	167	124	98	84	75
$6^1/_2$	589	344	256	192	159	140
$5^3/_4$	921	524	294	289	231	200

7.3.2.4 Cut Mud

Gas-contaminated mud will occasionally cause a kick. The mud-density reduction is usually caused by fluids from the core volume which are cut and released into the mud system. As the gas is circulated to the surface, it expands and reduces the overall hydrostatic pressure sufficiently to allow a kick to occur. Although the mud weight is cut severely at the surface, the hydrostatic pressure is not reduced significantly because most gas expansion occurs near the surface and not at the hole bottom.

7.3.2.5 Lost Circulation

Occasionally kicks are caused by lost circulation. A decreased hydrostatic pressure occurs because of a shorter mud column. When a kick occurs because of lost circulation, the problem may become se-vere. A large volume of kick fluid may enter the hole before the rising mud level is observed at the surface.

7.3.3 Kick Signs

A number of warning signs and possible kick indicators can be observed at the surface. It is the responsibility of each crew member to recognize and interpret these signs and to take proper action. Not all of these signs will positively identify a kick; some simply warn of potential kick situations. Key warning signs include the following: (1)Flow-rate increase; (2)Pit-volume increase; (3)Continuing flow in the well with the pumps off; (4)Pump-pressure decrease along with a pump-stroke increase; (5)Improper hole fill up on trips; (6)Change in string weight; (7)Drilling break; (8) Decrease in mud weight.

Each warning sign is identified in the following paragraphs as being of primary or secondary importance to kick detection.

7.3.3.1 Warnings Signs of Primary Importance to Kick Detection

1. Flow-Rate Increase

An increase in the flow rate leaving the well while

7.3.2.4 钻井液侵入

受气体污染的钻井液有时也会引起井涌。岩屑释放出的气体侵入钻井液循环系统会引起钻井液密度降低。气体随钻井液上返时，会膨胀并降低井内静液柱压力，从而导致井涌发生。需要说明的是，此时井筒的静液柱压力降低程度并不像钻井液密度在地面降低的那么严重，原因是大多数气体膨胀会发生在地面附近。

7.3.2.5 井漏

井漏偶尔会引起井涌。由于钻井液液柱高度降低导致静液压力减小。因此井漏引发的井涌问题可能会很严重。因为在地面观察到钻井液浆液面上升之前，可能已经有大量流体侵入了井筒。

7.3.3 井涌的征兆

可以在地面观测到的井涌指示和预警征兆有很多。每个作业人员都要会识别和解释这些征兆，采取恰当的措施。并非所有征兆都能直接识别出井涌，有一些只是表明存在潜在的井涌风险。井涌的主要征兆包括:(1)出口流速增加；(2)钻井液池内流体体积增加；(3)关泵情况下，井中依然存在持续流体流动；(4)泵压降低并伴随泵冲增加；(5)起下钻过程中灌浆不及时；(6)大钩载荷变化；(7)钻进放空；(8)钻井液密度降低。

下面分一级和二级重要指标来讨论各种预警征兆。

7.3.3.1 一级井涌监测预警征兆

1. 出口流速增加

以恒定泵速泵入，而井口返

pumping at a constant rate is a primary kick indicator. The increased flow rate can be interpreted to mean that the formation is helping the rig pumps to move fluid up the annulus by forcing formation fluids into the wellbore. This is a key indicator of a kick.

2. Pit-Volume Increase

If the pit volume has not been changed as a result of control actions from the surface, an increase indicates that a kick is occurring. Fluids entering the wellbore displace an equal volume of mud in the flowline and cause an increase in pit level. However, this change takes some time to manifest itself and does not provide an immediate indication of a kick.

3. Flowing Well

When the rig pumps are not moving the mud, continued flow from the well indicates that a kick is in progress. An exception is when the mud in the drillpipe is considerably heavier than that in the annulus (for example, in the case of a slug). Care must be taken to determine whether a slug is present. If so, the flow will decrease and eventually stop.

7.3.3.2 Warnings Signs of Secondary Importance to Kick Detection

1. Drilling Break

An abrupt increase in bit penetration rate, called a drilling break, is a warning sign of a possible kick. A gradual increase in penetration rate is an indicator of abnormal pressure and should not be misconstrued as an abrupt rate increase. When the rate suddenly increases, it can be assumed that the rock type has changed. It can also be assumed that the new rock type has the potential to kick (as in the case of a sand), even if the previously drilled rock did not have this potential (as in the case of shale). Although a drilling break may have been observed, it is not certain that a kick will then occur, but only that a new formation that may have kick potential is now being drilled.

出流速增加是一级井涌征兆。流速增加可以解释为地层压力推动地层流体侵入井筒，从而使钻井液在环空上返流速增加。这是发生井涌的重要标志。

2. 钻井液池内流体体积增加

如果没有在地面进行令钻井液池内流体体积改变的操作，那么钻井液池体积增大表示正在发生井涌。地层流体进入井筒会在钻井液出口管线中替代等体积的钻井液，进而引起钻井液池液面升高。但有时这种变化并不能立即标示出井涌，需要一段时间才能显现出来。

3. 井筒流动

停泵后，井中依旧存在持续流动的现象则表明正在发生井涌。也会存在一些例外的情况，如钻杆内的钻井液比环空中的钻井液密度大得多（如环空中存在段塞）。必须谨慎对待是否存在段塞的情况。如果存在，井内流体流动会逐渐减小并最终停止。

7.3.3.2 二级井涌监测预警征兆

1. 钻进放空

机械钻速的突然增加称为钻进放空，这是可能存在井涌的预警征兆。机械钻速逐渐增加表示存在异常压力，这种情况不应该归为机械钻速突然增加。机械钻速的突然增加可以认为是所钻的岩石类型发生了变化，也可以认为是之前所钻的地层（如页岩）不存在发生井涌的风险，而新钻入的地层（如砂岩）存在井涌风险。虽然已经观察到了钻进放空的现象，但还不能确定接下来就会发生井涌，只能确定正在钻进的新地层具有井涌的潜在风险。

When a drilling break occurs, it is a recommended practice that the driller should drill 3 to 5 ft (1 to 1.5 m) in to the new formation and then stop to check for flowing formation fluids. Flow checks are not always performed in top hole drilling or if drilling through a series of stringers where repetitive breaks are encountered; unfortunately many kicks and blowouts have occurred as a result of this failure to perform flow checks.

2. Pump-Pressure Decrease with Stroke Increase

A pump-pressure change may indicate a kick. Initial fluid entry into the borehole may cause the mud to flocculate, which may increase the pump pressure temporarily. As the flow continues, the low-density influx will displace the heavier drilling fluids, and the pump pressure may begin to decrease. As the fluid in the annulus becomes less dense, the mud in the drillpipe tends to drop, and the pump speed may increase. However, other drilling problems may cause these same signs. A hole in the pipe, called a *washout,* will cause pump pressure to decrease. A twistoff of the drillstring will give the same signs. It is proper procedure, however, to check for a kick if these signs are observed.

3. Reduced Mud Weight

Reduced mud weight in the flowline has occasionally caused a kick to occur. Fortunately, the lower mud weights generated by the cutting effect are found near the surface, occur generally as a result of gas expansion, and do not appreciably reduce mud density throughout the hole. Table 7.2 shows that gas cutting has a very small effect on bottomhole hydrostatic pressure. An important point to remember about gas cutting is that if the well did not kick during the time required to drill the gas zone and to circulate the gas to the surface, only a small possibility exists that it will kick later. Generally, gas cutting indicates that a gas-containing formation has been drilled; it does not mean that the mud weight must be increased.

发生钻进放空时，常用的做法是司钻在新地层中钻3～5ft（1～1.5m）后停钻，检查是否存在地层流体的侵入。当然这种检查并不能一直进行，在钻上部井段或钻穿一系列重复遭遇放空段时，往往不会一直进行溢流检查；但是这种漏检会引发井涌和井喷。

2. 泵压降低并伴随泵冲增加

泵压变化可能表示存在井涌。地层流体开始侵入井筒会导致钻井液絮凝，从而引起泵压暂时增大。随着地层流体持续侵入，侵入的低密度流体会替换密度较大的钻井液，泵压开始下降。随着环空中流体密度的降低，钻杆中的钻井液面下降，泵速可能增加。当然其他钻井事故也可能引起相同的征兆。比如，钻柱刺漏时，也会引起泵压下降。钻柱断裂也会有同样的征兆。如果观测到这些征兆应采用合理的程序检测是否发生了井涌。

3. 钻井液密度降低

管线内的钻井液密度降低偶尔会导致井涌发生。由气侵引起的钻井液密度降低在地面附近更明显，这是由于气体会发生膨胀，但并不会显著降低整个井筒的钻井液密度。表7.2表明，气侵对井底静液压力的影响很小。关于气侵重点要记住的是，如果在钻气层时没有发生井涌，并且气体已经循环到地面，那么之后存在井涌的可能性极低。一般而言，气侵表示已钻到含气地层，但这并不意味着必须增大钻井液密度。

Table 7.2 Examples of the pressure-reduction effect of gAS-cUT mud on bottomhole hydrostatic pressure

Depth	10 lbm/gal cut to 5 lbm/gal	18 lbm/gal cut to 16.2 lbm/gal	18.0 lbm/gal cut to 9 lbm/gal
1000	51	31	60

Continued

Depth	10 lbm/gal cut to 5 lbm/gal	18 lbm/gal cut to 16.2 lbm/gal	18.0 lbm/gal cut to 9 lbm/gal
5000	72	41	82
10000	86	48	95
20000	97	51	105

4. Improper Hole Fill-up

When the drillstring is pulled out of the hole, the mud level should decrease by a volume equivalent to that of the steel removed. If the hole does not fill with the volume of mud calculated to bring the mud level back to the surface, it can be assumed that a kick fluid has entered the hole and filled the displacement volume of the drillstring. Even though gas or salt water has entered the hole, the well may not flow until enough fluid has entered to reduce the hydrostatic pressure to a value lower than the formation pressure.

5. Change in String Weight

Drilling fluid provides a buoyant effect on the drillstring and reduces the actual pipe weight supported by the derrick. Heavier muds have a greater buoyant force than less dense muds. When a kick occurs and low-density formation fluids begin to enter the borehole, the buoyant force of the mud system is reduced. The string weight observed at the surface will increase. However, this change may be small and not readily observable.

7.3.4 Precautions Whilst Drilling

Whilst drilling, the drilling crew will be watching for the indicators described above. If one of the indicators are seen then an operation known as a flow check is carried out to confirm whether an influx is taking place or not. The procedure for conducting a flow check is as follows:

(1) Pick up the Kelly until a tool joint appears above the rotary table.

(2) Shut down the mud pumps.

(3) Set the slips to support the drillstring.

(4) Observe flowline and check for flow from the annulus.

4. 灌浆不当

起钻时，钻井液液面高度的减少应等于起出钻柱的体积。如果灌浆量少于理论灌注的计算值，则可以认为地层流体侵入井筒并填充了部分起出钻柱的体积。即使是气体或盐水进入了井筒，也要等到有足够的流体进入，使静液柱压力低于地层压力时才会发生溢流。

5. 大钩载荷变化

钻井液对钻柱会产生浮力作用，降低井架实际支撑的钻柱重量。高密度钻井液比低密度钻井液产生的浮力更大。井涌发生时，低密度的地层流体侵入井筒，使得钻井液对钻柱的浮力降低。在地面观测到的结果是大钩载荷增加。但是这种变化可能很小，且不易观察到。

7.3.4 钻进时注意事项

在钻井过程中，钻井人员需要密切关注上述征兆。一旦发现任何征兆，就需要进行溢流检查，来确认是否正在发生井侵。实施溢流检查的步骤如下：

（1）上提方钻杆直到钻具接头露出转盘面。

（2）关闭钻井泵。

（3）坐卡瓦支撑钻柱。

（4）观察钻井液出口管线，检查环空的流量。

(5) If the well is flowing, close the BOP. If the well is not flowing resume drilling, checking for further indications of a kick.

7.3.5 Precautions During Tripping

Since most blow-outs actually occur during trips, extra care must be taken during tripping. Before tripping out of the hole the following precautions are recommended:

(1) Circulate bottoms up to ensure that no influx has entered the wellbore.

(2) Make a flowcheck.

(3) Displace a heavy slug of mud down the drillstring. This is to prevent the string being pulled wet (i.e. mud still in the pipe when the connections are broken). The loss of this mud complicates the calculation of drillstring displacement.

It is important to check that an influx is not taking place and that the well is dead before pulling out of the hole since the well control operations become more complicated if a kick occurs during a trip. When the bit is off bottom it is not possible to circulate mud all the way to the bottom of well. If this happens the pipe must be run back to bottom the with the BOP's closed. This procedure is known as stripping-in.

As the pipe is tripped out of the hole the volume of mud added to the well, from the trip tank, should be monitored closely. To check for swabbing it is recommended that the drill bit is only pulled back to the previous casing shoe and then run back to bottom before pulling out of hole completely. This is known as a short trip. Early detection of swabbing or incomplete filling of the hole is very important.

7.4 Secondary Control

Audio 7.4

If a kick is detected and a pit gain has occurred on surface, it is clear that primary control over the well has been lost and all normal drilling or tripping operations must

（5）若存在流动，则关闭防喷器组合。若不存在流动，则恢复钻进，核查井涌的进一步征兆。

7.3.5 起下钻时注意事项

由于大多数的井喷都发生在起下钻过程中，因此在起下钻期间必须要格外注意。在钻柱起出井筒之前需遵循下列注意事项：

（1）充分循环以确保井筒中没有井侵流体。

（2）检查流量。

（3）向钻柱中注入一段加重钻井液，防止钻柱湿起出（即卸扣时钻柱内仍存留钻井液）。湿起出产生的钻井液损失使灌浆计算变得更加复杂。

起下钻过程中发生的井涌会使井控作业变得非常复杂，因此起钻前检查是否发生了井侵及井筒是否处于平衡状态至关重要。当钻头离开井底后，钻井液循环就无法到达井底。因此，如果发生溢流，必须在防喷器组合关闭的情况下将钻柱重新下入井底。这个过程称为强行下钻。

起钻时，应严密监测从计量罐灌入的钻井液体积。为了检查抽汲作用，建议将钻头上提至上一层套管鞋处，再下入井底，随后再完全从井筒中起出，这种操作称为短起下钻。尽早发现抽汲或灌浆不足非常重要。

7.4 二级井控

如果检测到井涌且地面钻井液池液面已经升高，那么这口井显然已经失去了一级井控。为了集中精力将这口井重新恢复至一级井控，必须停

cease in order to concentrate on bringing the well back under primary control.

The first step to take when primary control has been lost is to close the BOP valves, and seal off the drillstring to wellhead annulus at the surface. This is known as initiating secondary control over the well. It is not necessary to close off valves inside the drillpipe since the drillpipe is connected to the mudpumps and therefore the pressure on the drillpipe can be controlled.

Usually it is only necessary to close the uppermost annular preventer, but the lower pipe rams can also be used as a back up if required (Fig. 7.7). When the well is shut in, the choke should be fully open and then closed slowly so as to prevent sudden pressure surges. The surface pressure on the drillpipe and the annulus should then be monitored carefully. These pressures can be used to identify the nature of the influx and calculate the mud weight required to kill the well.

止所有正常钻进和起下钻作业。

一级井控失效时，首先要关闭防喷器组合的阀门，封闭钻柱在地面的井口环空，这种操作称为启动二级井控。由于钻杆连接到钻井泵且压力可控，因此不必关闭钻杆内部的阀门。

通常只需要关闭最上方的环形防喷器，但是下部的闸板防喷器在需要时也可作为备选（图7.7）。关井时应将节流阀完全打开，然后缓慢关闭以防止突然的压力波动。之后认真监测地面立管压力和套管压力。这些压力可用来识别井侵流体的性质，并计算压井所需的钻井液密度。

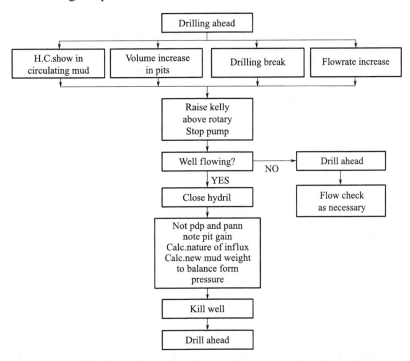

Fig. 7.7 Operational procedure following detection of a kick

7.4.1 Shut in Procedure

When one or more warning signs of a kick have been observed, steps should be taken to shut in the well. If there is any doubt as to whether the well is flowing, shut it in and check the pressures. Moreover, there is no

7.4.1 关井程序

当观测到一个或多个井涌征兆时，应立即采取关井措施。如果对井内是否存在溢流有任何疑问，应立即进行关井并检查压力。

difference in this context between "just a small flow" and a "full-flowing" well, because both can very quickly turn into a big blowout. There has been some hesitation in the past to close in a flowing well because of the possibility of sticking the pipe. It can be shown that for all types of pipe sticking, including differential pressure, heaving, or sloughing shale, it is better to close in the well quickly and reduce the kick influx.This approach in fact reduces the chances of pipe sticking. The primary concern at this point is to kill the kick safely; when feasible, the secondary concern is to avoid pipe sticking.

Some concern has been expressed about fracturing the well and creating an underground blowout as a result of shutting in a well when a kick occurs. If the well is allowed to flow, it will eventually become necessary to shut in the well, at which time the possibility of fracturing the well will be greater than if the well had been shut in im- mediately after the initial kick detection. Table 7.3 shows an example of the higher casing pressures that can result from continuous flow.

Table 7.3 Examples of the effect of continuous influx on casing pressure as a result of failure to close in the Well

Volume of gas gained, bbl	20	30	40
Casing pressure, psi	1468	1654	1796

7.4.1.1 Initial Shut-in

In a hard shut-in procedure, the annular preventer(s) are closed immediately after the pumps are shut down. In a soft shut-in procedure, the choke is opened before the preventers are closed, and the choke is closed afterward(Fig. 7.8). Arguments in favor of a soft shut-in procedure are (1) it avoids water hammer because fluid flow is not stopped abruptly and (2) it provides an alternate means of well control (the low-choke-pressure method) if the casing pressure becomes excessive. The water-hammer concern has been proved to be of no substance, and the low- choke-pressure method of well control is an unreliable procedure. It is best to use the hard shut-in procedure to minimize the kick volume.

出现溢流时,"小溢流量"和"大溢流量"之间并没有区别,因为它们都可以迅速恶化成为严重井喷。过去,一般不轻易做出关井的决定,那是因为关井会可能引起卡钻。但现在看来,对于所有类型的卡钻,包括压差卡钻、泥页岩膨胀或剥落造成的卡钻,最好迅速关井并减少井涌侵入量。这实际上降低了卡钻的风险。此时首先要关注的是安全压井,其次才是在可行条件下避免卡钻。

也有人对井涌时关井会造成的地层破裂和地下井喷表示担忧。但如果任其发展,最终还是需要进行关井作业,此时地层破裂的可能性会比刚开始检测出井涌就立即关井的可能性还要大。表7.3给出的例子表明,持续侵入的流体会引起套压升高。

7.4.1.1 初始关井

硬关井就是停泵后立即关闭环形防喷器。软关井是在关闭环形防喷器前先打开节流阀,然后再关闭节流阀(图7.8)。采用软关井的原因有:(1)避免水击,因为流体不会突然停止流动;(2)如果套压过大,软关井可以提供可行的井控方法(低节流压力法)。但事实已经证实,水击影响不大,而采用低节流压力进行井控也不可靠。所以选择硬关井来降低井涌量比较好。

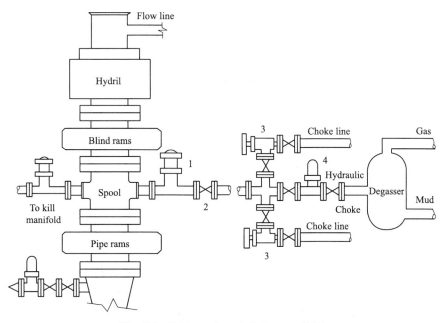

Fig. 7.8 BOP stack and choke manifold
1—Hydraulic valve; 2—Plug vale; 3—Adjustable choke; 4—Pressure guage

7.4.1.2 Drilling Kicks

Shut-in procedures:

(1) When a primary kick-warning sign has been observed, immediately raise the kelly or top drive until a tool joint is above the rotary table.

(2) Stop the mud pumps.

(3) Close the annular preventer.

(4) Notify company personnel.

(5) Read and record the shut-in drillpipe pressure, the shut-in casing pressure, and the pit gain(The shut-in drill- pipe pressure is referred to as the SIDPP and the shut-in casing pressure as the SICP).

Raising the kelly / top drive is an important part of this procedure. With the kelly/top drive out of the hole, the valve at the bottom of the kelly/top drive can be closed if necessary. Also, the annular preventer members can attain a more secure seal on pipe than against a kelly.

7.4.1.3 Tripping Kicks

A high percentage of well-control problems occur when a trip is in progress. Kick problems may be compounded when the rig crew is preoccupied with the trip mechanics and fails to observe the initial warning signs of a kick.

7.4.1.2 钻进井涌

关井程序如下：

（1）观测到井涌征兆时，应立即上提方钻杆或顶驱，直到钻柱接头起出转盘面。

（2）钻井泵停泵。

（3）关闭环形防喷器。

（4）通知作业人员。

（5）读取并记录关井立管压力、关井套管压力及钻井液池增量（关井立管压力称为SIDPP，关井套管压力称为SICP）。

上提方钻杆或顶驱是关井程序中的关键操作。当方钻杆或顶驱起出时，一般需要关闭方钻杆或顶驱下部的阀门。相对方钻杆而言，环形防喷器对钻杆的密封效果更好。

7.4.1.3 起下钻井涌

大部分井控问题都发生在起下钻过程中。如果作业人员只关注起下钻而忽略观测井涌征兆时，出现的井涌问题可能会更加复杂。

Shut-in procedures:

(1) When a primary warning sign of a kick has been observed, immediately set the top tool joint on the slips.

(2) Install and make up a full-opening, fully opened safety valve on the drillpipe.

(3) Close the safety valve and the annular preventer.

(4) Notify company personnel.

(5) Pick up and make up the kelly/ top drive.

(6) Open the safety valve.

(7) Read and record the SIDPP, the SICP, and the pit gain.

Installing a full-opening safety valve in preference to an inside BOP (float) valve is of prime importance because of the advantages offered by the full-opening valve. If flow occurs up the drillpipe as a result of a trip kick, the fully opened, full-opening valve is physically easier to stab. Also, a float-type inside BOP valve will close automatically when the upward-moving fluid contacts the valve. If wireline work such as drillpipe perforation or logging becomes necessary, the full-opening valve will accept logging tools approximately equal to its inside diameter, whereas the float valve may prohibit wireline work altogether. After the kick is shut in, an inside BOP float valve may be stabbed onto the full-opening valve to enable stripping operations.

7.4.1.4 Diverter Procedures

Because a shallow underground blowout is difficult to control and may cause the loss of the rig, an attempt is usually made to divert a surface blowout away from the rig. This is the common practice on land. Special attention must be paid to opening the diverter lines before shutting in the well.

(1) When a primary warning sign of a kick has been observed, immediately raise the kelly/top drive until a tool joint is above the rotary table.

(2) Increase the pump rate to maximum output.

(3) Open the diverter-line valve(s).

(4) Close the diverter unit (or annular preventer).

关井程序如下：

（1）观测到井涌主要征兆时，立即将上部钻具接头坐在卡瓦上。

（2）抢接全开式钻杆安全阀。

（3）关闭安全阀和环形防喷器。

（4）通知作业人员。

（5）上提并回接方钻杆或顶驱。

（6）打开安全阀。

（7）读取并记录关井立管压力、关井套管压力和钻井液池增量。

由于全开式安全阀有很多优点，因此，在钻柱上安装全开式安全阀比钻杆内防喷器（浮阀）会更好。如果起钻时井涌导致井内流体沿钻杆上返，安装完全打开的全开式安全阀更容易实现抢接作业。如果是钻杆内防喷器，当上返流体到达阀门时，浮阀会自动关闭。如果需要进行钻杆射孔或测井等电缆作业，全开式安全阀允许尺寸接近其内径大小的测井工具通过，而浮阀则无法实现电缆作业。关井之后，内防喷器浮阀可以接在全开式安全阀上部，以便强行起下钻作业。

7.4.1.4 放喷程序

浅层钻井的井喷最难控制，而且可能会导致钻机受损，这时通常会尽量将溢出的流体引到地面距离钻机一定安全距离的地方进行放喷。这是陆地钻机的常规演练。在关井之前必须打开放喷管线。

（1）观测到井涌主要征兆时，立即上提方钻杆或顶驱直至钻柱接头位于转盘面上方。

（2）提高泵速。

（3）打开放喷管线阀门。

（4）关闭放喷装置（或环形

(5) Notify company personnel.

Recent experiences show that shallow gas flows are difficult to control. Industry philosophy is improving, and new handling procedures are being developed.

7.4.2 Interpretation of Shut-in Pressures

Shut-in pressures are defined as pressures recorded on the drill-pipe and the casing when the well is closed. Although both pressures are important, the drillpipe pressure will be used almost exclusively in killing the well. If the SIDPP reads as zero, check to see whether a drillpipe float valve is installed.

During a kick, fluids flow from the formation into the wellbore. When the well is closed to prevent a blowout, pressure builds at the surface because of the entry of formation fluid into the annulus and the difference between the mud hydrostatic pressure and the formation pressure. Because this pressure imbalance cannot exist for long, the surface pressures will build so that, eventually, the surface pressure plus the mud and influx hydrostatic pressures in the well will be equal to the formation pressure. The equations below express this relationship for the drillpipe and the annular side, respectively.

$$p_{form} = SIDPP + p_{DP}$$
$$p_{form} = SICP + p_{ann} + p_{kick} \tag{7.1}$$

where, p_{form} is the formation pressure, p_{DP} is the drillstring hydrostatic pressure, p_{ann} is the annulus hydrostatic pressure, and p_{kick} is the kick pressure.

7.4.2.1 Interpretation of Recorded Pressures

Fig.7.9 illustrates an important basic principle. It can be observed that the formation pressure is greater than the drillpipe hydrostatic pressure by an amount equal to the shut-in drillpipe pressure. The drillpipe pressure gauge is a bottomhole pressure (BHP) gauge. The casing-pressure reading cannot be considered as a direct BHP gauge measurement because the amount of formation fluid in the annulus is generally unknown.

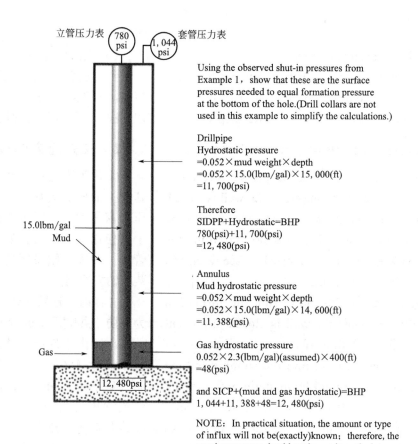

Fig.7.9 Pressure relationships at shut-in conditions

7.4.2.2 Constant BHP Concept

Fig. 7.9 can be used to illustrate another important basic principle. The 780psi (5.4MPa) reading observed on the drillpipe gauge is the amount of pressure that was necessary to balance the mud pressure at the hole bottom with the pressure in the gas sand at 15,000 ft (4,600 m). A basic law of physics states that formation fluids travel from areas of high pressure to areas of lower pressure only and that they do not travel between areas of equal pressures, assuming that gravity segregation can be neglected. If the drillpipe pressure is controlled so that the total mud pressure at the hole bottom is slightly greater than the formation pressure, then no additional kick influx will enter the well. The concept is the basis of the constant BHP method of well control, in which the pressure at the hole bottom is kept constant and at least equal to the formation pressure.

7.4.2.2 井底常压的概念

图 7.9 还可用来阐明另一个重要的基本原理。780psi（5.4MPa）的立管压力表读数表征的是为了平衡 15000ft（4600m）深度处含气砂岩的地层压力，需要在钻井液静液柱压力的基础上增加的附加压力。根据物理学的基本定律假设忽略重力分离作用，地层流体只会从高压区流向低压区，在等压区不会流动。如果控制立管压力使得井底的钻井液压力略大于地层压力，就不会有额外井涌侵入。这个概念是井底常压井控法的基础，即要保持井底压力恒定，且至少等于地层压力。

7.4.2.3 Effects of Time

After shut-in, a finite amount of time will elapse before both pressures stabilize. The kick flow rate will eventually drop to close to zero when the pressure in the wellbore is almost equal to that in the formation. The amount of time this takes varies with the difference between the wellbore and the formation pressures, the permeability, the fluid viscosity, and the length and diameter of hole in the kicking formation. Stabilization may take a few minutes to several hours depending on the conditions surrounding the kick. In general, 15 minutes are allowed to obtain shut-in pressures. The pressure will typically increase rapidly at first and then level off, although it will not necessarily become stable. The breakpoint in the pressure curve is taken as the SIDPP.

Several other factors affect the time needed for pressures to stabilize. One reason that they may not necessarily stabilize is differences in density between the kick fluid and the drilling fluid. Gas migration involves the movement of low-density fluids up the annulus. The kick may start to migrate up the hole, with an attendant increase in surface pressure (in the absence of any mitigation technique). It will tend to build pressure at the surface if time is allowed for migration. In addition, the influx may tend to degrade hole stability and to cause either stuck pipe or hole bridging. These problems must be considered when reading the shut-in pressures.

7.4.2.4 Trapped Pressure

Trapped pressure is any pressure recorded on the drillpipe or annulus that is greater than the amount needed to balance the BHP. Pressure can be trapped in the system in several ways. One common way is for gas to migrate up the annulus and expand; another is to shut in the well before the mud pumps have stopped running. Using a pressure reading that includes trapped pressure may result in erroneous kill calculations. There exist guidelines for checking for and releasing trapped pressure. If these are not properly executed, the well will be much more difficult to

7.4.2.3 时间的影响

关井后经过一定时间，套管压力和立管压力将会趋于稳定。当井筒压力几乎等于地层压力时，井涌流量最终会降到零。这一过程所需的时间会因地层压力、孔隙度、钻井液黏度、井涌层井筒长度和外径等不同井筒条件而有所不同。达到稳定状态少则数分钟，多则几小时。一般来说，15min后就可以读取关井压力。关井压力通常先迅速增加，之后趋于平缓。压力曲线中的转折点可作为立管压力值。

其他的因素也会影响压力稳定所需的时间。不稳定的原因之一是井涌流体和钻井液之间存在密度差。气体运移会使侵入的低密度流体沿环空向上移动。随着井涌流体向井筒上部运移，地面压力开始增加（没有采取任何缓解措施）。最后，会在地面附近形成附加压力。此外，井涌可能会使井壁稳定性降低，引起卡钻或砂桥。在读取关井压力时必须要考虑上述这些问题。

7.4.2.4 圈闭压力

圈闭压力是指立管压力或套管压力大于平衡井底压力的读数。形成圈闭压力的原因有许多，常见的有两种：一种是气体向上运移膨胀；另一种是关井后未及时停泵。使用包含圈闭压力的压力读数进行压井计算可能会产生误差。目前已经有一些监测和释放圈闭压力的准则。如果执行不当，就会增加压井操作的难度。这些

kill. These guidelines are as follows:

(1) When checking for trapped pressure, bleed from the casing side only. There are several reasons for this: ① the choke is located on the casing side; ② it avoids contamination of the mud in the drillpipe; and ③ it avoids the possibility of plugging the bit jets.

(2) Use the drillpipe pressure as a guide because it is a direct BHP indicator.

(3) Bleed small amounts (1/4 to 1/2 bbl) of mud at a time. Close the choke after bleeding and observe the pressure on the drillpipe.

(4) Continue to alternate the bleeding and subsequent pressure-observation procedures as long as the drillpipe pressure continues to decrease. When the drillpipe pressure ceases to drop, stop bleeding and record the true SIDPP and casing pressure.

(5) If the drillpipe pressure should decrease to zero during this procedure, continue to bleed and check pressures on the casing side as long as the casing pressure decreases (note: this step will normally not be necessary).

Because the trapped pressure is in excess of that needed to balance the BHP, it can be bled off without allowing any additional influx into the well. However, after the trapped pressure has been bled off, if bleeding is continued, more influx will be allowed into the well, and surface casing pressures will begin to increase. Although bleeding procedures can be implemented at any time, it is advisable to check for trapped pressure when the well is shut in initially and to recheck whether any pressure remains on the shut-in drillpipe after the drillpipe has been displaced with kill mud.

7.4.2.5 Drillpipe Floats

A kick can occur when a drillpipe float valve is used. Because a float valve prevents movement of fluid and pressure up the drillpipe, a drillpipe pressure reading will not be available after the well has been shut in. Several procedures are available for obtaining a drillpipe pressure value; the choice depends on the amount of information known when the kick occurs (see Fig. 7.10).

准则如下：

（1）检测圈闭压力时要仅从环空排气。原因有以下几点：①节流阀安装在套管一侧；②避免污染钻杆内的钻井液；③避免堵塞钻头喷嘴。

（2）要用立管压力作为计算依据，因为立管压力直接反应井底压力。

（3）排放时，每次排放少量（1/4～1/2 bbl）的钻井液。放喷后关闭节流阀，观察立管压力。

（4）如果立管压力持续下降，就要连续交替放喷并执行压力观测程序。立管压力停止下降时，停止放喷并记录真实的立管压力和套管压力。

（5）如果在此过程中立管压力降至零，那么只要套管压力下降，就继续放喷并检查套管压力。

由于圈闭压力超过了平衡井底所需的压力，因此可以通过放喷释放，且不会产生额外井侵。但当圈闭压力释放掉以后，如果继续放喷，就会引发进一步井侵，地面的套管压力开始上升。尽管可以随时进行放喷，但建议还是在最初关井时检查是否存在圈闭压力，用压井钻井液替掉钻杆内钻井液后，再次检查关井立管上是否仍存在压力。

7.4.2.5 钻杆浮阀

使用钻杆浮阀时也可能发生井涌。因为浮阀阻止了流体向上流动，阻碍了压力向上传播，关井后无法获得立管压力读数。要想获取立管压力值，有以下几种做法可以实现，怎样选择取决于从井涌地层获得的已知信息量（图7.10）。

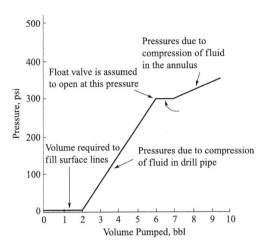

Fig.7.10 Procedure for establishing SIDPP with a float in the drillstring

The procedure to obtain a drillpipe pressure value if the slow pumping rate (kill rate) is known is as follows:

(1) Shut in the well, record the SICP, and obtain the kill rate either from the driller or from the daily tour report.

(2) Instruct the driller to start the pumps and maintain the pumping rate (measured in strokes) at the kill rate.

(3) As the driller starts the pumps, use the choke to regulate the casing pressure to the same pressure that was originally recorded at shut-in conditions.

(4) After the pumps are running at the kill rate with the casing pressure properly regulated at shut-in pressure, record the pressure on the drillpipe while pumping.

(5) Shut down the pumps and close the choke.

(6) The SIDPP equals the total pumping pressure minus the kill-rate pressure.

If the kill rate is not known, the procedure is as follows:

(1) Shut in the well.

(2) Line up a low-volume, high-pressure reciprocating pump on the standpipe.

(3) Start pumping and fill all the lines.

(4) Gradually increase the torque on the pumps until they begin to move fluid down the drillpipe.

(5) The SIDPP is the amount of pressure required to initiate fluid movement. This is assumed to be the

在低泵速（压井泵速）条件下，获取立管压力值的过程如下：

（1）关井，记录套压，并从司钻口中或日报中获得压井速率。

（2）通知司钻启动泵，保持泵速（以冲程计量）处于压井速率。

（3）司钻启动泵后，使用节流阀将套管压力调节到与初始关井套管压力相同。

（4）保持压井速率，将套管压力调节到关井时的套管压力值，记录此时的立管压力大小。

（5）停泵并关闭节流阀。

（6）关井立管压力等于总泵压减去压井泵速压力。

若压井泵速未知，则过程如下：

（1）关井。

（2）在立管上装一个小排量、高压力的往复泵。

（3）开始泵送并填充所有管线。

（4）逐渐增加泵上的压力，直到流体开始沿钻杆向下流动。

（5）假设关井立压大小就是克服井底浮阀所需的压力，那么

amount needed to overcome the pressure acting against the bottom side of the valve.

7.4.2.6 Kick Identification

When a kick occurs, it may prove useful to know the type of influx (gas, oil, or salt water) entering the wellbore. It must be remembered that the well-control procedures outlined here are designed to kill all types of kicks safely. The equation required to perform the kick-influx calculation is:

$$\rho_{kick} = g_{mud} - \frac{SICP - SIDPP}{h_{kick}} \qquad (7.2)$$

where, ρ_{mud} is mud density in psi/ft, h_{kick} is kick height in ft, and ρ_{kick} is kick density in psi/ft. The influx gradient can be evaluated using the guidelines in Table 7.4.

Table 7.4 Gradients for different types of influx

Gradient,psi/ft	0.05 ~ 0.2	0.2 ~ 0.4	> 0.5
Type of influx	Gas	Probable combination of gas, oil, and/or saltwater	Oil or saltwater

Although both the SIDPP and SICP can be determined accurately, it is difficult to determine the influx height. This requires knowledge of the pit gain and the exact hole size.

It is convenient to analyse the shut-in pressures by comparing the situation with that in a U-tube (Fig. 7.11). One arm of the U-tube represents the inner bore of the drillstring, while the other represents the annulus. A change of pressure in one arm will affect the pressure in the other arm so as to restore equilibrium.

The pressure at the bottom of the drillstring is due to the hydrostatic head of mud, while in the annulus the pressure is due to a combination of mud and the formation fluid influx (Fig. 7.12). Hence, when the system is in equilibrium, the bottom hole pressure will be equal to the drill pipe shut-in pressure plus the hydrostatic pressure exerted by the drilling mud in the drillstring. Hence:

关井立压就是流体开始流动时的压力。

7.4.2.6 井涌识别

发生井涌时，首先要了解侵入井眼的流体类型（天然气、原油或盐水）。本书讨论的井控是针对所有流体类型的压井。计算侵入流体的公式为：

式中，ρ_{mud} 是钻井液密度，h_{kick} 是井涌高度，ρ_{kick} 是井涌流体密度。可用表7.4中的数据评估井涌流体梯度。

尽管可以精确地确定立压和套压，但井涌深度很难确定，需要准确获取井筒尺寸和钻井液池增量信息。

通过U形管原理，可以方便地分析关井压力（图7.11）。U形管的一个分支代表钻柱内，另一个分支代表环空。压力平衡状态下，一个分支的压力变化会影响另一个分支。

钻柱底部的压力来自钻井液的静水压头，而环空中的压力则是钻井液和地层侵入流体共同作用的结果（图7.12）。因此当系统处于平衡状态时，井底压力将等于关井立管压力加上钻柱中钻井液施加的静水压力：

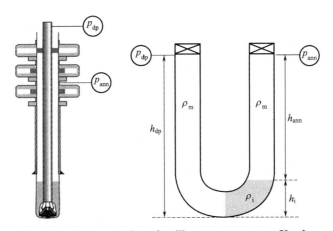

Fig. 7.11 Interpretation of wellbore pressures as a U-tube

hdp-vertical height of mud column in driupipe; hann-vertical height of mud column in annulus

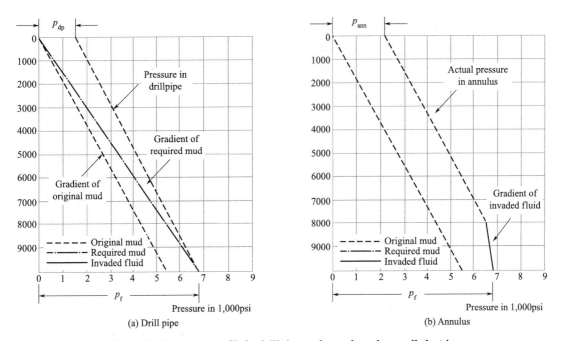

(a) Drill pipe (b) Annulus

Fig. 7.12 Pressure profile in drillpipe and annulus when well shut-in

$$p_{bh} = p_f$$
$$p_{dp} + \rho_m d = p_{bh} \tag{7.3}$$

where, p_{dp} = shut in drillpipe pressure (psi); ρ_m = mud pressure gradient (psi/ft); d = vertical height of mud column (ft); p_{bh} = bottomhole pressure (psi).

If the well is in equilibrium and there is no increase in the surface pressures the bottomhole pressure must be equal to the formation pore pressure.

式中，p_{dp} 为关井立管压力，psi；ρ_m 为钻井液压力梯度，psi/ft；d 为钻井液液柱高度，ft；p_{bh} 为井底压力，psi。

如果井内压力处于平衡状态且地面压力没有增加，则井底压力一定等于地层孔隙压力。

Since the mud weight in the drill pipe will be known throughout the well killing operation and p_{dp} can be used As a direct indication of bottom hole pressure (i.e. the drillpipe pressure gauge acts as a bottom hole pressure gauge). No further influx of formation fluids must be allowed during the well killing operation. In order to accomplish this the bottom hole pressure, p_{bh} must be kept equal to, or slightly above, the formation pressure.

This is an important concept of well control and the one on which everything else is based. This is the reason that this technique for well killing is sometimes referred to as the constant bottom hole pressure killing methods

On the annulus arm of the U-tube, the bottom hole pressure is equal to the surface annulus pressure and the combined hydrostatic pressure of the mud and influx:

$$p_{ann}+h_i\rho_i+(d-h_i)\rho_m=p_{bh} \tag{7.4}$$

where, p_{ann} = shut-in annulus pressure (psi); h_i = height of influx (ft); ρ_i = pressure gradient of influx (psi/ft). and to achieve equilibrium:

$$p_{bh}=p_f$$

One further piece of information can be inferred from the events observed at surface when the well has been shut-in. The vertical height of the influx (h_i) can be calculated from the displaced volume of mud measured at surface (i.e. the pit gain) and the cross-sectional area of the annulus.

$$h_i=V/A \tag{7.5}$$

where, V = pit gain (bbl); A = cross section area (bbl/ft).

Both V and A (if open hole) will not be known exactly, so h_i can only be taken as an estimate.

7.4.3 Kill Mud Weight

Kill calculations require the mud weight, which will be needed to balance the bottomhole formation pressure. The kill mud weight is defined as the weight of mud necessary to balance the formation pressure. It will be shown later in this chapter that using the exact required mud weight without variations reduces downhole stresses. Because the drillpipe pressure has been defined as the reading from a BHP gauge, the SIDPP can be used to calculate the mud weight necessary to kill the well. The equation for kill mud weight is:

$$KMW = 19.25 \frac{SIDPP}{TVD} + OMW \qquad (7.6)$$

where, TVD is true vertical depth in ft, OMW is original mud weight in lbm/gal, and KMW is kill mud weight in lbm/gal. Because the casing pressure does not appear in the above equation, a high casing pressure does not necessarily indicate a high kill mud weight. The same is true for a high pit gain.

7.4.4 Determination of the Type of Influx

By combining equations 7.3 and 7.4 the influx gradient, ρ_i can be found from:

$$\rho_i = \rho_m - (p_{ann} - p_{dp})/h_i$$

Note: The expression is given in this form since $p_{ann} > p_{dp}$, due to the lighter fluid being in the annulus.

From the gradient calculated from equation 3 the type of fluid can be identified as follows:

Gas: 0.075 ~ 0.150 psi/ft;
Oil: 0.3 ~ 0.4 psi/ft;
Seawater: 0.470 ~ 0.520 psi/ft.

If ρ_i was found to be about 0.25 this may indicate a mixture of gas and oil. If the nature of the influx is not known it is usually assumed to be gas, since this is the most severe type of kick.

7.4.5 Factors Affecting the Annulus Pressure, Pann

7.4.5.1 Size of Influx

The time taken to close in the well should be no more than 2 minutes. If the kick is not recognised quickly enough, or there is some delay in closing in the well, the influx continues to flow into the annulus. The effect of this is shown in Fig.7.13. As the volume of the influx allowed into the annulus increases the height of the influx increases and the higher the pressure on the annulus when the well is eventually shut-in. Not only will the eventual pressure at surface increase but as can be seen from Fig. 7.14, the pressure along the entire wellbore increases. There are two dangers here:

(1) At some point the fracture pressure of one of the formations in the openhole section may be exceeded. This may lead to an underground blow-out-formation fluid entering the wellbore and then leaving the wellbore at some shallower depth. Once a formation has been fractured it may be impossible to weight the mud up to control the flowing formation and there will be continuous crossflow between the zones. If an underground blow-out occurs at a shallow depth it may cause *cratering* (breakdown of surface sediment, forming a large hole into which the rig may collapse).

关井后的最终套压就越大。结果是不仅地面压力增加，整个井筒上的压力都会增加，如图 7.14 所示。会引发以下两种危险：

（1）井筒压力可能会超过裸眼井段中某个地层的破裂压力，导致地下井喷，即地层流体进入井眼后，又经过井眼进入较浅的地层。一旦地层破裂，就不能用加重钻井液来控制井侵，并且这时地层之间存在窜流。如果地下井喷发生在浅层，还可能引起塌陷（破坏地表沉积层，形成大孔洞可导致钻机塌陷）。

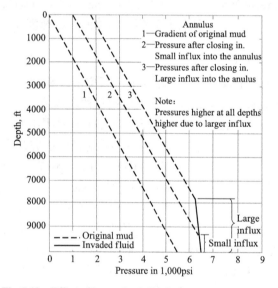

Fig. 7.13 Effect of increasing influx before the well is shut in

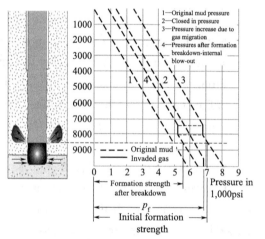

Fig.7.14 Underground blow-out conditions

(2) There is the possibility that p_{ann} will exceed the burst capacity of the casing at surface.

7.4.5.2 Gas Buoyancy Effect

An influx of gas into the wellbore can have a significant effect on the annulus pressure. Since there is such a large difference in density between the gas and the mud, a gas bubble entering the well will be subjected to a large buoyancy effect. The gas bubble will therefore rise up the annulus. As the gas rises it will expand and, if the well is

（2）地面的套压可能会超过套管的破裂强度，造成套管破裂。

7.4.5.2 气体浮力效应

气体侵入井筒会对套管压力产生重大影响。由于气体密度和钻井液密度差异很大，因此进入井筒的气泡很大程度上会受到浮力作用。如果井口是打开的，随着气泡上升，气体会膨胀并从环空排出钻

open, displace mud from the annulus. If, however, the well is shut in mud cannot be displaced and so the gas cannot expand. The gas influx will rise, due to buoyancy, but will maintain its high pressure since it cannot expand. As a result of this p_{ann} will increase and higher pressures will be exerted all down the wellbore (note the increase in bottom hole pressure). The situation is as shown in Fig. 7.15.

井液。反之，如果井口关闭，钻井液无法排出，气体也无法膨胀。在浮力作用下气体上升，但由于不能膨胀，气体会保持高压。结果是套管压力增加，且较高的压力会施加在整个井筒上（注意井底压力的增加），如图7.15所示。

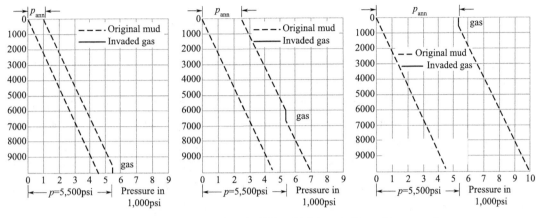

Fig. 7.15 Migration of gas bubble which is not allowed to expand

This increase in annulus, and therefore bottom hole, pressure will be reflected in the drillpipe pressure. This situation can, therefore, be identified by a simultaneous rise in drillpipe and annulus pressure. It is evident that this situation cannot be allowed to develop as it may lead to the problems mentioned earlier (casing bursting or underground blow-out). From the point at which the well is shut in the drillpipe and annulus pressures should be continuously monitored.

If p_{ann} and p_{dp} continue to rise simultaneously it must be assumed that a high pressure gas bubble is rising in the annulus. In this case, the pressure must be bled off from the annulus by opening the choke. Only small volumes (1/4 ～ 1/2 bbl) should be bled off at a time. By opening and closing the choke the gas is allowed to expand, and the pressure should gradually fall. The process should be continued until p_{dp} returns to its original shut in value (again p_{dp} is being used as a bottom hole pressure gauge). This procedure can be carried out until preparations to kill the well are complete. During this procedure no further influx of fluids will occur,

套管压力的增加及由此增加的井底压力也会反映在立管压力上。因此，可以通过立管压力和套管压力同步增长来确定是否发生气侵。显然，这种情况是不允许进一步发展的，因为会引发前面提到的事故（套管破裂或地下井喷）。从关井那一刻开始，就应该持续监测立管压力和套管压力。

若套管压力和立管压力继续同步上升，就表明高压气泡正在环空中上移。这种情况下，必须打开节流阀从环空释放压力，但一次只能释放少量（1/4 ～ 1/2 bbl）。不断打开、关闭节流阀，气体得以膨胀，压力逐渐下降。直到将立管压力恢复到其初始关井压力（可再次用来计量井底压力）。通常情况下，在压井准备工作完成前，可不断打开、关闭节流阀，保持 p_{dp} 大于初始关井压力，就不

provided p_{dp} remains above its original value.

7.4.6 MAASP

Another important parameter which must be calculated is the maximum allowable annular surface pressure (MAASP). The MAASP is the maximum pressure that can be allowed to develop at surface before the fracture pressure of the formation just below the casing shoe is exceeded. Remember that an increase in the annulus pressure at surface will mean that the pressures along the entire wellbore are increasing also. Normally the weakest point in a drilled well is the highest point in the open hole section (i.e. at the previous casing shoe). During the well control operation it is important that the pressure is not allowed to exceed the fracture gradient at this weakest point. The fracture pressure of the formation just below the casing shoe will be available from leak off tests carried out after the casing was set. If no leak-off test was carried out an estimate can be made by taking a percentage of the minimum geostatic gradient for that depth. If an influx occurs and the well is killed with a kill mud this calculation should be repeated to determine the new MAASP. The MAASP should not exceed 70% of the burst resistance of the casing.

7.5 Well Killing Procedures

The procedure used to kill the well depends primarily on whether the kick occurs whilst drilling (there is a drillstring in the well) or whilst tripping (there is no drillstring in the well).

7.5.1 Drill String out of the Well

Audio 7.5

One method of killing a well when there is no drill string in the hole is the Volumetric Method. The volumetric method uses the expansion of the gas to maintain bottom hole pressure greater than formation pressure. Pressures are adjusted by bleeding off at the choke in small amounts. This is a slow process which maintains constant bottom hole pressure

7.4.6 最大允许关井套压

另一个必须计算的重要参数是最大允许关井套压（MAASP）。最大允许关井套压是指确保井筒压力不超过上一层套管鞋下方（薄弱层）地层破裂压力的最大关井套压。地面套管压力的增加意味着整个井筒沿程压力也在增加。通常，正钻井的最薄弱点是裸眼段中的最高点（即在上一开套管鞋处）。在井控作业期间，不允许井筒内压力超过最薄弱点的破裂压力。上层套管封固后进行的漏失试验会提供套管鞋下部地层破裂压力。如果没有进行漏失实验，也可以通过对该深度的最小主应力梯度按一定百分比进行估算。在发生井侵并采用压井钻井液进行压井时，需要重新计算MAASP。MAASP不应超过套管抗内压强度的70%。

7.5 压井程序

压井程序的选择主要取决于井涌是发生在钻进时（井中有钻柱），还是发生在起下钻时（井中无钻柱）。

7.5.1 井内无钻柱

井内无钻柱时采用体积法压井。体积法是利用气体膨胀来保持井底压力大于地层压力。可通过节流阀少量放喷来调节压力。这是一个缓慢的过程，既可保持井底压力恒定，又允许气泡在浮力作用下运移到地面。当气体到

while allowing the gas bubble to migrate to surface under the effects of buoyancy. When the gas reaches surface it is gradually bled off whilst mud is pumped slowly into the well through the kill line. Once the gas is out of the well, heavier mud must be circulated. This can be done with a snubbing unit. This equipment allows a small diameter pipe to be into the hole through the closed BOPs.

7.5.2 Drill String in the Well

When the kick occurs during drilling, the well can be killed directly since:

(1) The formation fluids can be circulated out;

(2) The existing mud can be replaced with a mud with sufficient density to overbalance the formation pressure.

If a kick is detected during a trip the drill string must be stripped to bottom, otherwise the influx cannot be circulated out. Stripping is the process by which pipe is allowed to move through the closed BOPs under its own weight. Snubbing is where the pipe is forced through the BOP mechanically.

There are basically two methods of killing the well when the drill string is at the bottom of the borehole. These are: (1) The One Circulation Method; (2) The Drillers Method.

7.5.2.1 The One Circulation Method

The procedure used in this method is to circulate out the influx and circulate in the heavier mud simultaneously. The influx is circulated out by pumping kill mud down the drillstring displacing the influx up the annulus. The kill mud is pumped into the drill string at a constant pump rate and the pressure on the annulus is controlled on the choke so that the bottomhole pressure does not fall, allowing a further influx to occur.

The advantages of this method are:

(1) Since heavy mud will usually enter the annulus before the influx reaches surface the annulus pressure will be kept low. Thus there is less risk of fracturing the formation at the casing shoe.

(2) The maximum annulus pressure will only be exerted on the wellhead for a short time.

(3) It is easier to maintain a constant p_{bh} by adjusting the choke.

7.5.2.2 Driller's Method (Two Circulation Method)

In this method the influx is first of all removed with the original mud. Then the well is displaced to heavier mud during a second circulation. The one circulation method is generally considered better than the Drillers method since it is safer, simpler and quicker. Its main disadvantage is the time taken to mix the heavier mud, which may allow a gas bubble to migrate.

7.5.3 One Circulation Well Killing Method

When an influx has been detected the well must be shut in immediately. After the pressures have stabilised, the drillpipe pressure (p_{dp}) and the annulus pressure (p_{ann}) should be recorded. The required mud weight can then be calculated using Equation 7.8.

$$\rho_k = \rho_m + \frac{(p_{dp} + p_{ob})}{d}$$
$$\rho_k d = p_{bh} + p_{ob} \quad (7.8)$$
$$\rho_k d = p_{dp} + \rho_m d + p_{ob}$$

where, ρ_k=kill mudweight (psi/ft); p_{ob}=overbalance (psi).

These calculations can be conducted while the heavy, kill mud is being mixed. These are best done in the form of a worksheet (Fig.7.13). It is good practice to have a standard worksheet available in the event of such an emergency. Certain information should already be recorded (capacity of pipe, existing mud weight, pump output).

Notice on the worksheet that a slow pump rate is required. The higher the pump rate the higher the pressure drop, in the drill string and annulus, due to friction. A low pump rate should, therefore, be used to minimise the risk of fracturing the formation. (A kill rate of 1～4 bbl/min. is recommended). The pressure drop (p_{c1}) which occurs while pumping at the kill rate will be known from pump rate tests which are conducted at regular intervals during the drilling operation. It is assumed that this pressure drop applies only to the drillstring and does not include the annulus. Initially, the pressure at the top of the drillstring, known as the standpipe pressure will be the sum of $p_{dp} + p_{c1}$ (Fig. 7.16). The phrase standpipe pressure comes from the fact that the pressure gauge which is used to measure the pressure on the drillstring is connected to the standpipe. As the heavy mud is pumped down the drillstring, the standpipe pressure will change due to:

（1）Larger hydrostatic pressure from the heavy mud

（2）Changing circulating pressure drop due to the heavy mud

By the time the heavy mud reaches the bit the initial shut-in pressure p_{dp} should be reduced to zero psi. The standpipe pressure should then be equal to the pressure drop due to circulating the heavier mud

$$p_{c2} = p_{c1} \times \rho_k / \rho_m \qquad (7.9)$$

where, ρ_m = original mud gradient.

（p_{c1}）可以从钻井作业的低泵冲实验中得到。这种测试会在钻井作业期间进行，每隔一定时间实施一次。假设只关注钻柱内产生的压降，而不考虑环空。最初，钻柱顶部的压力（称为立管压力）为 $p_{dp} + p_{c1}$ 之和（图7.16）。之所以叫立管压力，是因为用于测量钻柱压力的压力计与立管相连。当加重钻井液泵入钻柱时，立管压力会由于以下原因而变化：

（1）加重钻井液产生了更大的静水压力；

（2）加重钻井液改变了循环压降。

当加重钻井液到达钻头时，初始关井压力 p_{dp} 应降至0psi。立管压力应该等于循环加重钻井液产生的压力降。

式中，ρ_m 为原始钻井液密度。

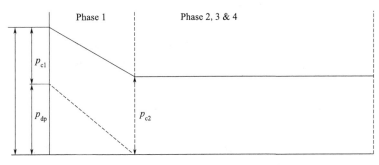

Fig. 7.16 Standpipe pressure versus time

The time taken (or strokes pumped) for the drillstring volume to be displaced to heavy mud can be calculated by dividing the volumetric capacity of the drillstring by the pump output. This information is plotted on a graph of standpipe pressure vs. time or number of pump strokes (volume pumped). This determines the profile of how the standpipe pressure varies with time and number of pump strokes, during the kill procedure.

用钻柱体积除以泵的排量，可以计算出将钻柱内原钻井液置换为加重钻井液所需的时间（或泵的冲数）。这些数据可绘制在立管压力与时间或泵冲数（泵容积）的图表上，从而确定在压井过程中立管压力如何随着时间和泵冲数而变化。

The one circulation method can be divided into 4 stages and these will be discussed separately. When circulating the influx out there will be a pressure drop across the choke, p_{choke}. The pressure drop through the choke plus the hydrostatic head in the annulus should be equal to the formation pressure, p_f. Thus p_{choke} is equivalent to p_{ann} when circulating through a choke.

7.5.3.1　Phase I (displacing drillstring to kill mud)

As the kill mud is pumped at a constant rate down the drillstring the choke is opened. The choke should be adjusted to keep the standpipe pressure decreasing according to the pressure vs. time plot discussed above. In fact the pressure is reduced in steps by maintaining the standpipe pressure constant for a period of time and opening the choke to allow the pressure to drop in regular increments. Once the heavy mud completely fills the drillstring the standpipe pressure should become equal to p_{c2}. The pressure on the annulus usually increases during phase I due to the reduction in hydrostatic pressure caused by gas expansion in the annulus.

7.5.3.2　Phase II (pumping heavy mud into the annulus until influx reaches the choke)

During this stage of the operation the choke is adjusted to keep the standpipe pressure constant (i.e. standpipe pressure = p_{c2}). The annulus pressure will vary more significantly than in phase I due to two effects:

(1) The increased hydrostatic pressure due to the heavy mud entering the annulus will tend to reduce p_{ann}.

(2) If the influx is gas, the expansion of the gas will tend to increase p_{ann} since some of the annular colom of mud is being replaced by gas, leading to a decrease in hydrostatic pressure in the annulus.

The profile of annulus pressure during phase II therefore depends on the nature of the influx (Fig.7.17).

一次循环法可以分为四个阶段，下面将依次进行讨论。当溢流循环排出时，节流阀处会有一个压降 p_{choke}。通过节流阀的压降加上环空静水压头应该等于地层压力 p_f。因此循环通过节流阀时压降 p_{choke} 相当于套压 p_{ann}。

7.5.3.1　第一阶段（钻柱内替成加重钻井液）

打开节流阀，将压井泥浆以恒定的速率向钻柱下方泵送。同时，调节节流阀，保持立管压力按之前讨论的压力—时间曲线降低。事实上，先维持一段时间内立管压力恒定，再打开节流阀，使立管压力按规律的增量下降，逐步降低。一旦加重钻井液完全充满钻柱，此时立管压力应该等于 p_{c2}。通常，在第一阶段由于气体膨胀引起环空静液压力降低，导致环空套压增加。

7.5.3.2　第二阶段（将加重钻井液泵入环空，直到井侵流体到达节流阀）

在此作业阶段调节节流阀来维持立管压力恒定（即立管压力 =p_{c2}）。综合考虑两方面因素，套管压力的变化会比第一阶段更明显：

（1）加重钻井液进入环空，静水压力增加，使 p_{ann} 减小。

（2）如果是气侵，环空中的一部分钻井液会被气体替代，而气体膨胀会使 p_{ann} 增大，引起环空的静液压力下降。

因此，第二阶段的环空压力剖面取决于井侵流体的性质（图7.17）。

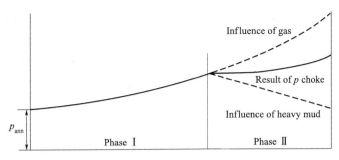

Fig. 7.17 Effect of different kick fluids on annulus pressure

7.5.3.3 Phase Ⅲ (all the influx removed from the annulus)

As the influx is allowed to escape, the hydrostatic pressure in the annulus will increase due to more heavy mud being pumped through the bit to replace the influx. Therefore, p_{ann} will reduce significantly. If the influx is gas this reduction may be very severe and cause vibrations which may damage the surface equipment (choke lines and choke manifold should be well secured). As in phase Ⅱ the standpipe pressure should remain constant.

7.5.3.4 Phase Ⅳ (stage between all the influx being expelled and heavy mud reaching surface)

During this phase all the original mud is circulated out of the annulus and is the annulus is completely full of heavy mud. If the mud weight has been calculated correctly, the annulus pressure will be equal to 0 (zero), and the choke should be fully open. The standpipe pressure should be equal to p_{c2}. To check that the well is finally dead the pumps can be stopped and the choke closed. The pressures on the drillpipe and the annulus should be 0 (zero). If the pressures are not zero continue circulating the heavy weight mud. When the well is dead, open the annular preventer, circulate, and condition the mud prior to resuming normal operations.

7.5.3.5 Summary of One Circulation Method

The underlying principle of the one circulation method is that bottom hole pressure is maintained at a level greater than the formation pressure throughout the operation, so that no further influx occurs. This is achieved by adjusting the choke, to keep the standpipe

7.5.3.3 第三阶段（从环空排出井侵）

随着井侵的排出，环空中的静液压力会增加，因为有更多的加重钻井液经钻头注入环空。此时，套管压力显著降低。如果是气侵，这种下降更为严重，还可能引起损害地面设备的震动（节流管线和节流管汇应固定牢靠）。和第二阶段一样，这一阶段立管压力应保持恒定。

7.5.3.4 第四阶段（所有井侵流体排出，加重钻井液到达地面）

在此阶段中，所有原钻井液都从环空循环排出，并且加重钻井液完全充满环空。如果钻井液密度计算正确，环空压力会等于零，节流阀应完全打开，立管压力应等于p_{c2}。为了检查是否压死油井，可以停泵并关闭节流阀。正常情况下，此时立压和套压应为零。如果不为零，则需要继续循环加重钻井液。当井压死时，打开环形防喷器、循环并处理钻井液，之后再恢复正常作业。

7.5.3.5 总结

工程师法的基本原理是在整个作业过程中，井底压力维持在高于地层压力的水平上，这样就不会再发生井侵。可以通过调节节流阀将立管压力保持在设计剖

pressure on a planned profile, whilst circulating the required mud weight into the well. A worksheet may be used to carry out the calculations in an orderly fashion and provide the required standpipe pressure profile. While the choke is being adjusted the operator must be able to see the standpipe pressure gauge and the annulus pressure gauge. Good communication between the choke operator and the pump operator is important.

Fig.7.18 shows the complete standpipe and annulus pressure profiles during the procedure. Notice that the maximum pressure occurs at the end of phase Ⅱ, just before the influx is expelled through the choke, in the case of a gas kick. Safety factors are sometimes built into the procedure by:

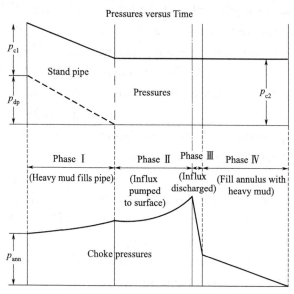

Fig. 7.18 Summary of standpipe and annulus pressure during the "one circulation" method

(1) Using extra back pressure (200 psi) on the choke to ensure no further influx occurs.

(2) Using a slightly higher mud weight. Due to the uncertainties in reading and calculating mud densities it is sometimes recommended to increase mud weight by 0.5 ppg more than the calculated kill weight. This will slightly increase the value of p_{c2}, and mean that the shut in drill pipe pressure at the end of phase Ⅰ will be negative. Whenever mud weight is increased care should be taken not to exceed the fracture pressure of the

formations in the openhole. (An increase of 0.5 lbm/gal mud weight means an increased hydrostatic pressure of 260 psi at 10000ft). Some so-called safety margins may lead to problems of overkill.

7.5.4 Drillers Method for Killing a Well

The Drillers Method for killing a well is an alternative to the One Circulation Method. In this method the influx is first circulated out of the well with the original mud. The heavy weight kill mud is then circulated into the well in a second stage of the operation. As with the one circulation method, the well will be closed in and the circulation pressures in the system are controlled by manipulation of the choke on the annulus. This procedure can also be divided conveniently into 4 stages:

7.5.4.1 Phase Ⅰ (circulation of influx to surface)

During this stage the well is circulated at a constant rate, with the original mud. Since the original mud weight is being circulated the standpipe pressure will equal $p_{dp} + p_{c1}$ throughout this phase of the operation. If the influx is gas then p_{ann} will increase significantly (Fig.7.19). If the influx is not gas the annulus pressure will remain fairly static.

7.5.4 司钻法压井

除了工程师法，还可以用司钻法进行压井。司钻法先用原钻井液将井侵流体从井中循环出来。然后用加重钻井液循环入井。和工程师法类似，关井的同时通过操作环空的节流阀来控制系统中的循环压力。司钻法压井也可以分为四个阶段：

7.5.4.1 第一阶段（井侵循环出地面）

在此阶段，用原钻井液以恒定的泵速循环。由于是用原钻井液进行循环，在整个操作阶段，立管压力将等于$p_{dp}+p_{c1}$。如果是气侵，那么套管压力p_{ann}将会显著增大（图7.19）。如果不是气侵，则环空压力将保持稳定。

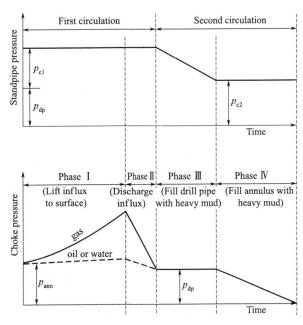

Fig. 7.19 Summary of standpipe and annulus pressure during the "Drillers" method

7.5.4.2 Phase Ⅱ (discharging the influx)

As the influx is discharged the choke will be progressively opened. When all the influx has been circulated out p_{ann} should reduce until it is equal to the original shut in drillpipe pressure p_{dp} so that $p_{ann} + \rho_m d = p_f$.

7.5.4.3 Phase Ⅲ (filling the drillstring with heavy mud)

At the beginning of the second circulation, the stand pipe pressure will still be $p_{dp} + p_{c1}$, but will be steadily reduced by adjusting the choke so that by the end of phase III the standpipe pressure = p_{c2} (as before).

7.5.4.4 Phase Ⅳ (filling the annulus with heavy mud)

In this phase p_{ann} will still be equal to the original p_{dp}, but as the heavy mud enters the annulus p_{ann} will reduce. By the time the heavy mud reaches surface p_{ann} = 0 and the choke will be fully opened.

7.6 Exercise

(1) Impact of Mudweight and Hole Fillup on Bottomhole Pressure:

① An 8½ in hole is drilled to 8,000ft using mud with a density of 12 lbm/gal. If the formation pore pressure at this depth was 4,700 psi what would be the mud pressure overbalance, above the pore pressure.

② If the mud density were 10 lbm/gal what would be the overbalance?

③ If the fluid level in the annulus in a. above dropped to 200 ft due to inadequate hole fill up during tripping, what would be the effect on bottom hole pressure?

(2) Whilst drilling the 8½ in hole of the well the mud pit level indicators suggest that the well is flowing.

① What action should the driller take?

② What action should the driller take if he was pulling out of hole at the time that the kick was recognised?

③ What other indicators of a kick would the driller check for?

(3) Killing Operation Calculations. Whilst drilling the 8½ in hole section of a well the mud pit level indicators indicate that the well is flowing. When the well is made safe the following information is collected:

① drillpipe pressure = 100 psi

② casing pressure = 110 psi

③ pit gain = 10 bbls

Using this and the information provided below carry out the necessary calculations to determine:

① the formation pressure and kill mudweight;

② the type of influx;

③ the time to kill the well;

④ the time to the end of stage 1, 2, and 3 of the killing operation;

⑤ the pump pressure during stages 2, 3, and 4 of the operation.

③ 司钻还应检查哪些溢流的其他征兆？

（3）压井作业计算。在钻 8½in 井段时，钻井液池液位指示器表明这口井正在发生溢流。关井稳定后，收集到以下信息：

① 立管压力为 100 psi；

② 套管压力为 110 psi；

③ 钻井液池增量为 10bbl。

通过上述信息下面给出的数据计算以下内容：

① 地层压力和压井液密度；

② 井侵的类型；

③ 压井时间；

④ 压井作业的第 1、2 和 3 阶段结束时间；

⑤ 压井时第 2、3 和 4 阶段操作期间的泵压。

Casing/Hole Data 套管 / 井眼尺寸

9⅝ in 53.5 lb/ft casing shoe	7000 ft
8½ in hole	9100 ft

Drillingstring Data 钻柱数据

5 in 19.5 lb/ft drillpipe in hole	capacity =0.0178 bbl/ft
BHA - 360 ft of 6.25 in×2 ¹³⁄₁₆ in collars	capacity =0.0077 bbl/ft

Pump Data 泵数据

Type - Triplex pump	Output = 0.1428 bbls/stk
kill rate/circ. press.	14 spm @ 600 psi circ. pressure

Mud Data 钻井液数据

Mud in hole	9.5 lbm/gal

Depth of kick：9100 ft.　　溢流井深：9100ft

Annular Capacities　　环空容积

Collar/Hole (6.25 in Collar×8½ in Hole)	0.0323 bbl/ft
D.P./Hole (5 in Drillipe×8½ in Hole)	0.0459 bbl/ft
D.P./Casing (5 in Drillpipe×9⅝ in Casing)	0.0465 bbl/ft

(4) Briefly explain the essential differences between the "one circulation method" and the "drillers method" for killing a well.

（4）简要描述工程师法和司钻法压井的区别。

301

Casing & Cementing 固井 8

8.1 Casing Program

8.1.1 Introduction

Audio 8.1

It is generally not possible to drill a well through all of the formations from surface (or the seabed) to the target depth in one hole section. The well is therefore drilled in sections, with each section of the well being sealed off by lining the inside of the borehole with steel pipe, known as casing and filling the annular space between this casing string and the borehole with cement, before drilling the subsequent hole section. This casing string is made up of joints of pipe, of approximately 40ft in length, with threaded connections. Depending on the conditions encountered, 3 or 4 casing strings may be required to reach the target depth. The cost of the casing can therefore constitute 20%～30% of the total cost of the well. Great care must therefore be taken when designing a casing programme which will meet the requirements of the well. There are many reasons for casing off formations:

(1) To prevent unstable formations from caving in;

(2) To protect weak formations from the high mudweights that may be required in subsequent hole sections, these high mudweights may fracture the weaker zones;

(3) To isolate zones with abnormally high pore pressure from deeper zones which may be normally

8.1 井身结构

8.1.1 简介

通常不可能只用一个单一井段就从地表（或海床）钻穿所有地层，到达目标深度。因此，一口井的钻井过程要分段进行。在每次钻下一井段之前，都需要将本井段用钢管（称为套管）进行封隔，并在套管和井眼之间的环空注入水泥浆封固。套管柱一般由长度约40ft的套管通过带螺纹的接箍连接组成。不同的钻遇条件，可能需要下3层或4层套管才能钻达目的井深。套管成本一般占油井总成本的20%～30%。所以必须格外重视符合钻井要求的井身结构设计。下套管的原因如下：

（1）避免不稳定地层发生坍塌；

（2）保护薄弱地层免受后续井段钻进时使用高密度钻井液造成的影响，这些高密度钻井液可能会压裂薄弱地层；

（3）将异常高压地层与深部的正常压力地层分隔开；

pressured;

(4) To seal off lost circulation zones;

(5) When set across the production interval: to allow selective access for production / injection/control the flow of fluids from, or into, the reservoir(s).

One of the casing strings will also be required to provide structural support for the wellhead and BOPs.

Each string of casing must be carefully designed to withstand the anticipated loads to which it will be exposed during installation, when drilling the next hole section, and when producing from the well. These loads will depend on parameters such as: the types of formation to be drilled; the formation pore pressures; the formation fracture pressures; the geothermal temperature profile; and the nature of the fluids in the formations which will be encountered. The designer must also bear in mind the costs of the casing, the availability of different casing types and the operational problems in running the casing string into the borehole.

Since the cost of the casing can represent up to 30% of the total cost of the well, the number of casing strings run into the well should be minimised. Ideally the drilling engineer would drill from surface to the target depth without setting casing at all. However, it is normally the case that several casing strings will have to be run into the well in order to reach the objective formations. These strings must be run concentrically with the largest diameter casing being run first and smaller casing strings being used as the well gets deeper. The sizes and setting depths of these casing strings depends almost entirely on the geological and pore pressure conditions in the particular location in which the well is being drilled. Some typical casing string configurations used throughout the world are shown in Fig. 8.1.

In view of the high cost of casing, each string must be carefully designed. This design will be based on the anticipated loads to which the casing will be exposed. When drilling a development well, these loads will have been encountered in previous wells and so the casing programme can be designed with a high degree of confidence, and

（4）封隔漏失层；

（5）当穿过整个生产层段时，允许进行选择性生产、注入和控制来自或进入储层的流体流动。

表层套管柱还需要为井口和防喷器组合提供结构支撑。

每层套管都必须仔细设计，从而可以承受安装、钻下一井段和生产过程中所面临的预期载荷。决定这些载荷大小的参数包括待钻地层的类型、地层孔隙压力、地层破裂压力、地热温度剖面和将遇到的地层流体的性质。设计人员还必须考虑套管的成本、不同类型套管的可用性及下套管时的操作问题。

由于套管的成本可占油井总成本的30%，因此应该尽量减少下入套管的层数。理想的情况是，从地面钻至目的井深，无须下任何套管。然而，通常情况下为了钻达目的层，必须要在井内下入几层套管。这些套管必须是同心下入，先下入直径最大的套管，随着井深的增加，再下入直径较小的套管。这些套管的尺寸和下深完全取决于具体钻井所在位置的地质条件和孔隙压力等因素。图8.1给出了世界各地常用的一些典型井身结构。

由于套管的成本较高，每层管柱都必须根据套管承受的预期荷载进行精心设计。对于开发井，这些载荷在以前所钻的井中都钻遇过，可以据此设计出高可信度、低成本的井身结构。而对于探井

minimal cost. In an exploration well, however, these loads can only be estimated and problems may be encountered which were not expected. The casing design must therefore be more conservative and include a higher safety margin when quantifying the design loads for which the casing must be designed. In addition, in the case of an exploration well, the casing configuration should be flexible enough to allow an extra string of casing to be run, if necessary. A well drilled in an area with high pressures or troublesome formations will usually require more casing strings than one in a normally pressured environment (Fig. 8.2).

而言，这些载荷只能预测，而且考虑到可能会遇到意想不到的情况，套管设计时要更加保守，留有更高的安全余量。此外，对于探井来说，井身结构应该足够灵活，以便在必要时能够多下一层套管。在高压或复杂地层地区进行钻井时，通常需要比在正常压力环境中下的井身结构复杂，下入更多的套管层次（图 8.2）。

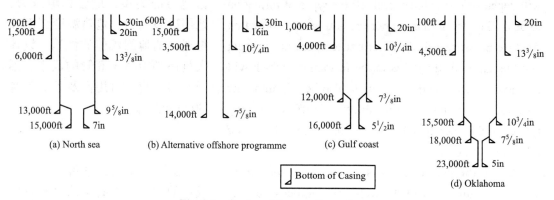

Fig. 8.1　Casing string configurations

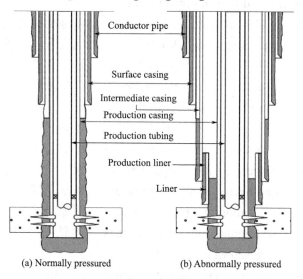

Fig. 8.2　Casing string terminology

A casing string consists of individual joints of steel pipe which are connected together by threaded connections. The joints of casing in a string generally have the same outer diameter and are approximately 40ft

套管是通过接头的螺纹连接在一起的。通常同一层套管的套管外径是相同的，单根套管长度一般为 40ft 左右。套管柱底部通

long. A bull-nose shaped device, known as a guide shoe or casing shoe, is attached to the bottom of the casing string and a casing hanger, which allows the casing to be suspended from the wellhead, is attached to the top of the casing. Various other items of equipment, associated with the cementing operation, may also be included in the casing string, or attached to the outside of the casing e.g. float collar and centralisers.

8.1.2 Casing Terminology

There are a set of generic terms used to describe casing strings. The classification system is based on the specific function of the casing string so, for instance, the function of the surface string shown in Fig. 8.2 is to support the wellhead and BOP stack. Although there is no direct relationship between the size of casing and its function, there is a great deal of similarity in the casing sizes used by operators. The chart in Fig. 8.3 shows the most common casing size and hole size configurations. The dotted lines represent less commonly used configurations. The terms which are generally used to classify casing strings are shown below.

常会安装一个具有内倒角的装置，称为引鞋或套管鞋，顶部会安装一个套管悬挂器，从而确保套管能够悬挂在井口。此外还有和固井作业相关的其他设备，包括安装在套管柱内的浮箍，或装在套管外的扶正器等。

8.1.2 套管类型

不同套管层次通常用一组通用术语用来描述。按每层套管的特定功能划分。表层套管的功能是支撑井口和防喷器组合。虽然套管尺寸和功能没有直接关系，但作业人员常用的套管尺寸却有很大的相似性。图8.3给出了最常见的套管和井眼尺寸搭配组合。其中，虚线表示不太常用的组合。常用的套管类型如下。

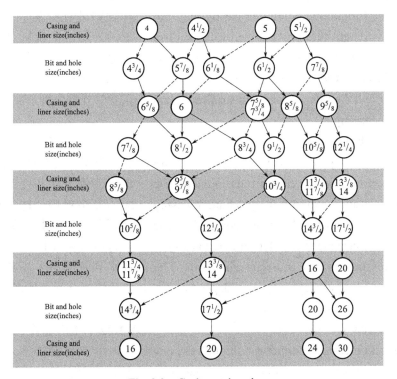

Fig. 8.3 Casing string sizes

8.1.2.1 Conductor Casing

The conductor is the first casing string to be run, and consequently has the largest diameter. It is generally set at approximately 100ft below the ground level or seabed. Its function is to seal off unconsolidated formations at shallow depths which, with continuous mud circulation, would be washed away. The surface formations may also have low fracture strengths which could easily be exceeded by the hydrostatic pressure exerted by the drilling fluid when drilling a deeper section of the hole. In areas where the surface formations are stronger and less likely to be eroded the conductor pipe may not be necessary.

8.1.2.2 Surface Casing

The surface casing is run after the conductor and is generally set at approximately 1,000 ～ 1,500 ft below the ground level or the seabed. The main functions of surface casing are to seal off any fresh water sands, and support the wellhead and BOP equipment. The setting depth of this casing string is important in an area where abnormally high pressures are expected. If the casing is set too high, the formations below the casing may not have sufficient strength to allow the well to be shut-in and killed if a gas influx occurs when drilling the next hole section. This can result in the formations around the casing cratering and the influx flowing to surface around the outside of the casing.

8.1.2.3 Intermediate Casing

Intermediate (or protection) casing strings are used to isolate troublesome formations between the surface casing setting depth and the production casing setting depth. The types of problems encountered in this interval include: unstable shales, lost circulation zones, abnormally pressured zones and squeezing salts. The number of intermediate casing strings will depend on the number of such problems encountered.

8.1.2.4 Production Casing

The production casing is either run through the pay zone, or set just above the pay zone (for an open hole completion or prior to running a liner). The main purpose

8.1.2.1 导管

导管是每口井下入的第一层套管，直径最大，通常下到地面或海床以下约100ft的深度。它的作用是封隔浅部的松散地层，因为连续的钻井液循环会对这些地层造成冲刷。此外，浅表地层的破裂压力也可能较低，当钻进下部较深井段时，钻井液产生的静水压力很容易超过该处的破裂压力，造成漏失。在浅表地层岩石强度较高且不易被侵蚀的地区，也可以不下导管。

8.1.2.2 表层套管

表层套管继导管之后下入，下入深度一般为1000～1500ft。表层套管的主要功能是封隔淡水砂岩，并支撑井口和防喷器组合。在预计会出现异常高压的地区，表层套管的下入深度非常重要。如果套管下入深度不够，在下一井段钻进时会发生气侵，套管以下的地层可能没有足够的强度来支撑关井和压井作业。这会导致套管周围的地层凹陷，地层流体从套管外侧溢到地面。

8.1.2.3 技术套管

技术（或中间）套管用来封隔表层套管下深和生产套管下深之间的复杂地层。这些复杂地层包括不稳定页岩层、漏失层、异常压力地层和具有挤压效应的盐层。技术套管的下入层数取决于可能要钻遇复杂问题的多少。

8.1.2.4 生产套管

生产套管要下入产层，或只下到产层上方（对于裸眼完井或筛管完井）。生产套管的主要用途

of this casing is to isolate the production interval from other formations (e.g. water bearing sands) and/or act as a conduct for the production tubing. Since it forms the conduct for the well completion, it should be thoroughly pressure tested before running the completion.

8.1.2.5　Liner

A liner is a short (usually less than 5,000ft) casing string which is suspended from the inside of the previous casing string by a device known as a liner hanger. The liner hanger is attached to the top joint of the casing in the string. The liner hanger consists of a collar which has hydraulically or mechanically set slips (teeth) which, when activated, grip the inside of the previous string of casing.These slips support the weight of the liner and therefore the liner does not have to extend back up to the wellhead. The overlap with the previous casing is usually 200 ～ 400ft. Liners may be used as an intermediate string or as a production string.

The advantages of running a liner, as opposed to a full string of casing, are that:

(1) A shorter length of casing string is required, and this results in a significant cost reduction.

(2) The liner is run on drillpipe, and therefore less rig time is required to run the string.

(3) The liner can be rotated during cementing operations. This will significantly improve the mud displacement process and the quality of the cement job.

After the liner has been run and cemented it may be necessary to run a casing string of the same diameter as the liner and connect onto the top of the liner hanger, effectively extending the liner back to surface. The casing string which is latched onto the top of the liner hanger is called a tie-back string. This tie-back string may be required to protect the previous casing string from the pressures that will be encountered when the well is in production.

In addition to being used as part of a production string, liners may also be used as an intermediate string to case off problem zones before reaching the production zone. In this case the liner would be known as a drilling

是将生产层段与其他地层（如含水砂岩）隔离，并充当生产油管的导管。由于它充当了完井导管，所以在进行完井作业之前，应该对其进行压力测试。

8.1.2.5　尾管

尾管是一种短的（通常小于5000ft）套管柱，通过尾管悬挂器从上一层的套管柱内部悬挂下来。尾管悬挂器连接到尾管顶部的接头上，由一个具有水力或机械坐挂卡瓦（齿）的接箍组成，当卡瓦（齿）激活时，卡瓦将坐封在上一层套管柱内部。这些卡瓦支撑着尾管的重量，因此尾管不需要延伸至井口。与上一层套管的重叠深度通常在200 ～ 400ft之间。尾管可用作中间套管或生产套管。

与下入完整的套管柱相比，下尾管的优点是：

（1）所需套管柱长度较短，从而大大降低成本。

（2）尾管通过钻杆下入，因此节省了下套管所需的时间。

（3）固井作业时，可以旋转尾管。这会大大改善钻井液顶替过程，提高固井质量。

下入尾管并固井后，可能需要下入和尾管直径相同的套管柱，并连接到尾管悬挂器顶部，有效地将尾管延伸到地面。挂接在尾管悬挂器顶部的套管柱称为回接管柱。这种回接管柱可以保护上一层套管柱，使其免受油井生产过程中的各种压力载荷冲击。

除了作为生产套管的一部分，尾管还可以作为技术套管，隔离产层以上的复杂地层。这时尾管被称为钻井尾管（图8.2）。尾管还

liner (Fig. 8.2). Liners may also be used as a patch over existing casing for repairing damaged casing or for extra protection against corrosion. In this case the liner is known as a stub liner.

8.1.3 Casing Program Selection and Design

The process of selecting casing-setting depths, hole sizes, number of casing strings, and related parameters is referred to as the casing-program selection process.

The process of casing-program selection begins with specification of the surface and bottomhole well locations and the size of the production casing that will be used if hydrocarbons are found in commercial quantities. The number and sizes of tubing strings and the type of subsurface artificial-lift equipment that may eventually be placed in the well determine the minimum ID of the production casing. These specifications are usually determined for the drilling engineer by other members of the engineering staff. In some cases, the possibility of exploratory drilling below an anticipated productive interval must also be considered. The drilling engineer then must design a program of bit sizes, casing sizes, grades, and setting depths that will enable the well to be drilled and completed safely in the desired producing configuration.

To obtain the most economical design, casing strings often consist of multiple sections of different steel grades, wall thicknesses, and coupling types. Such a casing string is called a combination string. Additional cost savings can sometimes be achieved using liner-tieback combination strings instead of a full string run from the surface to the bottom of the hole. When this is done, the reduced tension loads experienced in running the casing in stages often make it possible to use lighter weights or lower grades of casing. The potential savings from use of a liner-tieback combination rather than a full string must be weighed against the additional risks and costs of a successful, leak-free tieback operation. Rig and equipment limitations often make the use of a liner-

8.1.3 井身结构设计

井身结构设计是指合理地选择套管的下入深度、尺寸、下入层数和相关参数的过程。

井身结构设计首先要确定井口和井底的位置，以及生产套管的尺寸。生产套管的最小内径取决于油管柱的数量和尺寸、最终下入井内的人工举升设备类型等因素。这些具体要求通常由地质、油藏工程人员确定并提供给钻井工程师。在有些情况下，进行井身结构设计时还需要考虑在预期产层以下继续进行勘探钻进的可能性。综上所述，钻井工程师必须设计出钻头尺寸、套管尺寸、套管钢级及下深合理的井身结构，从而确保油井能够在预期的产层安全地钻井和完井。

基于既经济又安全的原则，套管柱通常会由不同钢级、壁厚和扣型的多段套管组成。这种套管柱称为复合管柱。有时，也用尾管回接复合管柱取代从地面一直下到井底的完整套管柱。这样可以通过分段下入，降低套管的受拉载荷，从而可以使用质量更轻或钢级更低的套管，节省套管成本。但使用尾管回接复合套管柱取代完整套管柱时，需要在这一技术带来的成本节约及回接作业产生的额外风险和花费之间进行权衡。尾管回接作业更多的是由于受钻机和设备能力限制所采取的技术手段，因为回接的拉

tieback combination necessary because the tension loads are usually lower than when running a full string.

8.1.3.1 Selection of Casing-Setting Depths

The selection of the number of casing strings and their respective setting depths is generally based on consideration of the pore-pressure gradients and fracture gradients of the formations to be penetrated. The example shown in Fig. 8.4 illustrates the relationship between casing setting depth and these gradients. The pore-pressure-gradient and fracture-gradient data are obtained by the methods presented in Chapter 2, expressed as equivalent densities, and plotted against depth. A line representing the planned mud-density program is also plotted. The mud densities are chosen to provide an acceptable trip margin above the anticipated formation pore pressures to allow for reductions in effective mud weight caused by upward pipe movement during tripping operations. A commonly used trip margin is 0.5 lbm/gal or one that will provide 200 to 500 psi of excess bottom hole pressure over the formation pore pressure.

To reach the depth objective, the effective drilling-fluid density shown at point a is chosen to prevent the flow of formation fluid into the well (i.e., to prevent a kick). However, to carry this drilling-fluid density without exceeding the fracture gradient of the weakest formation exposed within the borehole, the protective intermediate casing must extend at least to the depth of Point b, where the fracture gradient is equal to the mud density needed to drill to Point a. Similarly, to drill to Point b and to set intermediate casing, the drilling-fluid density shown at Point c will be needed and will require surface casing to be set at least to the depth at Point d. When possible, a kick margin is subtracted from the true fracture-gradient line to obtain a design fracture-gradient line. If no kick margin is provided, it is impossible to absorb a kick at the casing-setting depth without causing a hydrofracture and a possible underground blowout.

力载荷常小于下入完整套管柱的拉力载荷。

8.1.3.1 套管下深的选择

套管层次及每层套管的下深，一般根据待钻地层的孔隙压力梯度和破裂压力梯度来确定。图8.4所示的例子说明了套管下深与这些压力梯度之间的关系。孔隙压力梯度和破裂压力梯度根据第2章给出的方法进行计算，通常用当量密度表示，并按深度绘制成图。还可以画出设计钻井液密度的曲线。钻井液密度必须在待钻地层压力的基础上增加一个起下钻的安全余量，从而克服起下钻作业时上提钻柱引起的有效钻井液密度降低。常用的起下钻安全余量为0.5 lbm/gal，或者保证井底压力大于地层孔隙压力200～500psi。

为了能够钻到目的井深，需要选择图8.4中a点所示的有效钻井液密度，防止地层流体侵入井内（即防止井涌）。同时，为了保证钻井液密度不超过井内最薄弱地层破裂压力梯度，中间套管至少要下到b点深度处，b点的破裂压力梯度等于钻至a点所需的钻井液密度。同样，要钻至b点并下入中间套管，就需要用c点所示的钻井液密度，且要求至少将表层套管下到d点深度处。这其中要用实际破裂压力梯度减去井涌余量，作为设计破裂压力梯度来用。如果不考虑井涌余量，在套管下入深度处发生井涌时，可能会造成地层破裂和地下井喷。

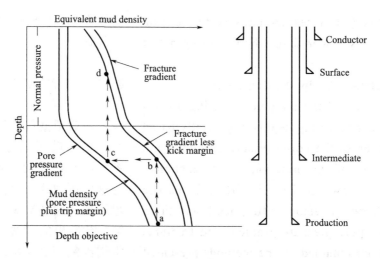

Fig. 8.4 Casing setting depths

Other factors, such as the need to protect freshwater aquifers, the presence of vugular lost-circulation zones, the presence of depleted low-pressure zones that tend to cause stuck pipe, the presence of salt beds that tend to flow plastically and close the borehole, and government regulations, can also affect casing-depth requirements. Moreover, experience in an area may show that it is easier to achieve a good casing-seat cement job in some formation types than in others, or that fracture gradients are generally higher in some formation types than in others. Under such conditions, a design must be found that simultaneously will meet these special requirements and the pore-pressure and fracture-gradient requirements outlined above.

The conductor-casing setting depth is based on the mud density required to prevent washout of the shallow borehole when drilling to the depth of the surface casing. The conductor casing must be able to sustain the pressures expected during diverter operations without washing around the outside of the conductor. The conductor casing is often driven into the ground, where soil resistance governs its length. The casing-driving operation is stopped when the number of blows per foot of depth exceeds some specified upper limit. Typically the conductor is not designed to support the weight of the surface casing or of subsequent strings.

8.1.3.2 Selection of Casing Sizes

The size of the casing strings is controlled by the necessary

套管设计时还要考虑其他因素，如存在需要保护的淡水层、存在洞穴型漏失层、易导致卡钻的枯竭低压层、易塑性流动导致井眼缩径的盐层及当地政府法规等，这些都会影响套管的下深。此外，根据地区经验，发现有些类型的地层比其他地层更容易进行下套管—固井作业；有些地层的破裂压力梯度比其他地层的更高。在这些情况下，进行井身结构设计时，既要满足这些特殊要求，又要满足上述的孔隙压力和破裂压力梯度范围。

导管的下入深度根据钻到表层套管深度时，防止浅层井眼冲蚀所需的钻井液密度来确定。导管必须能够承受放喷作业时的预期压力。导管下深由土壤的阻力决定。当打桩下入每英尺导管所需锤数超过一定上限时，停止下导管作业。通常情况下，导管不需要支撑表层套管或后续套管的重量。

8.1.3.2 套管尺寸的选择

套管的尺寸由生产套管的内

ID of the production string and the number of intermediate casing strings required to reach the depth objective. To enable the production casing to be placed in the well, the bit size used to drill the last interval of the well must be slightly larger than the OD of the casing connectors. The selected bit size should provide sufficient clearance beyond the OD of the coupling to allow for mudcake on the borehole wall and for casing appliances such as centralizers and scratchers. This, in turn, determines the minimum size of the second-deepest casing string. Using similar considerations, the bit size and casing size of successively more shallow well segments are selected.

Selection of casing sizes that permit the use of commonly used bits is advantageous because the bit manufacturers make readily available a much larger variety of bit types and features in these common sizes. However, additional bit sizes are available that can be used in special circumstances (Fig. 8.3).

8.2 Casing

8.2.1 Properties of Casing

When the casing configuration (casing size and setting depth) has been selected, the loads to which each string will be exposed will be computed. Casing, of the required size, and with adequate load bearing capacity will then be selected from manufacturer's catalogues or cementing company handbooks. Casing joints are manufactured in a wide variety of sizes, weights and material grades and a number of different types of connection are available. The detailed specification of the sizes, weights and grades of casing which are most commonly used has been standardised by the American Petroleum Institute - API. The majority of sizes, weights and grades of casing which are available can be found in manufacturer's catalogues and cementing company handbooks (e.g. Halliburton Cementing Tables). Casing is generally classified, in manufacturer's catalogues and handbooks, in terms of its

Audio 8.2

径和中间套管的层数决定。为了将生产套管顺利下入井中，最后一开井眼的钻头尺寸必须略大于生产套管接头的外径。钻头尺寸和套管接头外径的间隙要足够大，为在井壁上形成滤饼和使用套管辅助工具（如扶正器）留出空间。钻头尺寸又决定了上一层套管的最小尺寸，以此类推，可以逐层选取钻头尺寸和套管尺寸。

选择标准的钻头—套管尺寸有利于钻头制造商随时提供不同类型和特征的同尺寸钻头。然而，也可以在特殊情况下使用非标准钻头尺寸（图 8.3）。

8.2 套管柱

8.2.1 套管性能

井身结构（套管尺寸和下入深度）确定后，就需要计算出每层套管柱所承受的载荷，然后从制造商或固井公司提供的套管手册中，根据套管尺寸和承载能力选择合适的套管。套管单根的尺寸、单位长度重量和钢级多种多样，连接的扣型也有多种不同的方式。美国石油学会（API）给出了常用的套管尺寸、重量和钢级等属性的标准化规格。大部分的套管尺寸、重量和钢级都可以在制造商或固井公司提供的套管手册中找到。在制造商或固井公司提供的套管手册中，套管一般按尺寸（外径）、重量、钢级和扣型进行分类。

size (O.D.), weight, grade and connection type.

8.2.1.1 Casing Size (Outside Diameter - O.D.)

The size of the casing refers to the outside diameter (O.D.) of the main body of the tubular (not the connector). Casing sizes vary from 4.5 in to 36 in diameter. Tubulars with an O.D. of less than 4.5 in are called Tubing. The sizes of casing used for a particular well will generally be limited to the standard sizes that are shown in Fig. 8.3. The hole sizes required to accommodate these casing sizes are also shown in this diagram. The casing string configuration used in any given location e.g. 20 in×13⅜ in×9⅝ in×7 in× 4½ in is generally the result of local convention, and the availability of particular sizes.

8.2.1.2 Length of Joint

The length of a joint of casing has been standardised and classified by the API as follows (Table 8.1):

8.2.1.1 套管尺寸（外径）

套管尺寸是指套管本体（不是接头）的外径（O.D.）。套管直径从4.5in到36in不等。外径小于4.5in的称为油管。套管尺寸通常会在图8.3所示的标准尺寸范围内进行选择。这些套管尺寸所需要的井眼尺寸也在图8.3中给出。具体的井身结构，例如20in×13⅜in×9⅝in×7in×4½in，通常要结合当地惯例和每种尺寸的可用性来确定。

8.2.1.2 套管长度

API对套管的长度进行了标准化和分类，见表8.1。

Table 8.1　API length ranges

Length, ft	16～25	25～34	> 34
Average Length, ft	22	31	42

Although casing must meet the classification requirements of the API, set out above, it is not possible to manufacture it to a precise length. Therefore, when the casing is delivered to the rig, the precise length of each joint has to be measured and recorded on a tally sheet. The length is measured from the top of the connector to a reference point on the pin end of the connection at the far end of the casing joint. Range 2 is the most common length, although shorter lengths are useful as pup joints when attempting to assemble a precise length of string.

8.2.1.3 Casing Weight

For each casing size there are a range of casing weights available. The weight of the casing is in fact the weight per foot of the casing and is a representation of the wall thickness of the pipe. There are for instance four different weights of 9⅝ in casing (Table 8.2).

尽管套管要满足API的分类要求，但在套管制造时不可能完全达到相同的精确长度。因此，当套管送到井场时，必须丈量每根套管的准确长度并进行记录。丈量长度时要从接头顶部量到套管末端接头外螺纹上的参考点。表8.2中的第2行是最常用的套管长度，当然确定整个套管柱精确长度时，也会用到更短的套管作为短节来用。

8.2.1.3 套管重量

对于每种套管尺寸，都有一系列套管单位长度重量可供选择。单位长度重量是每英尺套管的重量，是套管壁厚的一种表示。例如，9⅝in套管有四种不同的单位长度重量（表8.2）。

Table 8.2 9⅝ in Casing weights

Weight, lbf/ft	OD, in	ID, in	Wall Thickness, in	Drift Diameter, in
53.5	9.625	8.535	0.545	8.379
47	9.625	8.681	0.472	8.525
43.5	9.625	8.755	0.435	8.599
40	9.625	8.835	0.395	8.679

Although there are strict tolerances on the dimensions of casing, set out by the API, the actual I.D. of the casing will vary slightly in the manufacturing process. For this reason the drift diameter of casing is quoted in the specifications for all casing. The drift diameter refers to the guaranteed minimum I.D. of the casing. This may be important when deciding whether certain drilling or completion tools will be able to pass through the casing e.g. the drift diameter of 9⅝ in 53.5 lbf/ft casing is less than 8½ in bit and therefore an 8½ in bit cannot be used below this casing setting depth. If the 47 lbf/ft casing is too weak for the particular application then a higher grade of casing would be used. The nominal I.D. of the casing is used for calculating the volumetric capacity of the casing.

8.2.1.4 Casing Grade

The chemical composition of casing varies widely, and a variety of compositions and treatment processes are used during the manufacturing process. This means that the physical properties of the steel varies widely. The materials which result from the manufacturing process have been classified by the API into a series of grades (Table 8.3). Each grade is designated by a letter, and a number. The letter refers to the chemical composition of the material and the number refers to the minimum yield strength of the material e.g. N-80 casing has a minimum yield strength of 80,000 psi and K-55 has a minimum yield strength of 55,000 psi. Hence the grade of the casing provides an indication of the strength of the casing. The higher the grade, the higher the strength of the casing.

尽管 API 规定的套管尺寸有严格的公差，但在制造过程中套管的实际内径会有轻微的变化。因此，在所有套管的规格中都引用了套管通径。通径是指必须保证的套管最小内径。这在决定某些钻井或完井工具能否穿过套管时非常重要。例如，9⅝in 53.5lbf/ft 的套管通径小于 8½in 钻头，因此在这种套管下入深度以下不能使用 8½in 钻头。如果改用单位长度重量为 47lbf/ft 的套管，有时可能套管强度不够，需要选用钢级更高的套管。套管的容积常用套管的公称内径来计算。

8.2.1.4 套管钢级

套管的化学成分差异很大，在制造过程中使用了多种成分和处理工艺，这意味着钢的物理性能差异很大。生产过程中产生的材料被 API 划分为一系列钢级（表 8.3）。每个钢级由一个字母和一个数字表示。字母表示材料的化学成分，数字表示材料的最小屈服强度。例如，N-80 套管的最小屈服强度为 80000 psi，K-55 的最小屈服强度为 55000 psi。因此，套管的钢级表明了套管的强度。钢级越高，套管强度越大。

Table 8.3 Casing grades and properties

Grade	Yield Strength psi		Tensile Strength psi
	min	max	
H-40	40,000	—	60,000
J-55	55,000	80,000	75,000
K-55	55,000	80,000	95,000
C-75	75,000	90,000	95,000
L-80	80,000	95,000	95,000
N-80	80,000	110,000	100,000
S-95*	95,000	—	110,000
P-110	110,000	140,000	125,000
V-150*	150,000	180,000	160,000

In addition to the API grades, certain manufacturers produce their own grades of material. Both seamless and welded tubulars are used as casing although seamless casing is the most common type of casing and only H and J grades are welded.

8.2.1.5 Connections

Individual joints of casing are connected together by a threaded connection. These connections are variously classified as: API; premium; gastight; and metal-to-metal seal. In the case of API connections, the casing joints are threaded externally at either end and each joint is connected to the next joint by a coupling which is threaded internally (Fig. 8.5). A coupling is already installed on one end of each joint when the casing is delivered to the rig. The connection must be leak proof but can have a higher or lower physical strength than the main body of the casing joint. A wide variety of threaded connections are available. The standard types of API threaded and coupled connection are: (1) Short thread connection (STC); (2) Long thread connection (LTC); (3) Buttress thread connection (BTC).

In addition to threaded and coupled connections there are also externally and internally upset connections such as that shown in Fig.8.6 A standard API upset connection is: Extreme line (EL).

除了 API 钢级外，有些制造商还制订了自己的材料等级。最常见的套管类型是无缝套管，但焊接套管只有 H 和 J 级，无论是无缝管还是焊接管都可以当作套管使用。

8.2.1.5 扣型

每根套管通过具有螺纹的连接扣衔接在一起。这些连接扣有不同的分类：API 型、附加型、气密型、金属对金属密封型。对于 API 连接扣，每根套管的两端都有外螺纹，通过具有内部螺纹的接箍相连（图 8.5）。当套管运送到井场时，已经在每根套管的一端安装好了接箍。连接扣必须防泄漏，但其物理强度可以比套管本体高或低。连接扣有不同的类型，API 螺纹和接箍连接扣的标准类型有：（1）短螺纹连接扣（STC）；（2）长螺纹连接扣（LTC）；（3）梯形螺纹连接扣（BTC）。

除了螺纹和接箍连接扣外，还有外部和内部加厚连接扣，如图 8.6 所示。标准 API 加厚连接扣为直连型（EL）。

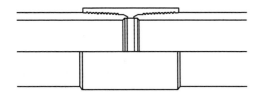

Fig. 8.5　Externally and internally upset casing connection

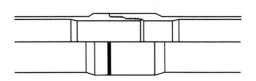

Fig.8.6　Threaded and coupled connection

The STC thread profile is rounded with 8 threads per inch. The LTC is similar but with a longer coupling, which provides better strength and sealing properties than the STC. The buttress thread profile has flat crests, with the front and back cut at different angles. Extreme line connections also have flat crests and have 5 or 6 threads per inch. The EL connection is the only API connection that has a metal to metal seal at the end of the pin and at the external shoulder of the connection, whereas all of the other API connections rely upon the thread compound, used to make up the connection, to seal off the leak path between the threads of the connection.

In addition to API connections, various manufacturers have developed and patented their own connections (e.g. Hydril, Vallourec, Mannesman). These connections are designed to contain high pressure gas and are often called gastight, premium and metal-to-metal seal connections. These connections are termed metal-to-metal seal because they have a specific surface machined into both the pin and box of the connection which are brought together and subjected to stress when the connection is made up.

Surveys have shown that over 80% of leaks in casing can be attributed to poor makeup of connections. This may be due to a variety of reasons: (1)Excessive torque used in making-up the connections; (2)Dirty threads; (3)Cross-threading; (4)Using the wrong thread compound.

The casing string should be tested for pressure integrity before drilling the subsequent hole section. Most of the causes of connection failure can be eliminated by good handling and running procedures on the rig.

The recommended make-up torque for API connections

短螺纹连接扣的螺纹牙形为每英寸8个螺纹。长螺纹连接扣与短螺纹连接扣相似，但其接箍较长，比短螺纹连接扣具有更好的强度和密封性能。梯形螺纹牙形具有平顶，前后切割角度不同。直连型连接扣也有平顶，每英寸有5或6个螺纹。直连型连接扣是唯一一种在内外螺纹处具有金属对金属密封的API连接扣，而所有其他API连接扣都依赖于接箍螺纹来密封以防泄漏。

除API连接扣外，许多制造商还开发了自己的连接扣并获得专利（如Hydril、Vallourec、Mannesman）。这些连接扣可用于容纳高压气体，通常称为气密、优质和金属对金属密封连接扣。之所以被称为金属对金属密封，是因为它们有一个特定的表面加工成外螺纹和内螺纹连接，当连接扣上紧后，内外螺纹咬合在一起，共同承担压力。

调查显示，超过80%的套管泄漏都是连接不当造成的。可能的原因有很多，主要包括：（1）连接时使用的扭矩过大；（2）螺纹不干净；（3）交叉螺纹；（4）使用错误的螺纹脂。

在下一井段钻井之前，应对套管柱进行压力完整性测试。通过现场良好的操作和运行程序，可以消除大多数连接扣故障。

API RP 5C1 中给出了API 连

is given in API RP 5C1. These recommended torques are based on an empirical equation obtained from tests using API modified thread compound on API connections. The recommended make up torque for other connections is available from manufacturers.

8.2.2 Casing Design

8.2.2.1 Casing Axial Forces

Audio 8.3

Having defined the size and setting depth for the casing strings, and defined the operational scenarios to be considered, the loads to which the casing will be exposed can be computed. The particular weight and grade of casing required to withstand these loads can then be determined. The uniaxial loads to which the casing is exposed are:

1. Collapse Load

The casing will experience a net collapse loading if the external radial load exceeds the internal radial load (Fig. 8.7). The greatest collapse load on the casing will occur if the casing is evacuated (empty) for any reason. The collapse load, p_c at any point along the casing can be calculated from:

接扣的推荐上扣扭矩。这些推荐扭矩是根据使用 API 改进螺纹脂对 API 连接扣进行测试时获得的经验公式得出的。对于其他连接扣，推荐的上扣扭矩可以从制造商处获得。

8.2.2 套管柱设计

8.2.2.1 套管轴向力

在确定了套管尺寸、套管下入深度和需要考虑的操作环境后，就可以计算出套管柱将要承受的载荷，以及承受这些荷载所需要的特定单位长度重量和钢级。套管承受的单轴载荷有以下几种。

1. 外挤载荷

当套管外的径向载荷超过管内径向载荷（图 8.7）时，套管会受到一个净外挤载荷。任何原因导致的套管全掏空会使套管柱承受最大的外挤载荷。套管柱上任一点的外挤载荷 p_c 计算如下：

$$p_c = p_e - p_i \tag{8.1}$$

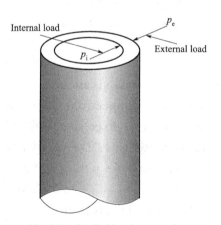

Fig. 8.7 Radial loads on casing

2. Burst Load

The casing will experience a net burst loading if the internal radial load exceeds the external radial load.

2. 内压载荷

当套管内的径向载荷超过管外径向载荷时，套管会受到一个

The burst load, p_b at any point along the casing can be calculated from:

$$p_b = p_i - p_e \qquad (8.2)$$

In designing the casing to resist burst loading the pressure rating of the wellhead and BOP stack should be considered since the casing is part of the well control system. The internal, p_i and external, p_e loads which are used in the determination of the burst and collapse loads on the casing are derived from an analysis of operational scenarios.

1) External Loads

The following issues are considered when deciding upon the external load to which the casing will be subjected:

(1) The pore pressure in the formation. If the engineer is satisfied that it will be possible to displace all of the mud from the annulus between the casing and borehole during the cementing operation, and that a satisfactory cement sheath can be achieved, the formation pore pressure is generally used to determine the load acting on the casing below the top of cement in the annulus, after the cement has hardened.

(2) The mud weight in which the casing was run. If a poor cement bond between the casing and cement or cement and borehole is anticipated then the pressure due to a column of mud in the annulus is generally used to determine the load acting on the casing below the top of cement in the annulus, after the cement has hardened. If the mud has been in place for more than 1 year the weighting material will probably have settled out and therefore the pressure experienced by the casing will be due to a column of mud mixwater.

(3) The pressure from a column of mud mixwater. The pressure due to the mud mixwater is often used to determine the external load on the casing during the producing life of the well. This pressure is equal to the density of fresh or seawater in the case of water-based mud and base oil in the case of oil based mud. The assumption is that the weighting material in the mud

净内压载荷。套管柱上任一点的内压载荷 p_b 计算如下：

$$(8.2)$$

由于套管是井控系统的一部分，因此在设计套管的抗内压强度时，需要同时考虑井口装置和防喷器组合的压力等级。用来确定套管内压和外挤载荷的管内载荷 p_i、管外载荷 p_e 应该根据实际作业情况进行分析。

1）管外载荷

在确定套管柱承受的管外荷载时，应考虑以下因素：

（1）地层中的孔隙压力。符合工程要求的固井作业需要将套管和井眼之间环空中的所有钻井液全部用水泥浆顶替出来，这样可以形成合格的水泥环。在这种情况下，水泥硬化后，通常用地层孔隙压力作为环空水泥返高以下套管的管外荷载。

（2）下套管时的钻井液密度。如果预期的套管和水泥环之间（第一界面）或水泥环和井眼之间（第二界面）胶结强度较差，则通常用下套管时钻井液的液柱压力作为水泥硬化后环空水泥返高下方套管柱的管外荷载。如果钻井液静止超过1年，加重材料可能会沉降，此时套管承受的管外载荷就是钻井液混合水产生的液柱压力。

（3）钻井液混合水液柱压力。在生产阶段，通常用钻井液混合水产生的压力作为套管的管外载荷。假设钻井液中的加重材料（通常为重晶石）已经从悬浮状态沉降下来。对于水基钻井液，管外载荷等于清水或盐水密度；对于油基钻井

(generally Barite) has settled from suspension.

(4) The pressure due to a column of cement slurry. The pressure exerted by a column of cement slurry will be experienced by the casing until the cement sets. It is assumed that hardened cement does not exert a hydrostatic pressure on the casing.

(5) Blockage in the annulus. If a blockage of the annulus occurs during a stinger cement operations. The excess pumping pressure on the cement will be transmitted to the annulus but not to the inside of the casing. This will result in an additional external load during stinger cementing. In the case of conventional cementing operations a blockage in the annulus will result in an equal and opposite pressures inside and outside the casing.

2) Internal Loads

It is commonplace to consider the internal loads due to the following:

(1) Mud to surface. This will be the predominant internal pressure during drilling operations. The casing designer must consider the possibility that the density of the drilling fluid may change during the drilling operation, due to for instance lost circulation or an influx.

(2) Pressure due to influx. The worst case scenario which can arise, from the point of view of burst loading, is if an influx of hydrocarbons occurs, that the well is completely evacuated to gas and simultaneously closed in at the BOP stack.

(3) Full evacuation. The worst case scenario which can arise, from the point of view of collapse loading, is if the casing is completely evacuated.

(4) Production tubing leak. In the case of production casing specifically a leak in the production tubing will result in the tubing pressure being exposed to the casing. The closed in tubing pressure is used as the basis of determining the pressure on the casing. This is calculated on the basis of a column of gas against the formation pressure. The pressure below surface is based on the combined effect of the tubing head pressure and the

液，管外载荷等于油基的密度。

（4）水泥浆液柱压力。在水泥浆凝固之前，套管会承受水泥浆液柱产生的管外载荷。水泥硬化后，通常认为不再对套管施加静水压力。

（5）环空堵塞。如果在插入式固井作业时发生环空堵塞，施加在水泥浆上的附加泵送压力将会传递到环空，但不会传递到套管内部，这会导致插入式固井时会有额外的管外荷载。常规固井作业，环空堵塞会在套管内外压力增加大小相等、方向相反的力。

2）管内载荷

在确定套管柱承受的管内荷载时，应考虑以下因素：

（1）返到地面的钻井液。钻井液产生的压力是钻井作业期间套管的主要管内载荷。套管设计时，必须要考虑井漏或井侵导致钻井作业期间钻井液密度发生变化的可能性。

（2）井侵引起的压力。从内压载荷的角度而言，最严重的情况可能就是在发生烃类井侵时，这时钻井液被完全排空，井内充满天然气，同时防喷器组合关闭。

（3）全掏空。从外挤载荷的角度而言，最严重的情况可能就是出现套管内完全掏空的现象。

（4）生产油管泄漏。对于生产套管而言，特别是当油管泄漏时，会导致生产套管承受油管压力。关井油管压力是确定生产套管所受管内载荷的依据，可以根据平衡地层压力的管内气柱进行计算。井内的压力可以通过油管头压力和封隔器流体柱（如果环

hydrostatic pressure due to a column of packer fluid (if there is any in the annulus).

(5) Fracture pressure of open formations. When considering the internal loads on a casing string the fracture pressure in any formations open to the internal pressures must be considered. The pressure in the open hole section cannot exceed the fracture pressure of the weakest formation. Hence, the pressures in the remaining portion of the borehole and the casing will be controlled by this fracture pressure. The formation just below the casing shoe is generally considered to be the weakest formation in the open hole section.

3) Net Radial Loading (Burst or Collapse Load)

When the internal and external loads have been quantified the maximum net radial loading on the casing is determined by quantifying the difference between the internal and external load at all points along the casing. If the net radial loading is outward then the casing is subjected to a burst load. If the net loading is inward then the casing is subjected to a collapse load.

3. Axial Load

The axial load on the casing can be either tensile or compressive, depending on the operating conditions (Fig. 8.8). The axial load on the casing will vary along the length of the casing. The casing is subjected to a wide range of axial loads during installation and subsequent drilling and production. The axial loads which will arise during any particular operation must be computed and added together to determine the total axial load on the casing.

（5）裸眼地层破裂压力。在确定套管承受的管内载荷时，必须考虑下一裸眼段地层的破裂压力。裸眼段的承载压力不能超过最薄弱地层的破裂压力。因此，整个裸眼段和套管内承受的压力将由最薄弱地层的破裂压力决定。通常认为上一层套管鞋下方的地层是整个裸眼井段中最薄弱的位置。

3）径向净载荷（内压或外挤载荷）

计算出管内载荷和管外载荷后，就可以利用套管柱上所有点的管内载荷和管外载荷差值确定套管上的最大径向净载荷。如果径向净载荷方向向外，则套管会承受内压载荷；如果径向净载荷方向向内，则套管会承受外挤载荷。

3. 轴向载荷

套管上的轴向载荷可以是拉伸载荷也可以是压缩载荷，具体取决于作业工况（图8.8）。轴向载荷会随着套管长度而变化。在下套管及随后的钻井和生产过程中，套管会承受各种轴向载荷。在任何特殊作业期间引起的轴向载荷增加，都必须要进行计算和叠加，从而确定套管总的轴向载荷。

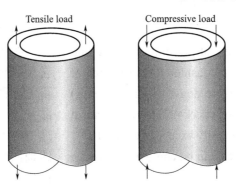

Fig. 8.8 Axial loads on casing

The sources of axial loads on the casing are a function of a number of variables:

(1) W the dry weight of the casing;

(2) φ the angle of the borehole;

(3) A_o the cross sectional area of the casing outside;

(4) A_i the cross sectional area of the casing inside;

(5) DLS the dogleg severity of the well at any point;

(6) p_i the pressure applied to the casing I.D.;

(7) A_s the cross sectional area of the pipe body;

(8) ΔT the change in temperature at any point in the well;

(9) Δp_i and Δp_e the change in internal and external pressure on the casing;

(10) μ the poissons ratio for the steel.

1) Dry weight of Casing

The suspension of a string of casing in a vertical or deviated well will result in an axial load. The total axial load on the casing (the weight of the casing) in air and can be computed from the following:

$$F_{wt} = W\cos\varphi \tag{8.3}$$

2) Buoyant Force on Casing

When submerged in a liquid the casing will be subjected to a compressive axial load. This is generally termed the buoyant force and can be computed from the following:

Open ended casing（套管开口）: $F_{buoy} = p_e(A_o - A_i)$ (8.4)

Closed ended casing（套管闭口）: $F_{buoy} = p_e A_o - p_i A_i$ (8.5)

3) Bending Stress

When designing a casing string in a deviated well the bending stresses must be considered. In sections of the hole where there are severe dog-legs (sharp bends) the bending stresses should be checked. The most critical sections are where dogleg severity exceeds 10° per 100′. The axial load due to bending can be computed from the following:

$$F_{\text{bend}} = 64 \times DLS \times D_{\text{o}} \times W \tag{8.6}$$

where, D_{o} is the casing outside diameter.

4) Plug Bumping Pressure

The casing will experience an axial load when the cement plug bumps during the cementation operation. This axial load can be computed from the following:

$$F_{\text{plug}} = p_{\text{surf}} \times A_{\text{i}} \tag{8.7}$$

5) Overpull when casing stuck

If the casing becomes stuck when being run in hole it may be necessary to apply an overpull' on the casing to get it free. This overpull can be added directly to the axial loads on the casing when it became stuck:

$$F_{\text{pt}} = F_{\text{dt}} \tag{8.8}$$

where, F_{dt} is direct ten sion.

6) Effects of Changes in Temperature

When the well has started to produce the casing will be subjected to an increase in temperature and will therefore expand. Since the casing is restrained at surface in the wellhead and at depth by the hardened cement it will experience a compressive (buckling) load. The axial load generated by an increase in temperature can be computed by the following:

$$F_{\text{temp}} = -200 \times A_{\text{s}} \times \Delta T \tag{8.9}$$

7) Overpull to Overcome Buckling Forces

When the well has started to produce the casing will be subjected to compressive (buckling) loads due to the increase in temperature and therefore expansion of the casing. Attempts are often made to compensate for these buckling loads by applying an overpull to the casing when the cement in the annulus has hardened. This tensile load (the overpull) is 'locked into' the string by using the slip type hanger. The overpull is added directly to the axial load on the casing when the overpull is applied.

$$F_{\text{op}} = F_{\text{do}} \tag{8.10}$$

式中，D_{o} 为套管外径。

4）注水泥碰压产生的轴向力

在注水泥作业时，上下胶塞相碰时，会对套管施加轴向载荷。该轴向载荷可以用以下公式进行计算：

5）卡套管上提拉力

如果套管在下入时遇卡，则可能需要通过上提套管来进行解卡。套管遇卡时的上提拉力直接作用在套管上，产生轴向载荷：

式中，F_{dt} 为上提拉力。

6）温度变化产生的轴向载荷

当油井开始生产时，受温度升高的影响套管将会膨胀。但由于套管在地面井口和井下深度受到硬化水泥的限制，套管不能自由伸缩，将会承受压缩（屈曲）荷载。温度升高产生的轴向载荷可以用以下公式进行计算：

7）克服屈曲施加的拉力

在油井生产时由于温度升高，套管将承受压缩（屈曲）载荷，从而导致套管膨胀。因此在环空中的水泥硬化时，通常会尝试通过对套管施加额外的拉力来补偿这些屈曲荷载。这个拉伸载荷（过拉）通过使用卡瓦式悬挂器"锁定"到管柱上。有过拉力施加时，过拉力会直接叠加到套管的轴向载荷上。

wher, F_{do} is divect overpull.

8) Axial Force Due to Ballooning (During Pressure Testing)

If the casing is subjected to a pressure test it will tend to 'balloon'. Since the casing is restrained at surface in the wellhead and at depth by the hardened cement, this ballooning will result in an axial load on the casing. This axial load can be computed from the following:

$$F_{Bal} = 2\mu(A_i \Delta p_i - A_o \Delta p_e) \tag{8.11}$$

9) Effect of Shock Loading

Whenever the casing is accelerated or decelerated, being run in hole, it will experience a shock loading. This acceleration and deceleration occurs when setting or unsetting the casing slips or at the end of the stroke when the casing is being reciprocated during cementing operations. This shock loading can be computed from the following:

$$F_{shock} = 1780\mu A_s \tag{8.12}$$

A velocity of 5cm/s is generally recommended for the computation of the shockloading.

During installation the total axial load F_t is some combination of the loads described above and depend on the operational scenarios. The objective is to determine the maximum axial load on the casing when all of the operational scenarios are considered.

(1) Free Running of Casing:

$$F_t = F_{wt} - F_{buoy} + F_{bend} \tag{8.13}$$

(2) Running Casing taking account of Shock Loading:

$$F_t = F_{wt} - F_{buoy} + F_{bend} + F_{shock} \tag{8.14}$$

(3) Stuck Casing:

$$F_t = F_{wt} - F_{buoy} + F_{bend} + F_{pt} \tag{8.15}$$

(4) Cementing Casing:

$$F_t = F_{wt} - F_{buoy} + F_{bend} + F_{plug} + F_{shock} \tag{8.16}$$

(5) When cemented and additional overpull is applied

$$F_{\text{tbase}} = F_{\text{wt}} - F_{\text{buoy}} + F_{\text{bend}} + F_{\text{plug}} + F_{\text{op}} \qquad (8.17)$$

(6) During Drilling and Production the total axial load F_t is

$$F_t = F_{\text{tbase}} + F_{\text{bal}} + F_{\text{temp}} \qquad (8.18)$$

4. Biaxial and Triaxial Loading

It can be demonstrated both theoretically and experimentally that the axial load on a casing can affect the burst and collapse ratings of that casing. This is represented in Fig. 8.9. As the tensile load imposed on a tubular increases, the collapse rating decreases and the burst rating increases. As the compressive loading increases the burst rating decreases and the collapse rating increases. The burst and collapse ratings for casing quoted by the API assume that the casing is experiencing zero axial load. However, since casing strings are very often subjected to a combination of tension and collapseoading simultaneously, the API has established a relationship between these loadings

The Ellipse shown in Fig. 8.10 is in fact a 2D representation of a 3D phenomenon. The casing will in reality experience a combination of three loads (Triaxial loading). These are Radial, Axial and Tangential loads. The latter being a resultant of the other two. Triaxial loading and failure of the casing due to the combination of these loads is very uncommon and therefore the computation of the triaxial loads on the casing are not frequently conducted. In the case of casing strings being run in extreme environment (>12,000 psi wells, high H_2S) triaxial analysis should be conducted.

5. Design Factors

The uncertainty associated with the conditions used in the calculation of the external, internal, compressive and tensile loads described above is accommodated by increasing the burst collapse and axial loads by a Design Factor. These factors are applied to increase the actual loading figures to obtain the design loadings. Design factors are determined largely through experience, and

（6）钻井和生产阶段总轴向载荷 F_t 为：

4. 双轴和三轴载荷

理论和实验证明套管上的轴向载荷会影响套管的抗内压和抗外挤强度。如图8.9所示。随着施加在套管上的拉伸载荷增加，套管的抗外挤强度会降低，而抗内压强度会增加。随着施加在套管上的压缩载荷增加，套管的抗内压强度会降低，而抗外挤强度会增加。API在假设套管承受零轴向载荷的情况下给出了套管抗内压和抗外挤强度。考虑到套管会经常同时承受拉伸载荷和外挤载荷的情况，API也建立了这两种载荷之间的关系。

图8.10所示的椭圆实际上是套管三维受力情况的二维表示。套管实际上将承受径向、轴向和周向三种载荷（三轴载荷）。周向载荷是径向载荷和轴向载荷的结果。由径向、轴向和周向载荷产生的套管三轴受力及失效的情况非常罕见，因此通常情况下，不计算套管的三轴载荷。如果套管在极端环境下应用（例如压力大于12000psi，高H_2S）需要进行三轴分析。

5. 套管设计系数

在计算套管的外挤载荷、内压载荷、轴向压缩载荷和拉伸荷载时，需要考虑到一些不确定性条件，这些不确定性可以通过在内压、外挤和轴向载荷上增加一个设计系数来进行调整。实际载荷值加上这些设计系数得到了设计载荷。

are influenced by the consequences of a casing failure. The degree of uncertainty must also be considered (e.g. an exploration well may require higher design factors than a development well), The following ranges of factors are commonly used: (1) Burst design factors 1.0 ～ 1.33; (2) Collapse design factors 1.0 ～ 1.125; (3) Tension design factors 1.0 ～ 2.0; (4) Triaxial Design Factors 1.25.

套管设计系数主要通过经验来确定，并受套管损坏结果的影响。设计系数必须要考虑不确定性的程度（如探井可能需要比开发井更高的设计系数），通常使用以下系数范围：（1）抗内压设计系数1.0～1.33；（2）抗外挤设计系数1.0～1.125；（3）抗拉设计系数1.0～2.0；（4）三轴设计系数1.25。

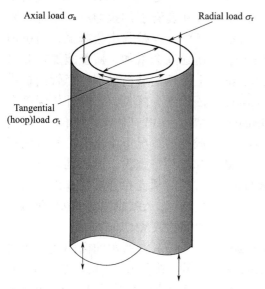

Fig. 8.9 Tri-axial loading on casing

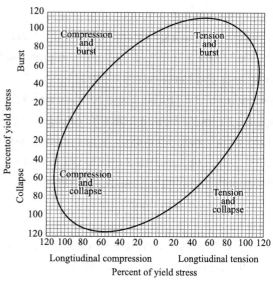

Fig. 8.10 Tri-axial loading ellipse

8.2.2.2 Casing Design Rules Base.

The loading scenarios to be used in the design of the casing string will be dictated by the operating company, on the basis of international and regional experience. These loading scenarios are generally classified on the basis of the casing string classification. The following rules base is presented as a typical example of a casing design rules base.

When the load case has been selected the internal and external loads are calculated on the basis of the rules below. These loads are then plotted on a common axis and the net loading (burst or collapse) is computed. An appropriate casing string can then be selected from the casing tables.

8.2.2.2 套管设计基本原则

作业公司通常会根据国际惯例和区域经验来确定套管设计中的载荷情况。这些载荷情况通常会由于套管类型不同而有所不同。下面通过一些典型的案例介绍套管设计时需要遵循的基本原则。

在确定载荷方案后，就可以根据以下原则计算管内载荷和管外载荷。将管内载荷和管外载荷绘制在常用坐标轴上，计算净载荷（内压或外挤）。然后从可选的套管数据表中选择合理的套管柱。

1. Conductor

The predominant concern in terms of failure of the conductor casing during installation is collapse of the casing. Whilst running the casing it is highly unlikely that the casing will be subjected to a differential pressure. When conducting the cement job the inside of the casing will generally contain the drilling fluid in which the casing was run into the well. The maximum external load will be due to the borehole-casing annulus being full of cement (assumes cement to surface). If a stab-in stinger cementation job is conducted there is the possibility that the annulus will bridge off during the cementing operation and since this pressure will be isolated from the annulus between the casing and the drill pipe stinger this pressure will not be experienced on the inside of the casing. Hence, very high collapse loads will be experienced by the casing below the point at which the bridging occurs.

2. Surface Casing

Once the surface casing has been set a BOP stack will be placed on the wellhead and in the event of a kick the well will be closed in at surface and the kick circulated out of the well. The surface casing must therefore be able to withstand the burst loads which will result from this operation. Some operators will require that the casing be designed to withstand the burst pressures which would result from internal pressures due to full evacuation of the well to gas.

The maximum collapse loads may be experienced during the cement operation or due to lost circulation whilst drilling ahead.

The design scenario to be used for collapse of surface casing in this course is when the casing is fully evacuated due to lost circulation whilst drilling. In this case the casing is empty on the inside and the pore pressure is acting on the outside.

The maximum burst load is experienced if the well is closed in after a gas kick has been experienced. The pressure inside the casing is due to formation pore pressure at the bottom of the well and a colom of gas which extends from the bottom of the well to surface. It is assumed that

1. 导管

导管安装过程中最主要的损坏形式是由外挤载荷导致的挤毁。下导管时，导管内外几乎不存在压差。但在注水泥作业时，套管内部为钻井液，管外最大荷载出现在环空充满水泥浆（假设水泥上返至地面）时。如果进行插入式固井作业，还有可能在固井作业期间发生环空桥堵。由于套管与井壁之间的环空和套管与钻杆之间的环空是隔离的，因此套管外由于桥堵引起的压力升高不会传递到套管内部。此时桥堵下方的套管将承受非常高的外挤载荷。

2. 表层套管

表层套管安装完成以后，需要在井口上装配防喷器组合，以便发生井涌时可以在地面进行关井作业并将井涌循环出井。因此，表层套管必须能够承受这种作业产生的内压载荷。有些套管设计要求设计人员考虑在井筒全掏空，井内充满天然气的情况下，套管能承受内压载荷。

在固井注水泥作业或钻进下一井段发生井漏时，可能会使套管承受最大的外挤载荷。

本书中考虑的表层套管抗外挤设计场景是在下一井段钻井时发生井漏，此时套管被完全掏空，套管外受到地层压力的作用。

如果发生气侵并关井，则表层套管将承受最大内压载荷。套管内的压力是由井底的地层孔隙压力和从井底蔓延至地表的气柱造成的。假设地层压力作用于套管外。

pore pressure is acting on the outside of the casing.

3. Intermediate Casing

The intermediate casing is subjected to similar loads to the surface casing. The design scenario to be used for collapse of intermediate casing in this course is when the casing is fully evacuated due to lost circulation whilst drilling. In this case the casing is empty on the inside and the pore pressure is acting on the outside.

The maximum burst load is experienced if the well is closed in after a gas kick has been experienced. The pressure inside the casing is due to formation pore pressure at the bottom of the well and a column of gas which extends from the bottom of the well to surface. It is assumed that pore pressure is acting on the outside of the casing.

4. Production Casing

The design scenarios for burst and collapse of the production casing are based on production operations.

The design scenario to be used for burst of production casing in this course is when a leak is experienced in the tubing just below the tubing hanger. In this event the pressure at the top of the casing will be the result of the reservoir pressure minus the pressure due to a column of gas. This pressure will the act on the fluid in the annulus of well and exert a very high internal pressure at the bottom of the casing.

The design scenario to be used for collapse of production casing in this course is when the annulus between the tubing and casing has been evacuated due to say the use of gas lift.

5. Other design considerations

In the previous sections the general approach to casing design has been explained. However, there are special circumstances which cannot be satisfied by this general procedure. When dealing with these cases a careful evaluation must be made and the design procedure modified accordingly. These special circumstances include:

(1) Temperature effects- high temperatures will tend to expand the pipe, causing buckling. This must be considered in geothermal wells.

(2) Casing through salt zones- massive salt formations can flow under temperature and pressure. This

3. 中间套管（技术套管）

中间套管承受与表层套管类似的载荷。本书中考虑的中间套管抗外挤设计场景也是在下一井段钻井时发生井漏，此时套管被完全掏空，套管内为空，套管外会受到孔隙压力的挤压载荷。

如果气侵后关井，则会存在最大内压荷载。假设作用在套管外的载荷是每一深度处的孔隙压力，套管内的载荷需要按照井内充满气体、井底处的气柱压力平衡地层孔隙压力这一条件来进行计算。

4. 生产套管

生产套管的内压和外挤设计要根据生产作业条件来确定。

本书中用于生产套管内压设计的场景是油管悬挂器下方的油管发生漏失。在这种情况下，套管顶部的压力等于储层压力减去气柱压力。这一压力作用在油管与生产套管之间环空的流体上，并在套管底部产生一个较高的内压。

本书中用于生产套管外挤设计的场景是，由于使用气举等措施使得油管和套管之间的环空已经被全掏空。

5. 其他设计注意事项

在前面的章节中，已经解释了套管设计的一般方法。然而，有一些特殊情况不能通过一般程序得到满足。在处理这些情况时，必须仔细评估并相应修改设计程序。这些特殊情况包括：

（1）温度效应：高温会使套管膨胀从而导致屈曲。在地热井中必须考虑这一点。

（2）穿越盐层的套管：盐层在温度和压力下大规模蠕动。这将对

will exert extra collapse pressure on the casing and cause it to shear. A collapse load of around 1 psi/ft (overburden stress) should be used for design purposes where such a formation is present.

(3) Casing through H_2S zones - if hydrogen sulphide is present in the formation it may cause casing failures due to hydrogen embrittlement.L-80 grade casing is specially manufactured for use in H_2S zones.

8.3 Cementing

8.3.1 Introduction

Cement is used primarily as an impermeable seal material in oil and gas well drilling. It is most widely used as a seal between casing and the borehole, bonding the casing to the formation and providing a barrier to the flow of fluids from, or into, the formations behind the casing and from, and into, the subsequent hole section(Fig. 8.11). Cement is also used for remedial or repair work on producing wells.

Audio 8.4

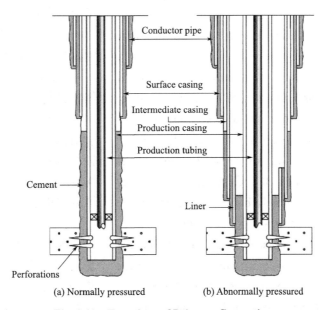

Fig. 8.11 Functions of Primary Cementing

It is used for instance to seal off perforated casing when a producing zone starts to produce large amounts of water and/or to repair casing leaks. This section will present: the reasons for using cement in oil and gas well drilling; the design of the cement slurry; and the operations involved in the placement of the cement slurry. The methods used to determine if the cementing operation has been successful will also be discussed.

8.3.1.1 Functions of Oilwell Cement

There are many reasons for using cement in oil and gaswell operations. As stated above, cement is most widely used as a seal between casing and the borehole, bonding the casing to the formation and providing a barrier to the flow of fluids from, or into, the formations behind the casing and from, and into, the subsequent hole section. However, when placed between the casing and borehole the cement may be required to perform some other tasks. The most important functions of a cement sheath between the casing and borehole are

(1) To prevent the movement of fluids from one formations to another or from the formations surface through the annulus between the casing and borehole;

(2) To support the casing string (specifically surface casing);

(3) To protect the casing from corrosive fluids in the formations.

However, the prevention of fluid migration is by far the most important function of the cement sheath between the casing and borehole. Cement is only required to support the casing in the case of the surface casing where the axial loads on the casing, due to the weight of the wellhead and BOP connected to the top of the casing string, are extremely high. The cement sheath in this case prevents the casing from buckling.

The techniques used to place the cement in the annular space will be discussed in detail later but basically the method of doing this is to pump cement down the inside of the casing and through the casing shoe into the annulus (Fig. 8.12). This operation is known as a primary cement job. A successful primary cement job is essential to allow further drilling and production operations to proceed.

套管，或修复泄漏的套管。下面介绍在油气钻井中使用水泥的原因、水泥浆设计及注水泥作业。此外，还会讨论评价注水泥作业的方法。

8.3.1.1 油井水泥的作用

在油气井作业中使用水泥有许多原因。水泥是一种使用最广泛的密封材料，它可以封隔套管与井壁之间的环空，将套管与地层胶结在一起，阻止封固地层和下一井段间的流体流通。同时，在套管和井壁之间注入水泥时，还会起到其他的作用。套管与井壁之间的水泥环具有以下功能：

（1）防止流体通过套管与井壁之间的环空从一个地层窜流到另一个地层，或从地层内窜流到地面；

（2）用来支撑套管柱（特别是表层套管）；

（3）保护套管免受地层中腐蚀性流体的侵蚀。

当然，到目前为止，防止流体窜槽仍是水泥环最重要的功能。由于连接在套管顶部的井口和防喷器组合的重量导致套管承受的轴向载荷非常大，需要水泥来支撑套管。在这种情况下，水泥环可以防止套管屈曲。

固井注水泥的基本方法是将水泥浆泵入套管内，并通过套管鞋进入环空（图8.12）。这一作业被称为一次注水泥。成功的一次注水泥作业对于进一步的钻井和生产作业至关重要。

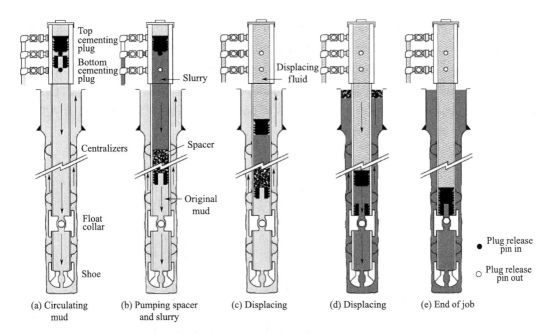

Fig. 8.12 Primary Cementing Operations

Another type of cement job that is performed in oil and gas well operations is called a secondary or squeeze cement job. This type of cement job may have to be done at a later stage in the life of the well. A secondary cement job may be performed for many reasons, but is usually carried out on wells which have been producing for some time. They are generally part of remedial work on the well (e.g. sealing off water producing zones or repairing casing leaks). These cement jobs are often called squeeze cement jobs because they involve cement being forced through holes or perforations in the casing into the annulus or the formation (Fig 8.13).

The specific properties of the cement slurry which is used in the primary and secondary cementing operations discussed above will depend on the particular reason for using the cement (e.g. to plug off the entire wellbore or simply to plug off perforations) and the conditions under which it will be used (e.g. the pressure and temperature at the bottom of the well).

The cement slurry which is used in the above operations is made up from: cement powder; water; and chemical additives. There are many different grades of cement powder manufactured and each has particular

在油气井作业中还有另一种注水泥作业，称为二次注水泥或挤水泥作业。这种作业可能在油井生产的后期进行。实施二次注水泥作业的原因有很多，但通常都是由于已经生产了一段时间的油井需要进行补救工作而采取的措施（如封堵产水层或修复泄漏套管）。这种注水泥作业通常被称为挤水泥作业，因为它们使用液体压力将水泥浆挤入套管的孔洞或射孔孔眼，并进入环空或地层（图8.13）。

上面讨论的一次和二次注水泥作业中使用的水泥浆性能取决于两方面因素，一方面是注水泥的特定原因（例如封堵整个井筒还是简单地封堵射孔），另一方面是使用水泥浆的条件（例如井底的压力和温度）。

用于上述作业的水泥浆由水泥干灰、水和化学添加剂组成。水泥干灰有许多不同的等级，每种等级都有特定的属性，适用于

attributes which make it suitable for a particular type of operation. These grades of cement powder will be discussed below. The water used may be fresh or salt water. The chemical additives (Fig. 8.14) which are mixed into the cement slurry alter the properties of both the cement slurry and the hardened cement.

特定的作业类型。这些水泥的等级将在后面讨论。配制水泥浆使用的水可以是淡水或盐水。加入水泥浆中的化学添加剂（图8.14）会改变水泥浆和水泥石的性能。

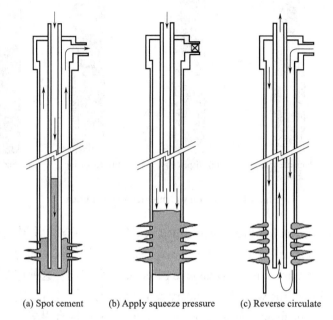

Fig. 8.13　Secondary or Squeeze Cementing Operation

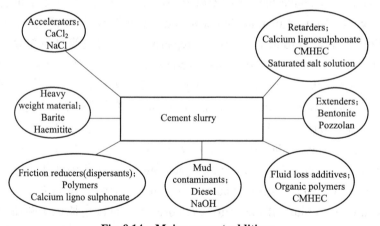

Fig. 8.14　Major cement additives

Each cement job must be carefully planned to ensure that the correct cement and additives are being used, and that a suitable placement technique is being employed for that particular application. In planning the cement job the engineer must ensure that:

(1) The cement can be placed correctly using the

每次注水泥作业都必须仔细规划，确保选用正确的水泥和添加剂，并针对具体的应用采用合适的注水泥技术。在注水泥设计时，工程师必须确保：

（1）可使用现有设备注水泥；

equipment available;

(2) The cement will achieve adequate compressive strength soon after it is placed;

(3) The cement will thereafter isolate zones and support the casing throughout the life of the well.

To assist the engineer in designing the cement slurry, the cement slurry is tested in the laboratory under the conditions to which it will be exposed in the wellbore. These tests are known as pilot tests and are carried out before the job goes ahead. These tests must simulate downhole conditions as closely as possible. They will help to assess the effect of different amounts of additives on the properties of the cement (e.g. thickening time, compressive strength development etc).

8.3.1.2 Classification of Cement Powders

There are several classes of cement powder which are approved for oilwell drilling applications, by the American Petroleum Institute- API. Each of these cement powders have different properties when mixed with water. The difference in properties produced by the cement powders is caused by the differences in the distribution of the four basic compounds which are used to make cement powder; C_3S, C_2S, C_3A, C_4AF (Table 8.4).

（2）水泥浆顶替不久就达到足够的抗压强度；

（3）在整个油井生产期内，水泥能够隔离不同层位并支撑套管。

为正确设计水泥浆，通常会按水泥浆在井筒中暴露的条件进行实验室水泥浆测试。这些先导试验会在注水泥作业之前进行，并且尽可能模拟井下条件。这有助于评价不同添加剂用量对水泥性能的影响（例如稠化时间、抗压强度等）。

8.3.1.2 水泥的分类

美国石油学会（API）推荐了几种常用的油井水泥。每种水泥与水混合时都具有不同的性能。制造水泥的四种基本化合物（C_3S、C_2S、C_3A、C_4AF）含量不同，水泥性能就会有差异（表8.4）。

Table 8.4　Composition of API Cements

API Class	Compounds*					Fineness cm^2/g
	C_3S	C_2S	C_3A	C_4AF	$CaSO_4$	
A	53	24	8	8	3.5	1600~1900
B	44	32	5	12	2.9	1500~1900
C	58	16	8	8	4.1	2000~2400
D&E	50	26	5	13	3	1200~1500
G	52	27	3	12	3.2	1400~1600
H	52	25	3	12	3.3	1400~1600

*Plus free lime, alkali, (Na, K, Mg)

Classes A and B- These cements are generally cheaper than other classes of cement and can only be used at shallow depths, where there are no special requirements. Class B has a higher resistance to sulphate

A级和B级水泥通常比其他类型的水泥便宜，只能用于没有特殊要求的浅地层。B级水泥比A级水泥具有更高的抗硫酸盐性。

than Class A.

Class C- This cement has a high C_3S content and therefore becomes hard relatively quickly.

Classes D, E and F- These are known as retarded cements since they take a much longer time to set hard than the other classes of cement powder. This retardation is due to a coarser grind. These cement powders are however more expensive than the other classes of cement and their increased cost must be justified by their ability to work satisfactorily in deep wells at higher temperatures and pressures.

Class G and H - These are general purpose cement powders which are compatible with most additives and can be used over a wide range of temperature and pressure. Class G is the most common type of cement and is used in most areas. Class H has a coarser grind than Class G and gives better retarding properties in deeper wells. There are other, non-API, terms used to classify cement. These include the following:

(1) Pozmix cement: This is formed by mixing Portland cement with pozzolan (ground volcanic ash) and 2% bentonite. This is a very lightweight but durable cement. Pozmix cement is less expensive than most other types of cement and due to its light weight is often used for shallow well casing cementation operations.

(2) Gypsum Cement: This type of cement is formed by mixing Portland cement with gypsum. These cements develop a high early strength and can be used for remedial work. They expand on setting and deteriorate in the presence of water and are therefore useful for sealing off lost circulation zones.

(3) Diesel oil cement: This is a mixture of one of the basic cement classes (A, B, G, H), diesel oil or kerosene and a surfactant. These cements have unlimited setting times and will only set in the presence of water. Consequently they are often used to seal off water producing zones, where they absorb and set to form a dense hard cement.

C级水泥的C_3S含量高，因此能够很快变硬。

D级、E级和F级油井水泥被称为缓凝水泥，因为与其他类型的水泥相比，它们需要更长的凝固时间。这种延迟是因为颗粒较粗。然而，这些水泥比其他类型的水泥更贵，虽然成本有所增加，但在高温高压深井固井中表现出更为良好的性能。

G级和H级水泥是通用水泥，和大多数添加剂相容，适用的温度和压力范围较广。G级水泥是最常见的水泥类型，用于大多数地区。H级水泥比G级水泥磨得粗一些，在较深的油井中具有更好的缓凝性能。还有其他非API水泥分类。其中包括：

（1）轻质水泥：由硅酸盐水泥与火山灰和2%膨润土混合而成。它是一种非常轻但耐用的水泥，比大多数其他类型的水泥便宜。由于密度小，通常用于浅井套管固井作业。

（2）石膏水泥：由硅酸盐水泥和石膏混合而成。这种水泥具有很高的早期强度，可用于补救工作。在凝固时膨胀，遇水时会水化，因此有助于封隔漏失层。

（3）柴油水泥：由一种基本水泥（A级、B级、G级、H级）、柴油或煤油，以及表面活性剂混合而成。这种水泥的凝固时间无限长，因为只有在有水的情况下它们才会凝固。所以，常用它们来封隔产水层，它们会在那里吸水并固化形成致密的硬水泥。

8.3.1.3 Mixwater Requirements

The water which is used to make up the cement slurry is known as the mixwater. The amount of mixwater used to make up the cement slurry is shown in Table 8.5.

8.3.1.3 混合水要求

用于配制水泥浆的水称为混合水。配制水泥浆的混合水用量见表8.5。

Table 8.5 API Mixwater requirements for API cements

API Class	A	B	C	D	E	F	G	H
Mix Water L/100kg	46.17	46.17	55.93	38.18	38.18	38.18	5.0	4.3
Slurry Weight lb/gal	15.6	15.6	14.8	16.4	16.4	16.2	15.8	16.4

These amounts are based on :

(1) The need to have a slurry that is easily pumped.

(2) The need to hydrate all of the cement powder so that a high quality hardened cement is produced.

(3) The need to ensure that all of the free water is used to hydrate the cement powder and that no free water is present in the hardened cement.

The amount of mixwater that is used to make up the cement slurry is carefully controlled. If too much mixwater is used the cement will not set into a strong, impermeable cement barrier. If not enough mixwater is used: The slurry density and viscosity will increase; The pumpability will decrease, Less volume of slurry will be obtained from each sack of cement.

The quantities of mixwater quoted in Table 8.5 are average values for the different classes of cement. Sometimes the amount of mixwater used will be changed to meet the specific temperature and pressure conditions which will be experienced during the cement job.

8.3.2 Properties of Cement

The properties of a specific cement slurry will depend on the particular reason for using the cement, as discussed above. However, there are fundamental properties which must be considered when designing any cement slurry.

8.3.2.1 Compressive Strength

The casing shoe should not be drilled out until the

混合水用量依据以下需求确定：

（1）满足水泥浆易于泵送的需要；

（2）满足将所有水泥灰水化的需要，以便产出高质量的硬化水泥；

（3）满足确保所有游离水用来进行水泥水化的需要，且硬化水泥中不存在游离水。

配制水泥浆所使用的混合水用量需要准确控制。如果混合水使用太多，水泥浆就不会凝结成坚固、不渗透的水泥石。如果混合水用量不足：水泥浆密度和黏度将会增加；水泥浆的可泵性将降低；用相同水泥混配的水泥浆体积会比较少。

表8.5中推荐的混合水用量是不同等级水泥的平均值。有时为了适合固井作业时遇到的特定温度和压力，需要改变混合水的用量。

8.3.2 水泥性能

特定的水泥浆性能取决于水泥的具体用途。然而，在设计任何水泥浆时都必须考虑一些基本特性。

8.3.2.1 抗压强度

在水泥环的抗压强度未达到

cement sheath has reached a compressive strength of about 500 psi. This is generally considered to be enough to support a casing string and to allow drilling hardened cement sheath, disintegrating, due to vibration. If the operation is delayed whilst waiting on the cement to set and develop this compressive strength the drilling rig is said to be "waiting on cement" (WOC). The development of compressive strength is a function of several variables, such as: temperature; pressure; amount of mixwater added; and elapsed time since mixing. The setting time of a cement slurry can be controlled with chemical additives, known as accelerators. Table 8.6 shows the compressive strengths for different cements under varying conditions.

500 psi 时,不得钻掉套管鞋。人们通常认为达到 500 psi 的水泥环足以支撑套管柱,并且不会因继续钻井产生振动而破裂。水泥浆凝固形成一定抗压强度水泥石的等待时间,称为候凝时间(WOC)。抗压强度的变化是温度、压力、混合水用量、混合后经过时间等变量的函数。水泥浆的凝固时间可以通过化学添加剂,即促凝剂来控制。表 8.6 给出了不同类型的水泥在不同条件下的抗压强度。

Table 8.6　Compressive strength of cements

		Portland	API Class G	API Class H
Water, gal/100kg		5.19	4.97	4.29
Slurry density, lbm/gal		15.9	15.8	16.5
Slurry Vol,L/100kg		1.8	1.14	1.05
Temp, °F	Pressure, psi	Typical comp. strength@ 12hrs, psi		
60	0	615	440	325
80	0	1,470	1,185	1,065
95	800	2,085	2,540	2,110
110	1,600	2,925	2,915	2,525
140	3,000	5,050	4,200	3,160
170	3,000	5,920	4,380	4,485
200	3,000	—	5,110	4,575
		Typical comp. strength @ 24hrs, psi		
60	0	2,870	—	—
80	0	4,130	—	—
95	800	4,130	—	—
110	1,600	5,840	—	—
140	3,000	6,550	—	7,125
170	3,000	6,210	5,865	7,310
200	3,000	—	7,360	9,900

8.3.2.2 Thickening Time (Pumpability)

The thickening time of a cement slurry is the time during which the cement slurry can be pumped and displaced into the annulus (i.e. the slurry is pumpable during this time). The slurry should have sufficient thickening time to allow it to be: (1)Mixed; (2)Pumped into the casing; (3)Displaced by drilling fluid until it is in the required place.

Generally 2～3 hours thickening time is enough to allow the above operations to be completed. This also allows enough time for any delays and interruptions in the cementing operation. The thickening time that is required for a particular operation will be carefully selected so that the following operational issues are satisfied: (1) The cement slurry does not set whilst it is being pumped; (2) The cement slurry is not sitting in position as a slurry for long periods, potentially being contaminated by the formation fluids or other contaminants; (3) The rig is not waiting on cement for long periods.

Wellbore conditions have a significant effect on thickening time. An increase in temperature, pressure or fluid loss will each reduce the thickening time and these conditions will be simulated when the cement slurry is being formulated and tested in the laboratory before the operation is performed.

8.3.2.3 Slurry Density

The standard slurry densities shown in Table 5 may have to be altered to meet specific operational requirements. The density can be altered by changing the amount of mixwater or using additives to the cement slurry. Most slurry densities vary between 11～18.5 lbm/gal. It should be noted that these densities are relatively high when the normal formation pore pressure gradient is generally considered to be equivalent to 8.9 lbm/gal. It is generally the case that cement slurries generally have a much higher density than the drilling fluids which are being used to drill the well. The high slurry densities are however unavoidable if a hardened cement with a high compressive strength is to be achieved.

8.3.2.2 稠化时间（可泵性）

水泥浆的稠化时间是指水泥浆可以泵送并替入环空的时间（即在此期间水泥浆可以泵送）。水泥浆应有足够的稠化时间，从而确保以下工作顺利进行：（1）混配；（2）泵入套管；（3）用钻井液将其顶替到所需位置。

一般来说稠化时间达到2～3小时就足以完成上述作业。同时也为注水泥作业中的任何延迟和中断提供了足够的时间。遇到以下情况时需要对水泥浆稠化时间做特殊处理：（1）泵送时水泥浆不会凝结；（2）受到地层流体或其他污染物的污染，水泥浆顶替到位后长时间不凝结；（3）钻机不能长时间等待候凝。

井筒条件对稠化时间有着显著的影响。温度、压力或滤失量的增加都会缩短稠化时间，固井作业前在实验室进行配制和水泥浆测试时，尽可能模拟实际条件。

8.3.2.3 水泥浆密度

为了满足具体的作业要求，可能需要改变表8.5的标准水泥浆密度。通过改变混合水用量，或加入添加剂可以改变水泥浆的密度。大多数水泥浆的密度在11～18.5lbm/gal之间。应该注意的是，相对于正常地层孔隙压力当量密度（人们通常认为8.9lbm/gal），这些水泥浆密度还是比较高的。通常情况下，水泥浆的密度比钻井液的密度要高得多。然而，如果要获得高抗压强度的硬化水泥，就必须要使用较高的水泥浆密度。

8.3.2.4 Water Loss

The slurry setting process is the result of the cement powder being hydrated by the mixwater. If water is lost from the cement slurry before it reaches its intended position in the annulus its pumpability will decrease and water sensitive formations may be adversely affected. The amount of water loss that can be tolerated depends on the type of cement job and the cement slurry formulation. Squeeze cementing requires a low water loss since the cement must be squeezed before the filter cake builds up and blocks the perforations. Primary cementing is not so critically dependent on fluid loss. The amount of fluid loss from a particular slurry should be determined from laboratory tests. Under standard laboratory conditions a slurry for a squeeze job should give a fluid loss of 50 ～ 200mL. For a primary cement job 250 ～ 400mL is adequate.

8.3.2.5 Corrosion Resistance

Formation water contains certain corrosive elements which may cause deterioration of the cement sheath. Two compounds which are commonly found in formation waters are sodium sulphate and magnesium sulphate. These will react with lime and C_3S to form large crystals of calcium sulphoaluminate. These crystals expand and cause cracks to develop in the cement structure. Lowering the C_3A content of the cement increases the sulphate resistance. For high sulphate resistant cement the C_3A content should be 0 ～ 3%.

8.3.2.6 Permeability

After the cement has hardened the permeability is very low (<0.1mD). This is much lower than most producing formations. However if the cement is disturbed during setting (e.g. by gas intrusion) higher permeability channels (5 ～ 10D) may be created during the placement operation.

8.3.3 Cement Additives

Audio 8.5

Most cement slurries will contain some additives, to modify the properties of the slurry and optimise the cement job. Most additives are known by the trade-

8.3.2.4 失水

水泥浆的凝固过程是水泥被混合水水化的过程。如果水泥浆在到达环空预定位置之前就发生了失水，那么水泥浆的可泵性会降低，水敏性地层会受到不利的影响。允许的失水量取决于注水泥作业的类型和水泥浆的配方。挤水泥作业要求低失水，因为必须在滤饼形成并堵塞射孔之前挤压水泥。一次注水泥对失水量的依赖不大。特定水泥浆的失水量应该事先通过实验室试验来确定。在标准的实验室条件下，挤水泥作业的水泥浆失水量应为 50 ～ 200mL。一次注水泥作业的水泥浆失水控制在 250 ～ 400mL 就可以了。

8.3.2.5 耐腐蚀性

地层水中含有一定的腐蚀性成分，可能导致水泥环恶化。地层水中常见的两种化合物是硫酸钠和硫酸镁。它们将与石灰和 C_3S 发生反应，形成硫铝酸钙的大晶体。这些晶体膨胀并导致水泥结构中出现裂缝。降低水泥中 C_3A 的含量可以提高水泥的抗硫酸盐侵蚀能力。对于高抗硫酸盐水泥，C_3A 含量应为 0 ～ 3%。

8.3.2.6 渗透性

水泥硬化后，渗透性非常低（<0.1mD），这比大多数产层渗透率要低得多。但是，如果水泥在凝固过程中受到干扰（例如气体侵入），那么在顶替作业中可能会产生更高的渗透率通道（5 ～ 10D）。

8.3.3 水泥添加剂

为了改变水泥浆的性能并优化注水泥作业，大多数水泥浆都会含有一些添加剂。大多数添加剂都用水泥服务公司使用的商品

names used by the cement service companies.Cement additives can be used to:

(1) Vary the slurry density;

(2) Change the compressive strength;

(3) Accelerate or retard the setting time;

(4) Control filtration and fluid loss;

(5) Reduce slurry viscosity.

Additives may be delivered to the rig in granular or liquid form and may be blended with the cement powder or added to the mixwater before the slurry is mixed. The amount of additive used is usually given in terms of a percentage by weight of the cement powder (based on each sack of cement weighing 94 lb). Several additives will affect more than one property and so care must be taken as to how they are used.

It should be remembered that the slurry is mixed up and tested in the laboratory before the actual cement job.

8.3.3.1 Accelerators

Accelerators are added to the cement slurry to shorten the time taken for the cement to set. These are used when the setting time for the cement would be much longer than that required to mix and place the slurry, and the drilling rig would incur WOC time. Accelerators are especially important in shallow wells where temperatures are low and therefore the slurry may take a long time to set. In deeper wells the higher temperatures promote the setting process, and accelerators may not be necessary.

The most common types of accelerator are:

(1) Calcium chloride ($CaCl_2$) 1.5%～2.0%;

(2) Sodium chloride (NaCl) 2.0%～2.5%;

(3) Seawater.

It should be noted that at higher concentrations these additives will act as retarders.

8.3.3.2 Retarders

In deep wells the higher temperatures will reduce

名而命名。水泥添加剂可的作用包括：

（1）改变水泥浆密度；

（2）改变抗压强度；

（3）加速或延迟凝固时间；

（4）控制滤失和失水；

（5）降低水泥浆的黏度。

水泥添加剂可以按颗粒状或液态的形式运到现场，之后与干水泥灰混合，或是在混合水泥浆之前添加到混合水中。添加剂用量通常以干水泥灰的质量分数表示（以每袋94lb的水泥计）。几种添加剂混合使用会影响多种水泥浆性能，因此必须正确使用添加剂。

注意，在实际固井作业之前，应在实验室先对水泥浆进行混合和测试。

8.3.3.1 促凝剂

在水泥浆中加入促凝剂，可以缩短水泥凝固所需的时间。当水泥浆的凝固时间比混合顶替所需的时间长很多，且引起钻机的额外候凝时间时，就会使用促凝剂。促凝剂在温度较低的浅井中尤其重要，因为水泥浆在低温下需要很长的凝固时间。在深井中，较高的温度促进了凝固过程，因此不需要使用促凝剂。

最常见的促凝剂有：

（1）氯化钙（$CaCl_2$）1.5%～2.0%；

（2）氯化钠（NaCl）2.0%～2.5%；

（3）海水。

应当注意的是，在较高浓度下，这些添加剂将起到缓凝剂的作用。

8.3.3.2 缓凝剂

在深井中，较高的温度会缩

the cement slurry's thickening time. Retarders are used to prolong the thickening time and avoid the risk of the cement setting in the casing prematurely. The bottom hole temperature is the critical factor which influences slurry setting times and therefore for determining the need for retarders. Above a static temperature of 260 ～ 275 degrees F the effect of retarders should be measured in pilot tests.

The most common types of retarders are:

(1) Calcium lignosulphanate (sometimes with organic acids) 0.1% ～ 1.5%;

(2) Saturated Salt Solutions.

8.3.3.3 Lightweight Additives

Extenders are used to reduce slurry density for jobs where the hydrostatic head of the cement slurry may exceed the fracture strength of certain formations. In reducing the slurry density the ultimate compressive strength is also reduced and the thickening time increased. The use of these additives allows more mixwater to be added, and hence increases the amount of slurry which is produced by each sack of cement powder (the yield of the slurry). Such additives are therefore sometimes called extenders.

The most common types of lightweight additives are:

(1) Bentonite (2% ～ 16%) - This is by far the most common type of additive used to lower slurry density. The bentonite material absorbs water, and therefore allows more mixwater to be added. Bentonite will also however reduce compressive strength and sulphate resistance. The increased yield due to the bentonite added is shown in Table 8.7.

短水泥浆的稠化时间。缓凝剂用于延长稠化时间，避免套管内水泥发生过早凝结的风险。井底温度是影响水泥浆凝结时间的关键因素，决定是否需要缓凝剂。在静态温度达到260～275 ℉以上时，应在先导试验中测量缓凝剂的效果。

最常见的缓凝剂有：

（1）木质素磺酸钙（有时含有机酸）0.1%～1.5%；

（2）饱和盐溶液。

8.3.3.3 减轻剂

当水泥浆的静水压头超过某些地层的破裂压力时，可用减轻剂降低水泥浆的密度。在降低水泥浆密度的同时，极限抗压强度也会降低，稠化时间将延长。使用这些添加剂可以添加更多的混合水，从而增加了每袋水泥灰制备的水泥浆量（水泥浆产量）。因此，这种添加剂有时被称为混合剂。

最常见的减轻剂有：

（1）膨润土（2%～16%），这是目前最常见的降低水泥浆密度的添加剂。膨润土易吸水，因此可以添加更多的混合水。然而，膨润土也会降低抗压强度和抗硫酸盐性。加入膨润土而增加的水泥浆量见表8.7。

Table 8.7 Cements with bentonite

Slurry Composition					
Cement Class	Gel %	Mixwater gal/sk	Slurry Density		Slurry Volume L/100kg
			ppg	pcf	
G	0	4.96	15.9	118.70	1.14
G	4	7.35	14.3	107.00	1.49

Continued

Slurry Composition					
Cement Class	Gel %	Mixwater gal/sk	Slurry Density		Slurry Volume L/100kg
			ppg	pcf	
G	8	9.74	13.3	99.77	1.83
G	12	12.10	12.7	94.83	2.18
G	16	14.50	12.2	91.24	2.52

Thickening Time						
Cement Class	Gel %	Casing Schedules				
		2,000ft	4,000ft	6,000ft	8,000ft	10,000ft
		91℉	103℉	113℉	126℉	144℉
G	0	4:30	2:50	2:24	1:50	1:20
G	4	4:10	2:18	1:51	1:27	0:57
G	8	5:00	2:43	2:06	1:38	1:04

Compressive Strength, psi							
Cement Class	Gel %	Time h	80℉	100℉	120℉	140℉	160℉
G	0	24	1,800	3,050	4,150	5,020	6,700
G	4	24	860	1,250	1,830	1,950	2,210
G	8	24	410	670	890	1,090	1,340

(2) Pozzolan - This may be used in a 50/50 mix with the Portland cement. The result is a slight decrease in compressive strength, and increased sulphate resistance.

(3) Diatomaceous earth (10% ~ 40%) - The large surface area of diatomaceous earth allows more water absorption, and produces low density slurries (down to 11 lbm/gal).

8.3.3.4 Heavyweight Additives

Heavyweight additives are used when cementing through overpressured zones. The most common types of additive are:

(1) Barite (barium sulphate)-this can be used to attain slurry densities of up to 18lbm/gal. It also causes a reduction in strength and pumpability.

（2）火山灰，可与硅酸盐水泥按1：1比例混合使用。其结果是抗压强度略有下降，但抗硫酸盐能力增强。

（3）硅藻土（10% ~ 40%），表面积大，可以吸收更多的水，并形成低密度水泥浆（密度可低至11 lbm/gal）。

8.3.3.4 加重剂

超压地层固井时需要使用加重剂。最常见的加重剂有：

（1）重晶石（硫酸钡）：使用时可获得高达18 lbm/gal的水泥浆密度。但同时会降低水泥石强度和水泥浆的可泵性。

(2) Hematite (Fe_2O_3)-The high specific gravity of hematite can be used to raise slurry densities to 22lbm/gal. Hematite significantly reduces the pumpability of slurries and therefore friction reducing additives may be required when using hematite.

(3) Sand-graded sand (40~60 mesh) can give a 2lbm/gal increase in slurry density.

8.3.3.5 Fluid loss Additives

Fluid loss additives are used to prevent dehydration of the cement slurry and premature setting. The most common additives are:

(1) Organic polymers (cellulose) 0.5%~1.5%;

(2) Carboxymethyl hydroxyethyl cellulose (CMHEC) 0.3%~1.0%.

8.3.3.6 Friction Reducing Additives (Dispersants)

Dispersants are added to improve the flow properties of the slurry. In particular they will lower the viscosity of the slurry so that turbulence will occur at a lower circulating pressure, thereby reducing the risk of breaking down formations. The most commonly used are:

(1) Polymers 0.3~0.5 lb/sx of cement;

(2) Salt 1~16 lb/sx;

(3) Calcium lignosulphanate 0.5~1.5 lb/sx).

8.3.3.7 Mud Contaminates

As well as the compounds deliberately added to the slurry on surface, to improve the slurry properties, the cement slurry will also come into contact with, and be contaminated by, drilling mud when it is pumped downhole. The chemicals in the mud may react with the cement to give undesirable side effects. Some of these are listed below (Table 8.8).

The mixture of mud and cement causes a sharp increase in viscosity. The major effect of a highly viscous fluid in the annulus is that it forms channels which are not easily displaced. These channels prevent a good cement bond all round the casing. To prevent mud contamination of the cement a spacer fluid is pumped ahead of the cement slurry.

需要在水泥浆前加入隔离液。

Table. 8.8 Mud additive and its effect on cement

Mud additive	Effect on cement
barite	increases density and reduces compressive strength
caustic	acts as an accelerator
calcium compounds	decrease density
diesel oil	decrease density
thinners	act as retarders

8.3.4 Primary Cementing

The objective of a primary cement job is to place the cement slurry in the annulus behind the casing. In most cases this can be done in a single operation, by pumping cement down the casing, through the casing shoe and up into the annulus. However, in longer casing strings and in particular where the formations are weak and may not be able to support the hydrostatic pressure generated by a very long column of cement slurry, the cement job may be carried out in two stages. The first stage is completed in the manner described above, with the exception that the cement slurry does not fill the entire annulus, but reaches only a pre-determined height above the shoe. The second stage is carried out by including a special tool in the casing string which can be opened, allowing cement to be pumped from the casing and into the annulus. This tool is called a multi stage cementing tool and is placed in the casing string at the point at which the bottom of the second stage is required. When the second stage slurry is ready to be pumped the multi stage tool is opened and the second stage slurry is pumped down the casing, through the stage cementing tool and into the annulus, as in the first stage. When the required amount of slurry has been pumped, the multi stage tool is closed. This is known as a two stage cementing operation and will be discussed in more detail later.

8.3.4 一次注水泥

一次注水泥作业的目的是将水泥浆注入套管外的环空中。大多数情况下，这可以在一次作业中完成，方法是将水泥浆泵入套管，通过套管鞋，向上进入环空。然而，当套管柱较长，存在地层破裂压力较低，且无法支撑长水泥浆液柱产生的静水压力时，注水泥作业可以分成两级进行。第一级按上述方式完成，但水泥浆不填满整个环空，只上返至套管鞋上方预定高度。第二级通过打开安装在套管柱上的一个专用工具进行，它可以将泵入套管的水泥返入环空。该工具称为分级注水泥工具（分级箍），它安装在需要二级注水泥位置的套管柱底部。当二级水泥浆准备好泵入时，打开分级箍，将二级水泥浆泵入套管，通过分级箍进入环空，与一级注水泥一样。当已泵入所需的水泥浆量后，关闭分级箍。该操作被称为双级注水泥作业，将在后面详细讨论。

The height of the cement sheath, above the casing shoe, in the annulus depends on the particular objectives of the cementing operations. In the case of conductor and surface casing the whole annulus is generally cemented so that the casing is prevented from buckling under the very high axial loads produced by the weight of the wellhead and BOP. In the case of the intermediate and production casing the top of the cement sheath (Top of Cement - TOC) is generally selected to be approximately 300 ～ 500 ft. above any formation that could cause problems in the annulus of the casing string being cemented. For instance, formations that contain gas which could migrate to surface in the annulus would be covered by the cement. Liners are generally cemented over their entire length, all the way from the liner shoe to the liner hanger.

8.3.4.1 Single Stage Cementing Operation

The single stage primary cementing operation is the most common type of cementing operation that is conducted when drilling a well. The procedure for performing a single stage cementing operation (Fig 8.12) will be discussed first and then the procedure for conducting a multiple stage and stinger cementing operations will be discussed.

In the case of the single stage operation, the casing with all of the required cementing accessories such as the float collar, centralisers etc. is run in the hole until the shoe is just a few feet off the bottom of the hole and the casing head is connected to the top of the casing. It is essential that the cement plugs are correctly placed in the cement head. The casing is then circulated clean before the cementing operation begins (at least one casing volume should be circulated). The first cement plug (wiper plug) shown in Fig.8.15, is pumped down ahead of the cement to wipe the inside of the casing clean. The spacer is then pumped into the casing. The spacer is followed by the cement slurry and this is followed by the second plug (shut-off plug) shown in Fig.8.16. When the wiper plug reaches the float collar its rubber diaphragm is ruptured, allowing the cement slurry to flow through the plug, around the shoe, and up into the annulus. At

环空中套管鞋上方水泥环的高度取决于注水泥作业的特定目标。对于导管和表层套管，通常要对整个环空进行胶结，以防止套管在受到井口和防喷器组合的重量引起的高轴向拉力下发生变形。对于中间套管和生产套管，水泥环顶部（TOC）通常选在可能出现问题的地层上方约300～500ft处的环空中。例如，含有气体的地层要用水泥封固，因为其中的气体可能从环空上窜到地表。对于尾管，一般从尾管鞋到尾管悬挂器的整个长度都要注水泥。

8.3.4.1 单级注水泥作业

单级一次注水泥作业是钻井时最常见的注水泥作业。我们将首先讨论单级一次注水泥作业的流程（图8.12），然后讨论多级注水泥和内管法注水泥流程。

在单级作业的情况下，套管与所有必需的套管附件（如浮箍、扶正器等）一起下入井眼中，直到套管鞋距离井底只有几英尺，套管头与套管顶部相连。水泥塞必须正确放置在水泥头中。注水泥作业开始前，先循环清洗套管（至少循环一个套管体积）。如图8.15所示，第一个水泥塞（下胶塞）在水泥之前向下泵送，将套管内部擦拭干净。然后将隔离液泵入套管。在隔离液之后是水泥浆，紧接着是第二个水泥塞（上胶塞），如图8.16所示。当下胶塞到达浮箍时，橡胶隔膜破裂，使水泥浆通过桥塞，绕过引鞋，向上返入环空。在这一阶段，隔

this stage the spacer is providing a barrier to mixing of the cement and mud. When the solid, shut-off plug reaches the float collar it lands on the wiper plug and stops the displacement process. The pumping rate should be slowed down as the shut-off plug approaches the float collar and the shut-off plug should be gently bumped into the bottom, wiper plug. The casing is often pressure tested at this point in the operation. The pressure is then bled off slowly to ensure that the float valves, in the float collar and/or casing shoe, are holding.

离液避免了水泥浆和钻井液的混浆。当上胶塞到达浮箍时，它落在下胶塞上，顶替过程停止。随着上胶塞接近浮箍，泵送速度应减慢，上胶塞应轻轻撞到底部的下胶塞。此时要经常对套管进行压力测试。然后缓慢释放压力，以确保浮箍和套管鞋中的浮阀保持稳定。

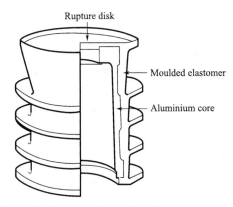

Fig. 8.15 Bottom Plug (wiper plug)

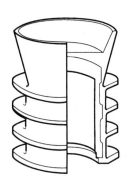

Fig.8.16 Top Plug (shut off plug)

The displacement of the top plug is closely monitored. The volume of displacing fluid necessary to bump the plug should be calculated before the job begins. When the pre-determined volume has almost been completely pumped, the pumps should be slowed down to avoid excessive pressure when the plug is bumped. If the top plug does not bump at the calculated volume (allowing for compression of the mud) this may be because the top, shut-off plug has not been released. If this is the case, no more fluid should be pumped, since this would displace the cement around the casing shoe and up the annulus. Throughout the cement job the mud returns from the annulus should be monitored to ensure that the formation has not been broken down. If formation breakdown does occur then mud returns would slow down or stop during the displacement operation.

The single stage procedure can be summarised as follows:

上胶塞的顶替受到严密监控。应该在作业开始前计算出挤胶塞所需的顶替液体积。当预先确定的顶替液几乎完全泵送时，应放慢泵速，以免上下胶塞相碰时压力过大。如果上胶塞未在计算体积（允许钻井液压缩）内碰压，这可能是因为尚未释放上胶塞。如果是这种情况，就不应该再泵入更多的液体，因为这将顶替套管鞋周围的水泥，并向上进入环空。在整个注水泥作业过程中，应监测水泥浆从环空返出，确保地层没有被压裂。如果确实发生地层破裂，那么在顶替作业期间，钻井液的回流速度会减慢或停止。

单级注水泥作业程序可总结如下：

(1) Circulate the casing and annulus clean with mud (one casing volume pumped);

(2) Release wiper plug;

(3) Pump spacer;

(4) Pump cement;

(5) Release shut-off plug;

(6) Displace with displacing fluid (generally mud) until the shut-off plug lands on the float collar;

(7) Pressure test the casing.

8.3.4.2　Multi Stage Cementing Operation

When a long intermediate string of casing is to be cemented it is sometimes necessary to split the cement sheath in the annulus into two, with one sheath extending from the casing shoe to some point above potentially troublesome formations at the bottom of the hole, and the second sheath covering shallower troublesome formations. The placement of these cement sheaths is known as a multi-stage cementing operation (Fig. 8.17). The reasons for using a multi-stage operation are to reduce:

(1) Long pumping times;

(2) High pump pressures;

(3) Excessive hydrostatic pressure on weak formations due to the relatively high density of cement slurries.

The procedure for conducting a multi-stage operation is as follows:

1. First Stage

The procedure for the first stage of the operation is similar to that described in Section above, except that a wiper plug is not used and only a liquid spacer is pumped ahead of the cement slurry. The conventional shut-off plug is replaced by a plug with flexible blades. This type of shut-off plug is used because it has to pass through the stage cementing collar which will be discussed below. It is worth noting that a smaller volume of cement slurry is used, since only the lower part of the annulus is to be cemented. The height of this cemented part of the annulus will depend on the fracture gradient of the formations which are exposed in the annulus (a height of 3000′ - 4000′ above the shoe is common).

（1）循环泥浆清洗套管和环空（泵入一个套管体积）；

（2）释放下胶塞；

（3）注入隔离液；

（4）注入水泥浆；

（5）释放上胶塞；

（6）用顶替液（通常是泥浆）将上胶塞挤到浮箍上，碰压；

（7）套管压力测试。

8.3.4.2　多级注水泥作业

在对长井段的中间套管进行固井时，有时需要将环空中的水泥环分成两部分，其中一部分水泥环从套管鞋延伸至有潜在复杂情况地层上方的某个点，另一部分水泥环覆盖较浅的复杂地层。这些水泥环的注入称为多级注水泥作业（图8.17）。使用多级注水泥作业的原因是为了避免以下情况：

（1）较长的泵送时间；

（2）高泵送压力；

（3）由水泥浆密度相对较高导致薄弱地层承受的过大静液压力。

多级注水泥作业程序如下：

1. 第一级

一级注水泥作业程序与上述相似，只是不使用下胶塞，在水泥浆之前只泵入隔离液。而常规的上胶塞被带有柔性叶片的水泥塞所取代。使用这种类型的上胶塞是因为它必须通过分级注水泥的分级箍（将在后续内容中讨论）。此时使用的水泥浆体积较小，因为只有环空的下部需要注水泥。环空水泥上返的高度取决于暴露在该段环空中的地层破裂压力梯度（通常在套管鞋上方3000～4000ft高度处）。

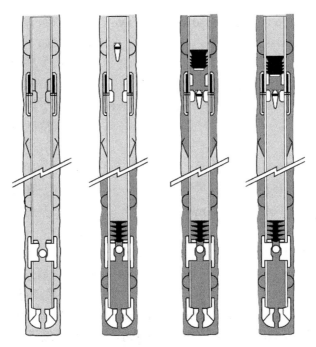

Fig. 8.17 Multi-Stage Cementing Operation

2. Second Stage

The second stage of the operation involves the use of a special tool known as a stage collar (Fig.8.18), which is made up into the casing string at a pre-determined position. The position often corresponds to the depth of the previous casing shoe. The ports in the stage collar are initially sealed off by the inner sleeve. This sleeve is held in place by retaining pins. After the first stage is complete a special dart is released form surface which lands in the inner sleeve of the stage collar. When a pressure of 1000 ～ 1500 psi is applied to the casing above the dart, and therefore to the dart, the retaining pins on the inner sleeve are sheared and the sleeve moves down, uncovering the ports in the outer mandrel. Circulation is established through the stage collar before the second stage slurry is pumped.

2. 第二级

二级注水泥作业要使用一种称为分级箍（图 8.18）的特殊工具，它安装在套管柱上的预定位置。该位置相当于之前的套管鞋深度。分级箍上的循环孔最初由内套密封，内套由销钉固定。一级注水泥完成后，从地面释放重力塞，落在分级箍的内套中。对重力塞上方的套管施加 1000 ～ 1500 psi 的压力，将分级箍打开套上的销钉剪断，使打开套下行露出循环孔。二级水泥浆泵入之前，先通过分级箍建立循环。

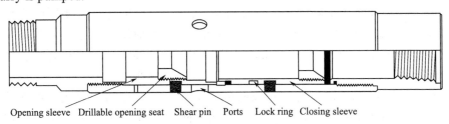

Fig. 8.18 Multi-Stage Cementing Collar

The normal procedure for the second stage of a two stage operation is as follows:

(1) Drop opening dart;

(2) Pressure up to shear pins;

(3) Circulate though stage collar whilst the first stage cement is setting;

(4) Pump spacer;

(5) Pump second stage slurry;

(6) Release closing plug;

(7) Displace plug and cement with mud;

(8) Pressure up on plug to close ports in stage collar and pressure test the casing.

To prevent cement falling down the annulus a cement basket or packer may be run on the casing below the stage collar. If necessary, more than one stage collar can be run on the casing so that various sections of the annulus can be cemented. One disadvantage of stage cementing is that the casing cannot be moved after the first stage cement has set in the lower part of the annulus. This increases the risk of channelling and a poor cement bond.

8.3.4.3 Inner String Cementing

For large diameter casing, such as conductors and surface casing, conventional cementing techniques result in:

(1) The potential for cement contamination during pumping and displacement;

(2) The use of large cement plugs which can get stuck in the casing;

(3) Large displacement volumes;

(4) Long pumping times;

(5) Large volume of cement left inside the casing between float collar and shoe.

An alternative technique, known as a stinger cement job, is to cement the casing through a tubing or drillpipe string, known as a cement stinger, rather than through the casing itself.

In the case of a stinger cement job the casing is run as before, but with a special float shoe (Fig. 8.19) rather than the conventional shoe and float collar. A special sealing

双级注水泥作业的第二级正常程序如下：

（1）释放重力塞；

（2）憋压剪断销钉；

（3）在一级水泥凝固时，通过分级箍循环钻井液；

（4）注入前置液；

（5）注入二级水泥浆；

（6）释放关闭塞；

（7）用泥浆顶替关闭塞和水泥；

（8）憋压，依靠关闭塞剪断关闭套销钉，关闭套下行关闭分级箍的循环孔，并对套管试压。

为防止水泥浆在环空中下沉，可在分级箍下方的套管上安装水泥伞或封隔器。如果有必要，可以在套管上安装多个分级箍，实现环空不同井段的注水泥。分级注水泥的缺点是环空下部一级注水泥凝结后，套管就不能移动了，这增加了窜槽和水泥胶结不良的风险。

8.3.4.3 内管法注水泥

对于导管或表层套管等大尺寸套管，常规固井会有以下问题：

（1）泵送和顶替过程中会存在水泥污染的可能性；

（2）使用大型胶塞可能会发生胶塞卡在套管中的复杂情况；

（3）需要进行大排量顶替；

（4）泵送的时间较长；

（5）浮箍和套管鞋之间的套管内会残留大量水泥。

因此，需要用内管法注水泥作业取而代之。内管法注水泥通过油管或钻杆进行注水泥，而不是通过套管本身。

在内管法注水泥作业中，套管按之前的方式下入，但是使用了特殊的浮鞋（图8.19），取代传

adapter, which can seal in the seal bore of the seal float shoe, is attached to the cement stinger. Once the casing has been run, the cementing string (generally tubing or drillpipe), with the seal adapter attached, is run and stabbed into the float shoe. Drilling mud is then circulated around the system to ensure that the stinger and annulus are clear of any debris. The cement slurry is then pumped with liquid spacers ahead and behind the cement slurry. No plugs are used in this type of cementing operation since the diameter of the stinger is generally so small that contamination of the cement is unlikely if a large enough liquid spacer is used. The cement slurry is generally under-displaced so that when the seal adapter on the stinger is pulled from the shoe the excess cement falls down on top of the shoe. This can be subsequently drilled out when the next hole section is being drilled. Under-displacement however ensures that the cement slurry is not displaced up above the casing shoe, leaving spacer and drilling mud across the shoe. After the cement has been displaced, and the float has been checked for backflow, the cement stinger can be retrieved. This method is suitable for casing diameters of 13⅜ in and larger. The main disadvantage of this method is that for long casing strings rig time is lost in running and retrieving the inner string.

统的浮鞋和浮箍。套管下入后，下入连接有注水泥插头的注水泥管柱（通常为油管或钻杆）并插入浮鞋中。然后，循环钻井液，确保插头和环空没有任何岩屑。随后依次注入隔离液和水泥浆。在这种注水泥作业中不使用胶塞，因为插头的直径通常很小，只要使用足够多的隔离液，就不太可能污染水泥。内管法注水泥水泥浆通常会顶替不足，因此当注水泥管柱上的插头从套管鞋中拔出时，多余的水泥会落到套管鞋顶部。这可以在下一个井段钻进时将其钻掉。然而，顶替不足可以确保水泥浆不会被顶替到套管鞋以上，避免出现隔离液和钻井液被顶替到套管鞋处的环空段。水泥顶替完之后，检查浮鞋是否回流，然后回收注水泥管柱。这种方法适用于直径为13⅜ in 及以上的套管。其主要缺点是，对于较长的套管柱，下入和回收内管柱会需要消耗钻机时间。

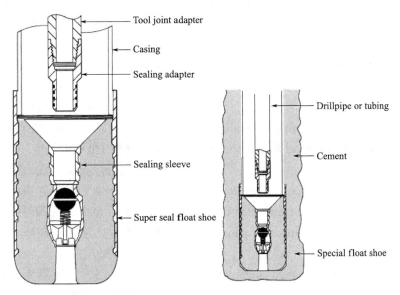

Fig. 8.19 Stinger Cementing Operation

8.3.4.4 Recommendations for a Good Cement Job

The main cause for poor isolation after a cement job is the presence of mud channels in the cement sheath in the annulus. These channels of gelled mud exist because the mud in the annulus has not been displaced by the cement slurry. This can occur for many reasons. The main reason for this is poor centralisation of the casing in the borehole, during the cementing operation. When mud is being displaced from the annulus the cement will follow the least path of resistance. If the pipe is not properly centralised the highest resistance to flow occurs where the clearance is least. This is where mud channels are most likely to occur (Fig. 8.20).

In addition, field tests have shown that for a good cement bond to develop the formation should be in contact with the cement slurry for a certain time period while the cement is being displaced. The recommended contact time (pump past time) is about 10 minutes for most cement jobs. To improve mud displacement and obtain a good cement bond the following practices are recommended:

(1) Use centralisers, especially at critical points in the casing string.

(2) Move the casing during the cement job. In general, rotation is preferred to reciprocation, since the latter may cause surging against the formation. A specially designed swivel may be installed between the cementing head and the casing to allow rotation(centralisers remain static and allow the casing to rotate within them).

(3) Before doing the cement job, condition the mud (low PV, low YP) to ensure good flow properties, so that it can be easily displaced.

(4) Displace the spacer is in turbulent flow. This may not be practicable in large diameter casing where the high pump rates and pressures may cause erosion or formation breakdown.

(5) Use spacers to prevent mud contamination in the annulus.

8.3.4.4 注水泥作业建议

注水泥作业后，封隔效果差的主要原因是环空水泥环中存在窜槽。这些窜槽的出现是由于环空中的钻井液没有完全被水泥浆所取代。发生这种情况的原因有很多，主要原因是注水泥作业时套管在井眼内不居中。当钻井液从环空中被顶替出来时，水泥浆将沿着阻力最小的路径流动。如果套管没有居中，那么间隙较小的地方就会产生较大的流动阻力。这就是窜槽最可能出现的地方（图 8.20）。

此外，现场试验表明，为了形成良好的水泥胶结，在顶替水泥浆时地层应该与水泥浆接触一段时间。大多数注水泥作业建议的接触时间约为 10 分钟。为提高泥浆的顶替效果并获得良好的水泥胶结，建议采用以下方法：

（1）使用扶正器，尤其是在套管柱的关键点上要用扶正器。

（2）注水泥作业期间活动套管。一般旋转比上下往复运动更可取，因为往复运动可能会对地层产生激动压力。在水泥头和套管之间可以安装一个专门设计的水龙头，以便旋转（扶正器保持静止，允许套管在扶正器内进行旋转）。

（3）在注水泥作业之前调整钻井液（低塑性黏度、低动切力），确保良好的流动性能，从而易于顶替。

（4）用湍流顶替隔离液。但这在大直径套管中是不可行的，因为高泵速和高压力可能导致侵蚀或地层破裂。

（5）使用隔离液，防止环空中的钻井液污染。

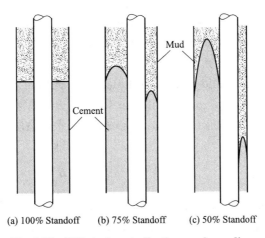

(a) 100% Standoff　　(b) 75% Standoff　　(c) 50% Standoff

Fig. 8.20　Effect of centralisation on channeling

8.3.5　Squeeze Cementing

Squeeze cementing is the process by which hydraulic pressure is used to force cement slurry through holes in the casing and into the annulus and/or the formation. Squeeze cement jobs are often used to carry out remedial operations during a workover on the well. The main applications of squeeze cementing are:

(1) To seal off gas or water producing zones, and thus maximise oil production from the completion interval;

(2) To repair casing failures by squeezing cement through leaking joints or corrosion hole;

(3) To seal off lost circulation zones;

(4) To carry out remedial work on a poor primary cement job (e.g. to fill up the annulus);

(5) To prevent vertical reservoir fluid migration into producing zones (block squeeze);

(6) To prevent fluids escaping from abandoned zones.

During squeeze cementing the pores in the rock rarely allow whole cement to enter the formation since a permeability of about 500 darcies would be required for this to happen. There are two processes by which cement can be squeezed:

(1) High pressure squeeze. This technique requires that the formation be fractured. which then allows the cement slurry to be pumped into the fractured zone.

8.3.5　挤水泥

挤水泥是利用液体压力将水泥浆挤入套管孔眼，进入环空和地层的作业。修井期间，挤水泥是常用的补救作业。主要应用于以下情况：

（1）封堵气层或水层，使完井井段的产油量最大化；

（2）通过将水泥挤入泄漏的接头或腐蚀孔眼，修复失效的套管；

（3）封隔漏失层；

（4）对一次注水泥作业不合格的情况进行补救；

（5）防止储层流体垂直流动至生产层（分段挤水泥）；

（6）防止流体从废弃层逸出。

挤水泥期间，水泥浆几乎不可能全部进入地层岩石的孔隙中，因为要实现这一点，岩石的渗透率需达到500D以上。挤水泥有两种方法：

（1）高压挤水泥。该技术要求地层破裂，然后将水泥浆压入裂缝中。

(2) Low pressure squeeze. During this technique the fracture gradient of the formation is not exceeded. Cement slurry is placed against the formation, and when pressure is applied the fluid content (filtrate) of the cement is squeezed into the rock, while the solid cement material (filter cake) builds up on the face of the formation.

8.3.6　Cement Plugs

At some stage during the life of a well a cement plug may have to be placed in the wellbore. A cement plug is designed to fill a length of casing or open hole to prevent vertical fluid movement. Cement plugs may be used for:

(1) Abandoning depleted zones;

(2) Seal off lost circulation zones;

(3) Providing a kick off point for directional drilling (eg side-tracking around fish);

(4) Isolating a zone for formation testing;

(5) Abandoning an entire well.

The major problem when setting cement plugs is avoiding mud contamination during placement of the cement. Certain precautions should be taken to reduce contamination.

(1) Select a section of clean hole which is in gauge, and calculate the volume required (add on a certain amount of excess). The plug should be long enough to allow for some contamination (500ft plugs are common). The top of the plug should be 250ft above the productive zone.

(2) Condition the mud prior to placing the plug.

(3) Use a preflush fluid ahead of the cement.

(4) Use densified cement slurry (ie less mixwater than normal).

After the cement has hardened the final position of the plug should be checked by running in and tagging the cement.

（2）低压挤水泥。使用该技术期间压力不会超过地层的破裂压力梯度。将水泥浆注入地层，施加压力时，水泥浆中的液体成分（滤液）被挤入岩石中，而固相水泥物质（滤饼）在地层剖面形成堆积。

8.3.6　水泥塞

在油井生产的特定阶段，可能需要在井筒中打水泥塞。水泥塞用于填充一段套管或裸眼井段，以防止流体垂直流动。水泥塞的作用如下：

（1）封隔废弃枯竭的产层；

（2）封隔漏失层；

（3）为定向钻井提供造斜点（例如在落鱼周围进行侧钻）；

（4）隔离测试地层；

（5）弃井作业。

打水泥塞的主要问题是在水泥注入期间避免钻井液污染。应采取以下预防措施减少污染。

（1）选择一段尺寸合乎要求的干净井眼，并计算打水泥塞所需的水泥浆体积（加上一定的余量）。水泥塞应足够长，以容许一些污染（常见的水泥塞长度是500ft）。水泥塞顶部应高出产层250ft以上。

（2）打水泥塞之前要调整钻井液性能。

（3）注水泥前使用冲洗液冲洗。

（4）使用较高密度的水泥浆（即混合水用量比正常情况下少）。

水泥硬化后，应通过下入探测标尺来检查水泥塞的最终位置。

8.3.7 Evaluation of Cement Jobs

A primary cement job can be considered a failure if the cement does not isolate undesirable zones. This will occur if:

(1) The cement does not fill the annulus to the required height between the casing and the borehole.

(2) The cement does not provide a good seal between the casing and borehole and fluids leak through the cement sheath to surface.

(3) The cement does not provide a good seal at the casing shoe and a poor leak off test is achieved

When any such failures occur some remedial work must be carried out. A number of methods can be used to assess the effectiveness of the cement job. These include:

8.3.7.1 Detecting Top of Cement (TOC)
1. Temperature Surveys (Fig.8.21)

This involves running a thermometer inside the casing just after the cement job. The thermometer responds to the heat generated by the cement hydration, and so can be used to detect the top of the cement column in the annulus.

2. Radioactive Surveys (Fig.8.22)

Radioactive tracers can be added to the cement slurry before it is pumped (Carnolite is commonly used). A logging tool is then run when the cement job is complete. This tool detects the top of the cement in the annulus, by identifying where the radioactivity decreases to the background natural radioactivity of the formation.

8.3.7.2 Detecting Top of Cement (TOC) and the Measuring the Quality of the Cement Bond

The cement bond logging(CBL) tools have become the standard method of evaluating cement jobs since they not only detect the top of cement, but also indicate how good the cement bond is. The CBL tool is basically a sonic tool which is run on wireline. The distance between transmitter and receiver is about 3 ft (Fig. 8.23). The logging tool must be centralised in the hole to give accurate results.

8.3.7 注水泥作业评价

如果水泥没有封隔需要隔离的地层，则认为一级注水泥作业失败，这可能在以下情况中出现：

（1）水泥没有填充到套管和井壁之间环空的预定高度。

（2）套管和井壁之间水泥密封不好，流体通过水泥环泄漏到地面。

（3）套管鞋处水泥的密封效果不佳，漏失实验也不理想。

上述任何故障发生时，必须要进行补救工作。评价注水泥作业效果的方法有很多，主要包括以下几类。

8.3.7.1 检测水泥返高（TOC）
1. 温度测量（图 8.21）

注水泥作业后在套管内下入测温计。测温计对水泥水化产生的热量作出反应，由此可以检测环空中水泥返高。

2. 放射性测量（图 8.22）

注入水泥浆之前，向水泥浆中添加放射性示踪剂。注水泥作业完成后，下入测井工具。通过检测放射性强度降低到地层天然放射性的位置，就可以确定环空中水泥的返高。

8.3.7.2 检测水泥返高（TOC）并测定水泥胶结质量

水泥胶结测井（CBL）不仅能检测水泥的返高，还能反映出水泥胶结的好坏，已经成为固井质量评价的标准方法。CBL工具本质上是一种声波测井工具，通过电缆下入，发射器和接收器之间的距离约为3英尺（图8.23）。测井工具必须在井筒中处于居中位置，才能得到准确的结果。

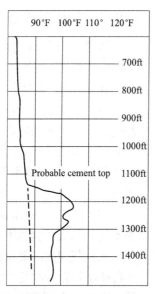

Fig. 8.21 Estimating top of cement in annulus by running a temperature log

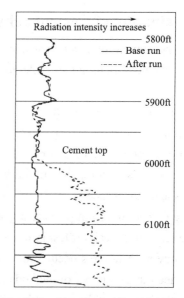

Fig. 8.22 Running radioactivity log

Both the time taken for the signal to reach the receiver, and the amplitude of the returning signal, give an indication of the cement bond. Since the speed of sound is greater in casing than in the formation or mud the first signals which are received at the receiver are those which travelled through the casing (Fig.8.24). If the amplitude (E1) is large (strong signal) this indicates that the pipe is free (poor bond). When cement is firmly bonded to the casing and the formation the signal is attenuated, and is characteristic of the formation behind the casing.

信号到达接收器所用的时间和返回信号的振幅都可以表明水泥胶结情况。由于声波在套管中的传播速度大于在地层或钻井液中的速度，因此接收器接收到的第一个信号是穿过套管的信号（图 8.24）。如果振幅（E1）较大（强信号），则表明套管未固定（胶结不良）。如果水泥将套管和地层牢牢胶结在一起时，信号会减弱，这是套管外地层的特征。

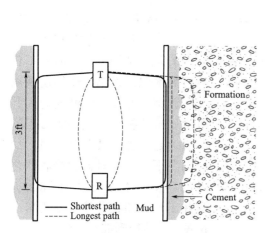

Fig.8.23 Schematic of CBL tool

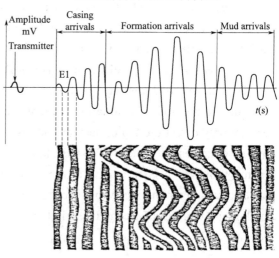

Fig. 8.24 Signals picked up by receiver

The CBL log usually gives an amplitude curve and provides an indication of the quality of the bond between the casing and cement. A VDL (variable density log), provides the wavetrain of the received signal (Fig.8.25), and can indicate the quality of the cement bond between the casing and cement, and the cement and the formation. The signals which pass directly through the casing show up as parallel, straight lines to the left of the VDL plot. A good bond between the casing and cement and cement and formation is shown by wavy lines to the right of the VDL plot. If the bonding is poor the signals will not reach the formation and parallel lines will be recorded all across the VDL plot.

CBL 通常给出振幅曲线，测量出套管和水泥之间的胶结质量。VDL（变密度测井）提供接收信号的波列（图 8.25），测量出套管与水泥、水泥与地层之间的水泥胶结质量。通过套管直接传来的信号在 VDL 图左侧显示为平行直线。VDL 图右侧出现波浪线表明套管与水泥、水泥与地层之间的胶结情况良好。如果胶结质量不好，信号就无法到达地层，在 VDL 图上显示的就是所有的曲线展现为平行线。

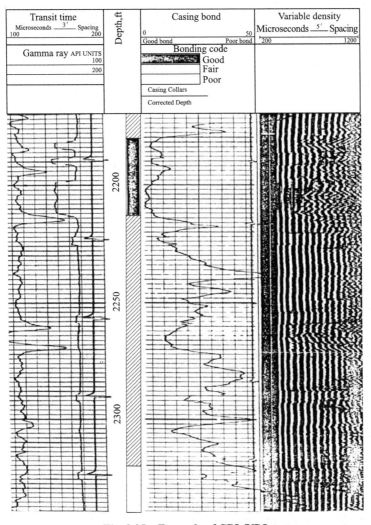

Fig. 8.25　Example of CBL/VDL

The interpretation of CBL logs is still controversial. There is no standard API scale to measure the effectiveness of the cement bond. There are many factors which can lead to false interpretation:

(1) During the setting process the velocity and amplitude of the signals varies significantly. It is recommended that the CBL log is not run until 24～36 hours after the cement job to give realistic results.

(2) Cement composition affects signal transmission.

(3) The thickness of the cement sheath will cause changes in the attenuation of the signal.

(4) The CBL will react to the presence of a microannulus (a small gap between casing and cement).The microannulus usually heals with time and is not a critical factor. Some operators recommend running the CBL under pressure to eliminate the microannulus effect.

8.3.8 Cementing Calculations

Audio 8.6

The following calculations must be undertaken prior to a cementation operation:

(1) Slurry Requirements;

(2) No. of sacks of Cement;

(3) Volume of Mixwater;

(4) Volume of Additives;

(5) Displacement Volume Duration of Operation.

These calculations will form the basis of the cementing programme. They should be performed in this sequence as will be seen below.

8.3.8.1 Cement Slurry Requirements

Sufficient cement slurry must be mixed and pumped to fill up the following (see Fig 8.26):

(1) the annular space between the casing and the borehole wall;

(2) the annular space between the casings (in the case of a two stage cementation operation);

(3) the openhole below the casing (rathole);

(4) the shoetrack.

CBL 解释还存在着争议。目前还没有一种权威的 API 标准来衡量水泥胶结的有效性。导致解释错误的因素有很多：

（1）在水泥凝结过程中，信号传播的速度和幅度变化很大，建议在注水泥作业 24～36 小时后再进行 CBL，以获得真实的结果。

（2）水泥的成分会影响信号传输。

（3）水泥环的厚度会引起信号衰减的变化。

（4）CBL 会对微环隙（套管和水泥之间的小间隙）有所反应。微环隙通常会随时间愈合，且不是一个关键因素。一些作业者建议在一定压力下下入 CBL 工具，以消除微环隙的影响。

8.3.8 注水泥计算

注水泥作业之前，必须完成以下计算：

（1）水泥浆用量；

（2）干水泥袋数；

（3）混合水体积；

（4）添加剂用量；

（5）顶替量。

这些计算将构成固井程序的基础。注水泥计算时应按照如下顺序进行。

8.3.8.1 水泥浆用量

固井时要混合足够的水泥浆，以便能填满以下所述的井内空间（图 8.26）：

（1）套管与井壁之间的环空；

（2）套管之间的环空（在双级注水泥作业时）；

（3）套管下方的裸眼（口袋）；

（4）引鞋。

The volume of slurry that is required will dictate the amount of dry cement, mixwater and additives that will be required for the operation.

In addition to the calculated volumes an excess of slurry will generally be mixed and pumped to accommodate any errors in the calculated volumes. These errors may arise due to inaccuracies in the size of the borehole (due to washouts etc.). It is common to mix an extra 10% ~ 20% of the calculated openhole volumes to accommodate these inaccuracies.

The volumetric capacities of the annual, casings, and open hole are available from service company cementing tables. These volumetric capacities can be calculated directly but the cementing tables are simple to use and include a more accurate assessment of the displacement of the casing for instance and the capacities based on nominal diameters.

In the case of a two stage operation (Fig. 8.27) the volume of slurry used in the first stage of the operation is the same as that for a single stage operation. The second stage slurry volume is the slurry required to fill the annulus between the casing and hole (or casing/casing if the multi-stage collar is inside the previous shoe) annular space.

所需水泥浆的体积将决定固井作业所需的干水泥、混合水和添加剂的用量。

为了应对水泥浆用量计算时可能出现的任何误差，一般会混合并泵入超过计算用量的水泥浆。计算的误差可能是井眼尺寸不准确（冲刷等）所致的。通常水泥附加量是裸眼井段体积的10% ~ 20%。

环空、套管和裸眼的体积容量可以从服务公司的固井台账中获得。虽然这些体积容量可以直接计算，但固井台账使用起来更简单，并包含了更准确的套管体积和基于公称直径的容量。

在双级注水泥作业中（图8.27），一级水泥浆体积与单级注水泥水泥浆体积相同。二级水泥浆体积将用于填充套管与井壁之间环空，或套管与套管之间的环空（如果分级箍位于上一层在套管鞋上方）。

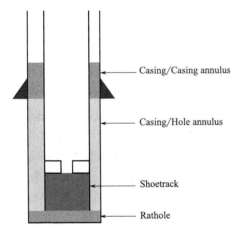

Fig. 8.26 Single stage cementing operation

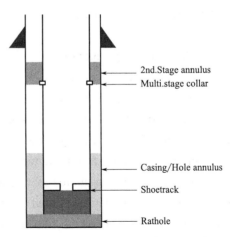

Fig. 8.27 Two-stage cementing operation

8.3.8.2 Number of Sacks of Cement

Although cement and other dry chemicals are delivered to the rig site in bulk tanks the amount of

8.3.8.2 干水泥袋数

尽管水泥和其他干化学剂通过散装罐输送到钻井现场，但干

dry cement powder is generally quoted in terms of the number of sacks of cement required. Each sack of cement is equivalent to 1 cu. ft of cement.

The number of sacks of cement required for the cement operation will depend on the amount of slurry required for the operation (calculated above) and the amount of cement slurry that can be produced from a sack of cement. The amount of cement slurry that can be produced from a sack of cement, known as the yield of the cement, will depend on the type of cement powder (API classification) and the amount of mixwater mixed with the cement powder. The latter will also depend on the type of cement and will vary with pressure and temperature. The number of sacks of cement required for the operation can be calculated from the following

No. of Sacks = Total Volume of Slurry /Yield of Cement

8.3.8.3　Mixwater Requirements

The mixwater required to hydrate the cement powder will be prepared and stored in specially cleaned mud tanks. The amount of mixwater required for the operation will depend on the type of cement powder used. The volume of mixwater required for the cement slurry can be calculated from:

Mixwater Vol. = Mixwater per sack × No. sxs

8.3.8.4　Additive Requirements

They are a variety of additives which may be added to cement. These additives may be delivered to the rigsite as liquid or dry additives. The amount of additive is generally quoted as a percentage of the cement powder used. Since each sack of cement weighs 94 lb, the amount of additive can be quoted in weight (lb) rather than volume. This can then be related to the number of sacks of additive. The number of sacks of additive can be calculated from:

Number of sacks of additive = No. of Sacks × % Additive

Weight of additive = No. sxs of Additive × 94

水泥灰的数量通常按所需水泥袋数计算。每袋水泥相当于 1ft^3 水泥浆。

注水泥作业所需的水泥袋数取决于作业中所需的水泥浆用量（如以上计算）及一袋水泥可产生的水泥浆量。一袋水泥可生产的水泥浆量，即水泥造浆量，将取决于水泥干灰的类型（API 分类）和与水泥混合的水量。混合水量也会取决于水泥的类型，并将随压力和温度而变化。注水泥作业所需的水泥袋数可用以下公式计算：

水泥袋数 = 总的水泥浆体积 / 水泥造浆量

8.3.8.3　混合水体积

干水泥灰所需的混合水会提前准备并储存在特别清洗的泥浆罐中。注水泥作业所需的混合水用量将取决于所使用的水泥类型。水泥浆所需的混合水体积可通过以下方法计算：

混合水体积 = 每袋水泥的混合水用量 × 水泥袋数

8.3.8.4　添加剂用量

添加到水泥中的各种添加剂可以是液体，也可以是干燥固体。添加剂的用量通常以所用水泥的百分比表示。由于每袋水泥质量为94lb，因此添加剂的用量可以用重量（磅）而不是体积来计算，这样就可与添加剂的袋数相关联。添加剂的袋数可由以下公式计算：

添加剂袋数 = 水泥袋数 × 添加剂百分比

添加剂质量 = 添加剂袋数 × 94

The amount of additive is always based on the volume of cement to be used.

8.3.8.5 Displacement Volume

The volume of mud used to displace the cement from the cement stinger or the casing during the cementing operation is commonly known as the displacement volume. The displacement volume is dependant on the way in which the operation is conducted.

1. Stinger Operation

The displacement volume can be calculated from the volumetric capacity of the cement stinger and the depth of the casing shoe. The cement is generally under displaced by 1～2 bbl of liquid.

Displacement Vol. = Volumetric capacity of stinger × Depth of Casing−1

2. Conventional Operation

In a conventional cementing operation the displacement volume is calculated from the volumetric capacity of the casing and the depth of the float collar in the casing.

Displacement Vol. = Volumetric Capacity of Casing × Depth of Float Collar

3. Two-stage Cementing Operation

In a two stage operation the first stage is firstly displaced by a volume of mud, calculated in the same way as a single stage cement operation described above. The second stage displacement is then calculated on the basis of the volumetric capacity of the casing and the depth of the second stage collar.

(1) 1st Stage :

Displacement Vol. = Volumetric Capacity of Casing × Depth of Float Collar

(2) 2nd stage :

Displacement Vol. = Volumetric Capacity of Casing × Depth of Multi stage collar

The amount of mud to be pumped during the displacement operation may be quoted in terms of a volume (bbl, cuft etc.) or in terms of the number of strokes of the mud pump required to pump the mud volume. It will therefore be necessary to determine the volume of fluid pumped with each stoke of the pumps (vol./stroke). The number of strokes required to

添加剂用量总是基于水泥用量来计算的。

8.3.8.5 顶替量

在注水泥作业期间，用于将水泥从注水泥插入管或套管中顶替出来的泥浆体积通常称为顶替量。顶替量取决于注水泥作业的实施方式。

1. 内管法注水泥作业

顶替量可以根据注水泥插入管的体积容量和套管鞋深度来计算。水泥浆一般会少顶替1～2bbl。

顶替量 = 水泥插入管的体积容量 × 套管深度 −1

2. 常规注水泥作业

在常规注水泥作业中，顶替量是根据套管的体积容量和套管内浮箍的深度计算出来的。

顶替量 = 套管的体积容量 × 浮箍深度

3. 双级注水泥作业

在双级注水泥作业中，一级的顶替量计算方法与上述单级注水泥作业相同。二级的顶替量则根据套管的体积容量和第二级接箍的深度来计算。

（1）一级：

顶替量 = 套管的体积容量 × 浮箍深度

（2）二级：

顶替量 = 套管的体积容量 × 分级接箍深度

顶替作业期间，泵送的水泥浆量可按体积（bbl、ft³等）表示，或者以泵入泥浆体积所需的泵冲数表示。所以，需要确定钻井泵一个冲程的泵出量（体积/冲程）。由此，顶替水泥浆所需的冲数可

displace the cement will therefore be calculated from:

Number of strokes = Volume of displacement fluid/ Vol. of fluid per stroke

8.3.8.6 Duration of Operation

The duration of the operation will be used to determine the required setting time for the cement formulation. The duration of the operation will be calculated on the basis of the mixing rate for the cement, the pumping rate for the cement slurry and the pumping rate for the displacing mud. An additional period of time, known as a contingency time, is added to the calculated duration to account for any operational problems during the operation. This contingency is generally 1 hour in duration. The duration of the operation can be calculated from:

Duration = Vol. of Slurry/ Mixing Rate + Vol. of Slurry/ Pumping Rate + Displacement Vol. / Displacement Rate + Contingency Time (1h)

8.3.8.7 Example of Cement Volume Calculations

The 9⅝ in Casing of a well is to be cemented in place with a single stage cementing operation. The appropriate calculations are to be conducted prior to the operation. The details of the operation are as follows (Fig. 8.28):

(1) 9⅝ in casing set at: 13800ft,

(2) 12¼ in hole: 13810ft;

(3) 13⅜ in casing set at: 6200ft;

(4) TOC outside 9⅝ in casing: 3000ft above shoe;

(5) Assume gauge hole, add 20% excess in open hole.

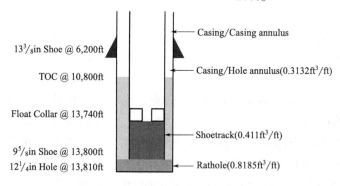

Fig. 8.28 Example of cementing calculation

The casing is to be cemented with class G cement with the following additives: (1) 0.2% D13R (retarder); (2) 1% D65 (friction reducer); (3) Slurry density=15.9lbm/gal.

剂); ② 1% D65（减阻剂）; ③水泥浆密度 15.9 lbm/gal。

(1) Slurry Volume Between The Casing and Hole:　　（1）环空水泥浆用量：

9⁵/₈in casing, 12¹/₄₈in hole capacity (9⁵/₈₈in 套管和 12¹/₄₈in 井眼间环空单位长度容积)	= 0.3132ft³/ft
annular volume（环空体积）	=3000×0.3132= 939.6ft³
plus 20% excess（20% 附加量）	=187.9ft³
Total（合计）	= 1127.5ft³ => 1128ft³

(2) Slurry Volume Below The Float Collar:　　（2）浮箍以下的水泥浆用量：

Cap. of 9⁵/₈₈, 47lb/ft csg （单位重量为 47lb/ft 的 9⁵/₈₈ in 套管单位长度容积）	= 0.4110ft³/ft
Shoe track volume（浮箍至浮鞋间容积）	= 60×0.411=24.66ft³
Total（合计）	= 25ft³

(3) Slurry volume in the rathole:　　（3）口袋处水泥塞水泥浆用量：

Cap. of 12¼ in hole （12¼ in 井眼单位长度容积）	=0.8185ft³/ft
Rat hole volume （口袋容积）	=10×0.8185=8.2ft³
plus 20% （20% 附加量）	=1.6ft³
Total（合计）	=9.8ft³ =>10ft³

(4) Total cement slurry volume:　　（4）总水泥浆用量：

$$V=1128 + 25 + 10= 1163\text{ft}^3$$

(5) Amount of cement and mixwater:　　（5）水泥和混合水的用量：

Yield of class G cement for density of 15.9 lbm/gal（密度为 15.9lbm/gal 水泥浆配浆率）	=1.14ft³/sk
mixwater requirements（单位混合水量）	=4.96gal/sk
No. of sks of cement（干水泥用量）	=1163/1.14=1020sx
Mixwater required（混合水量）	=1020×4.96gal= 5059gal = 120bbl

(6) Amount of Additives:　　（6）添加剂用量：

Retarder D13R (0.2% by weight)（缓凝剂，0.2%）	=(0.2/100)×1020×94 (lb/sk)= 192lb
Friction reducer (1.0% D65 by weight)（减摩剂，1.0%）	= (1/100)×1020×94(lb/sk)= 959lb

（7）Displacement Volume: （7）顶替量：

Displacement volume（顶替量）	= vol between cement head and float collar = 0.4110 × 13740 = 5647ft^3 = 1006bbl
add 2 bbl for surface line（地面管线增加 2bbl）	= 1008bbl

For Nat. pump 12-P-160, 7in liner 97% eff, 0.138bbl/stk

No. of strokes = 1008/0.138= 7300 strokes

8.4　Exercise

(1) List the functions of oilwell cement.

(2) List the major cement additives and their functions.

(3) Describe the process of the primary cementing opterations.

(4) When we need a secondary or squeeze cement job?

(5) How can we improve the efficiency of cementing?

(6) The 20 in casing of a well is to be cemented to surface with class C high early strength cement + 6% Bentonite using a stinger type cementation technique. Calculate the following for the 20" casing cementation:

① The number of sacks of cement required (allow 100% excess in open hole);

② The volume of mixwater required;

③ An estimate of the time taken to carry out the job.

(Note: use an average pumping time of 5 bbl/min.)

8.4　习题

（1）简述油井水泥的功能。

（2）简述水泥添加剂主要类型及其功能。

（3）描述初级固井作业的过程。

（4）何时需要二次固井或挤水泥作业？

（5）如何提高固井效率？

（6）20 in 套管需要用 C 级高早强度水泥 +6% 膨润土通过插入式固井方式封固到地面。计算下列固井过程：

① 所需的水泥袋量（允许裸眼段 100% 附加量）；

② 所需的混合水量；

③ 估算固井作业时间。

（注：平均泵送时间为 5bbl/min。）

30 in Casing:	0 ~ 400ft
20 in Casing 94lb/ft:	0 ~ 500ft
20 in Casing 133lb/ft:	500 ~ 1500ft
26 in Open hole Depth:	1530ft
Stinger:	5 in 19.5 in drillpipe
Class C Cement + 6% Bentonite	
Density:	13.1lbm/gal
Yield:	1.88ft^3/sk
Mixwater Requirements:	1.36 ft^3/sk

Drilling Risks 9
钻井事故与复杂井况

During drilling operations, some type of a drilling problem will almost certainly occur, even in very carefully planned wells. The reason is that geological conditions for two wells that are near each other may differ (nonhomogeneous formation); therefore, different problems can be encountered.

Audio 9.1

The key to success in achieving well objectives is to design drilling programs based on anticipation of potential hole problems. Drilling problems, when encountered, can be very costly, the most prevalent of which are: (1)Pipe sticking; (2)Lost circulation; (3) Hole deviation; (4)Pipe failures; (5)Borehole instability; (6)Mud contamination; (7)Formation damage; (8) Hydrogen sulfide–bearing formations and shallow gas; (9)Equipment and personnel related problems.

An understanding of these problems, their causes, their anticipation and planning for solutions is essential to control overall well cost control and succeed in reaching the intended target zone. This chapter addresses some of these problems definition, classification, possible solutions, preventive measures and when applicable.

9.1 Pipe Sticking

9.1.1 Definition and Classification

In drilling operations, a pipe is considered stuck if it cannot be freed (pulled out of the hole) without damage to the pipe and without exceeding the maximum allowed hook load of the drilling rig.

在钻井作业过程中，即便是进行过精心设计，仍不可避免地会发生一些钻井复杂情况。这主要是地层存在非均质性所引起的，即便是相邻的两口井，面临的地质条件也可能不同，因此不同的井会遇到不同的问题。

提前预测钻井潜在的复杂情况，并以此为基础进行钻井设计是确保钻井成功的关键。钻井过程中，一旦遇到钻井复杂情况，钻井成本就会增加，常见的钻井复杂情况有：（1）卡钻；（2）井漏；（3）井斜；（4）钻杆失效；（5）井壁失稳；（6）钻井液污染；（7）地层伤害；（8）含硫化氢地层和浅层气；（9）设备和人员相关的复杂情况。

了解上述钻井复杂情况及其发生原因和预防措施，对于控制钻井总成本、成功钻达目的层至关重要。本章详细论述了一些钻井复杂情况的定义、分类、预防措施及其适用条件。

9.1 卡钻

9.1.1 定义和分类

在钻井作业中，如果无法在不损坏钻具和不超过钻机最大允许钩载的情况下令钻具自由上下行（或从井眼中提出），则认为发

There are more ways to get stuck in a hole than there are words to describe the emotions of the driller after this happens. Just about any item that goes into a hole—including drillpipe, drill collars, casing, tubing, and downhole production equipment—can get stuck. This section reviews the most common ways of getting stuck in both open and cased holes. There are mainly the following types of pipe sticking problems. Differential pressure pipe sticking; Undergauge Hole Sticking; Sloughing-Hole Sticking; Keyseat Sticking; Sand Sticking and Mud Sticking; Cemented Sticking; Mechanical Sticking.

9.1.1.1 Differential Pressure Pipe Sticking

Differential-pressure sticking, often called differential sticking, is very prevalent in the drilling industry. Differential sticking causes most of the fishing operations that occur in the Gulf of Mexico. Differential pressure pipe sticking occurs when a portion of the drill string becomes embedded into a mud cake that forms on the wall of a permeable formation during the drilling process. Basically, the string is stuck against the side of the well because of a large pressure differential between the fluid in the borehole and that in the formation. Differential pressure pipe sticking occurs when friction forces in the wellbore acting on the drillstring in a normal direction exceed either the rig's ability to move the pipe or the strength of the pipe. Hydrostatic pressure creates a differential that forces the pipe into a filter cake across a permeable zone.

Formation-pressure increases above normal pressure—usually called abnormal pressure—require increased mud weights to control the high-pressure formations. Lower-pressure, uncased formations higher in the hole will also be exposed to these higher mud weights and, consequently, to increased pressure differentials. Pressure regressions can sometimes occur in deeper drilling intervals. At these levels, the formation pressure is receding, while the mud weight remains constant to control the high-pressure formations that have already been penetrated and remain uncased. With each newly drilled section, the tendency toward

生了卡钻。

卡钻对于司钻来讲是一件司空见惯却又十分麻烦的事情，几乎进入井筒的任何设备，包括钻杆、钻铤、套管、油管和井下生产设备，都有可能被卡住。发生在裸眼或套管井眼内最常见的卡钻方式，主要有以下几种类型：压差卡钻、缩径卡钻、坍塌卡钻、键槽卡钻、砂桥卡钻、水泥卡钻和机械卡钻。

9.1.1.1 压差卡钻

压差卡钻，又称黏吸卡钻，在钻井作业中十分普遍，是墨西哥湾地区大多数钻具打捞作业的主要原因。钻井作业过程中，当一部分钻柱嵌入在渗透性地层井壁上形成的滤饼中时，就会发生压差卡钻。这基本上是由于井筒内流体压力和地层孔隙压力之间的压差较大，井筒中沿法向作用在钻柱上的摩擦力会超过钻机提升能力或钻柱强度，导致钻柱被卡在井壁一侧。井筒内静液压力会产生压差，迫使钻柱进入可渗透区域的滤饼内。

高于正常压力的地层压力称为异常高压，对于高压地层需要提高钻井液密度来确保安全钻进。而此时上部的低压、裸眼地层也会同样暴露在高密度钻井液下，从而导致压差增大。低压地层不仅会在上部地层出现，有时也会发生在深部井段，在这一深度上地层压力正在下降，而为了平衡已钻井段和未下套管井段的异常高压，钻井液密度应保持不变，这样每钻进一段，压差和卡钻的

differential pressure and sticking can increase. Properly designed casing programs can significantly reduce stuck-pipe occurrences during changing pressure regimes.

Differential sticking occurs only across a permeable formation and, in fact, the higher the permeability, the higher the probability of differential sticking. As the drilling fluid moves across the permeable zone, it tends to lose its fluid phase to the permeable formation, leaving behind the solid phase. These remaining solids often settle out onto the side of the borehole. This nearly impermeable filter cake can become very thick. Meanwhile, if the hydrostatic pressure of the mud in the permeable zone is much higher than the formation pressure in the permeable zone, there will be a pressure gradient toward the formation across the borehole wall. If, by chance, the drillpipe or collars are lying in the filter cake (which is likely because all boreholes have some degree of deviation), a hydraulic seal can form. Now the pressure gradient lies across the string. Because filter cake has a high friction coefficient, the force required to pull the string tangentially across the filter cake is high. In many cases, the rig is not powerful enough to pull the string or the string is not strong enough to handle the load. Differential sticking is usually the problem if the drillstring cannot be moved up or down or rotated, yet circulation can be maintained.

If the mud pressure, p_m, that acts on the outside wall of the pipe is greater than the formation fluid pressure, p_p, as is generally the case except in underbalanced drilling, then the pipe is said to be differentially stuck, as shown in Fig. 9.1.

This sticking usually occurs, initially, only across a permeable zone such as sand, where friction resistance is a function of several variables. Forward operations are halted until the stuck pipe can be removed from the wellbore or a sidetrack hole can be drilled. Both of these options are costly in terms of both time and money.

The differential pressure acting on the portion of the drill pipe that is embedded into the mud cake can be expressed as follows: $\Delta p = p_m - p_p$. The pull force,

趋势都会增加。这种情况下，合理的井身结构设计可以显著地减少压力变化引起的卡钻事故。

压差卡钻仅发生在可渗透地层上，事实上，渗透率越高，压差卡钻的概率越大。当穿过渗透性地层时，钻井液的液相往往会滤失到渗透性地层内，留下固相附着在井壁上形成滤饼，这种几乎不透水的滤饼会变得很厚。同时，如果钻井液的静液压力远高于渗透性地层中的地层孔隙压力，则会在井壁上产生一个朝向地层的压力梯度。此时，如果碰巧钻杆或钻铤位于滤饼中（这可能是因为所有井眼都有一定程度的井斜），则可能形成液压密封。压力梯度施加在钻柱上，而滤饼的摩擦系数又很高，就会使将钻柱切向拉过滤饼所需的拉力变得很大。在很多情况下，钻机的提升能力或钻柱本身的强度不足，难以处理如此大的负荷。如果钻柱不能上下移动或旋转，但仍能保持循环，则通常意味着出现了压差卡钻。

如果作用在钻柱外壁上的钻井液压力 p_m 大于地层孔隙压力 p_p（欠平衡钻井除外），此时钻柱发生压差卡钻，如图9.1所示。

最初，这种黏吸通常只发生在砂岩等渗透性地层，且摩擦阻力是多个变量的函数。发生卡钻事故后，如果不通过解卡作业或侧钻新井眼来解除，钻井作业将无法继续进行。无论是解卡作业还是侧钻新井眼都需要大量的时间和金钱。

作用在嵌入滤饼钻杆上的压差可表示为：$\Delta p = p_m - p_p$。解卡所需的拉力 F_{pull} 是压差 Δp、摩擦系

F_{pull}, that is required in order to free the stuck pipe is a function of the differential pressure, Δp; the coefficient of friction, μ; and the total contact area, A_c, between the pipe and mud cake surfaces:

$$F_{pull} = \mu A_c \Delta p \tag{9.1}$$

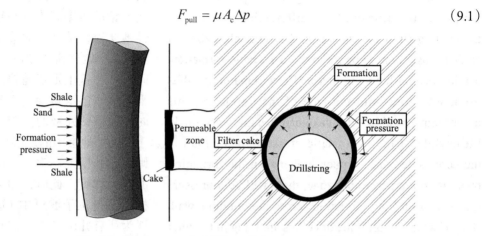

Fig. 9.1 Basic concept of differential pressure sticking

The coefficient of friction, μ, can vary from less than 0.04, for oil-based muds, to as much as 0.35, for weighted water-based muds with no lubricants added. The contact area, A_c, can be expressed in terms of the arc length, Ψ_{arc}, and length of the pipe body portion, L_{ep} (Fig.9.2) that are embedded in mud cake. The arc length, Ψ_{arc}, is given as follows:

$$\Psi_{arc} = 2\sqrt{\left(\frac{D_h}{2} - t_{mc}\right)^2 - \left(\frac{D_h}{2} - t_{mc}\frac{D_h - t_{mc}}{D_h - D_{op}}\right)^2} \tag{9.2}$$

where, D_h is the hole diameter; D_{op} is the outer pipe diameter; t_{mc} is the mud cake thickness.

In equation (9.2), D_{op} must be equal to or greater than $2t_{mc}$ and equal to or less than $D_h - t_{mc}$.

Thus, the contact area, A_c, can be expressed as follows:

$$A_c = \Psi_{arc} L_{ep} \tag{9.3}$$

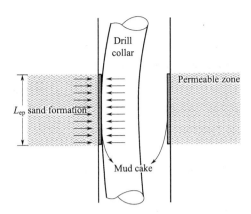

Fig. 9.2 Differential pressure pipe sticking with embedded pipe length

Equations (9.1)–(9.3) show controllable parameters that will cause higher pipe sticking force and the potential inability of freeing the stuck pipe. These are:

(1) Unnecessary high differential pressure;

(2) Thick mud cake (high continuous fluid loss to formation);

(3) Low-lubricity mud cake (high coefficient of friction);

(4) Excessive embedded pipe length in mud cake (time delay in freeing operations).

Although hole and pipe diameters and hole angle play a role in the pipe-sticking force, they are uncontrollable variables. Once they are selected to meet well design objectives, they cannot be altered. However, the shape of drill collars, as in square drill collars and drill collars with spiral grooves and external upset tool joints, can minimize the sticking force. Some of the indicators of differential pressure stuck pipe while drilling permeable zones or known depleted pressure zones are

(1) Increase in torque and drag;

(2) Inability to reciprocate the drill string and, in some cases, difficulty in rotating it;

(3) Drilling-fluid circulation is not interrupted.

Differential pressure pipe sticking can be prevented—or its occurrence can at least be mitigated—if some or all of the following precautions are taken:

(1) Maintain the lowest continuous fluid loss while adhering to the project economic objectives (e.g.,

式（9.1）至式（9.3）给出了导致粘滞力增大、解卡困难的可控参数：

（1）不必要的高压差；

（2）厚滤饼（地层持续高滤失量）；

（3）低润滑性滤饼（高摩擦系数）；

（4）卡入滤饼中的钻杆长度过大（解卡操作延时）。

虽然井眼直径、钻柱外径和井斜角对黏滞力也有影响，但它们是不可控的变量。一旦钻井设计选定，这些参数就不能改变。然而，钻铤的形状（如方形钻铤、螺旋钻铤和外加粗工具接头）可以减小黏滞力。在渗透性地层或压力枯竭地层钻进过程中，压差卡钻可以通过以下特征来识别：

（1）摩阻扭矩增大；

（2）钻柱无法往复移动，在某些情况下，钻柱难以旋转；

（3）钻井液循环没有中断。

采取以下预防措施，可以防止或至少可以减轻压差卡钻的发生：

（1）在坚持项目经济目标的同时，允许钻井液持续低量的漏

controlling the mud cake thickness);

(2) Maintain the lowest amount of drilled solids in the mud system or, if economical, remove all drilled solids;

(3) Use lowest differential pressure, with allowance for swab and surge pressures during tripping operations;

(4) Select a mud system that will yield smooth mud cake (low coefficient of friction);

(5) Maintain drill string rotation at all times, if possible.

Unsticking requires the reduction of the normal force, the coefficient of friction of the filter cake, or the hydraulically sealed area—or a combination of these. The sooner these methods can be undertaken, the greater the chance of success. One method used to unstick the string is to spot a lightweight fluid with a filter-cake-destroying chemical and then jar on the string. As stated earlier, differential pressure pipe sticking problems may not be totally prevented. Therefore, when the problem does occur, common field practices for freeing the stuck pipe include.

(1) Mud hydrostatic pressure reduction in the annulus;

(2) Oil spotting around the stuck portion of the drill string;

(3) Washing over the stuck pipe.

Some of the methods used to reduce the hydrostatic pressure in the annulus are

(1) Reduction of mud weight by dilution;

(2) Reduction of mud weight by gasifying with nitrogen;

(3) Placement of a packer in the hole above the stuck point.

9.1.1.2 Undergauge Hole Sticking

An undergauge hole is any hole that has a smaller diameter than the bit that drilled that section of hole. One potential cause of an undergauge condition is drilling a high-clay-content plastic shale with a freshwater mud.

失（如控制滤饼厚度）；

（2）如果经济可行，控制钻井液固相，确保钻井液中的固相含量最低；

（3）在考虑到起下钻时抽汲压力和激动压力的前提下，尽可能降低压差；

（4）选择合适的钻井液体系，产生光滑滤饼（降低摩擦系数）；

（5）如果可能，保持钻柱始终处于旋转状态。

解卡要求减小法向力、降低摩擦系数、减少液压密封面积，甚至要同时考虑这些方法，越早采用，成功的机会就越大。可以通过注入带有破坏滤饼化学剂的低密度钻井液，然后用震击器震击来进行解卡。如前所述，压差卡钻可能无法完全避免。因此，一旦发生卡钻，可以通过以下方式进行解卡：

（1）降低环空中钻井液的静液压力；

（2）在卡点位置注入油基钻井液；

（3）冲刷被卡的钻柱。

降低环空静液压力的方法有：

（1）通过稀释降低钻井液密度；

（2）通过注氮气降低钻井液密度；

（3）在卡点上方下一个管外封隔器。

9.1.1.2 缩径卡钻

缩径是指井眼直径小于本井段钻头直径的一种现象。产生缩径的一种情况是用清水钻井液钻进黏土含量高的塑性页岩地层，

If an oil-based mud is used, a plastic salt formation can "flow" into the wellbore. If the wellbore fluid has a hydrostatic pressure less than the formation pressure, shale or salt will slowly ooze into the wellbore. This process is slow, but can stick the drilling tools of the unwary.

An undergauge hole can also occur because a drill bit has been worn smaller while drilling through an abrasive formation. In this case, the hole is undergauge because the bit drilled it that way. If a new bit is run, it can jam into the undergauge section of the hole and become stuck. This is often called tapered-hole sticking (Fig.9.3). The presence of a thick filter cake as described above can also cause an undergauge hole. The filter cake can become so thick that tools cannot drag through it. The filter cake shows as a drag load on the weight indicator.

9.1.1.3 Sloughing-Hole Sticking

Sloughing-hole sticking occurs after a piece of the hole wall sloughs off. For example, water-sensitive shales that have been invaded by water will swell and break. If circulation is stopped, the broken pieces will collect around the drillstring and eventually pack the drillstring in place (Fig.9.4).

另一种情况是用油基钻井液钻进容易蠕变的盐膏层。如果井筒内的静液压力小于地层孔隙压力，页岩或盐将缓慢地挤入井筒。这个过程虽然缓慢，但可以在毫无警觉的情况下发生卡钻。

钻研磨性地层时，钻头由于磨损直径变小，会出现缩径。这种情况下，产生缩径是因为磨损后的钻头钻出的井眼变小。如果下入一个新钻头，可能会在井眼的缩径部分，发生卡钻，通常称为锥形孔卡钻（图9.3）。前面所描述的厚滤饼也会引起缩径卡钻。滤饼太厚，钻井工具无法通行，表现出来的是指重表上摩阻增加。

9.1.1.3 坍塌卡钻

坍塌卡钻是井壁坍塌崩落所引起的。例如，水敏页岩遇水会膨胀或破裂，如果此时钻井液停止循环，崩落的页岩碎片将聚集在钻柱周围，包住钻柱（图9.4）。

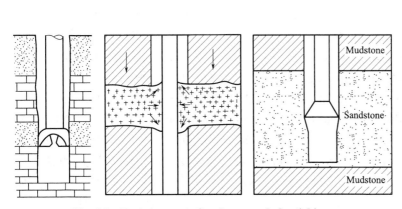

Fig. 9.3 Basic concept of undergauge hole sticking

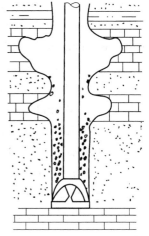

Fig. 9.4 Basic concept of sloughing-hole sticking

Shales under high formation pressure can slough as well. In this case, the formation pressure is greater

高压地层的页岩也会剥落。在这种情况下，地层孔隙压力大于井

than the wellbore hydrostatic pressure. Because shale has very low permeability, no flow is observed. The rock, which has a high pressure differential toward the wellbore, shears off the hole wall. The result can be seen as large cuttings on the shale shaker screen. If too much sloughing occurs, or if the wellbore is not cleaned properly, the drillstring can become stuck. More than likely, circulation will cease, and no string movement will be possible.

Steeply dipping and fractured formations can also slough into the hole. Drilling sites in over thrust belts are notorious for this problem. If there are cavities in the wellbore, cuttings can collect there. After circulation stops, the cuttings in the cavities may fall back into the hole.

9.1.1.4　Keyseat Sticking

In a deviated hole, or if ledges are present, the drillpipe can wear a slot into the borehole wall. This slot, called a keyseat, has essentially the same diameter as the drillpipe. While the drillstring is being pulled, the drill collars or bit will try to run through the keyseat. Because the diameter of the keyseat is smaller than that of the drill collars or bit, these tools become wedged in the keyseat. Circulation can be maintained in this situation. Of course, the usual response of a driller who sees the string start to stick is to pull harder. This exacerbates the situation, sticking the string even more solidly. Keyseat sticking usually occurs while the drillstring is being moved up the hole during a trip (Fig.9.5).

9.1.1.5　Sand Sticking and Mud Sticking

Sand sticking and mud sticking are similar. The sand particles or the solids in the mud can settle out of suspension. If there is little or no circulation, the rain of particles settles around the string, sticking the string in place. Sand sticking usually occurs in cased holes, although it can also occur in open holes. In cased holes, a leak can develop in the casing, enabling sand particles to flow into the well. The sand particles will then fall down and eventually pile up either on a packer or on some other restriction in the hole

筒内静液压力，由于页岩渗透率很低，流体流动受阻，由地层指向井筒方向的高压差作用在井壁岩石上，使得井壁产生剪切破坏，结果是在钻井液出口的振动筛上可以发现崩落的大岩屑。如果坍塌过多，或者井眼清洁不够，就可能会出现卡钻。很可能会出现钻井液失去循环，钻柱无法移动的现象。

高陡构造或裂缝性地层也会引起井塌。在逆冲带进行钻井作业更容易遭遇井塌。如果井筒中有空腔，岩屑会聚集在空腔里。当钻井液停止循环时，空腔内的岩屑就可能脱落到井筒中。

9.1.1.4　键槽卡钻

在斜井钻井作业过程中，或者在存在岩石台阶的地质情况下，钻杆可能会在井壁上磨出一个槽，这个槽称为键槽，其直径基本上与钻杆相同。当钻柱提起时，钻铤或钻头试图穿过键槽，但由于钻铤或钻头的直径大于键槽直径，便会楔入键槽。在这种情况下，钻井液依然可以循环，对司钻而言，发现开始卡钻的通常反应是上提遇阻。用力上提会加剧卡钻的进程，卡得更加牢固。键槽卡钻通常发生在起钻过程中（图9.5）。

9.1.1.5　砂桥卡钻

砂桥卡钻也叫沉砂卡钻，其中出砂引起的卡钻和钻井液固相沉积引起的卡钻相似，出砂的砂粒或钻井液中的固相颗粒从悬浮状态中沉积下来，如果钻井液失去循环或循环不畅，就会在钻柱周围沉淀，造成卡钻。砂桥卡钻通常发生在套管井中，也会在裸眼井中发生。套管可能存在割缝，砂粒会流入井内并下落，最终堆积在封隔器或

(Fig.9.6).

井筒的其他限制物上，造成砂桥（图 9.6）。

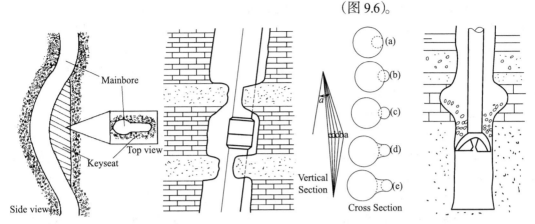

Fig. 9.5　Basic concept of keyseat sticking

Fig. 9.6　Basic concept of sand sticking and mud sticking

Mud sticking is similar. For whatever reason, the solids that form part of the mud can settle out of suspension. These solids can be barite particles or cuttings. In a high-temperature well, the mud can lose its fluid phase (filtrate), leaving the solids packed around the string. In addition, sometimes contaminants such as acids or salts can alter the mud properties leading to loss of the suspension properties of the mud.

Inadequate-hole-cleaning sticking can occur if the flow rate of the circulation fluid slows to the point that the solids carrying capacity of the circulation fluid is exceeded by the force of gravity. If the fluid is not viscous enough or flowing fast enough, the drag forces on the solids are less than the gravity forces. This means that the solids flow down the hole, instead of up and out of the hole.

The hole fills up with solids that build up around the string, eventually sticking the string. The flow rate can slow down for a number of reasons, including

(1) The driller may not be running the pumps fast enough;

(2) There could be a hole enlargement that slows the flow rate;

(3) The amount of solids may become overwhelming as a result of sloughing shales, unconsolidated formations,

钻井液固相沉积与此类似，不管出于什么原因，钻井液中的重晶石颗粒或岩屑等固相有时会从悬浮液中沉淀出来。在高温井中，钻井液会失去液相（滤失），使固相堆积在钻柱周围。此外，有时酸或盐等污染物会改变钻井液性能，导致钻井液失去悬浮性能。

如果钻井液的排量小于携岩能力的临界点，岩屑在重力作用下沉降会导致由于井眼清洁不充分产生的卡钻。如果流体黏度不够或流速过慢，作用在岩屑上的拖曳力小于重力，这意味着岩屑不会随钻井液向上流出井眼，而是向下沉积，致使井筒内钻柱周围堆满了岩屑，造成卡钻。排量降低的原因有很多，主要包括：

（1）钻井泵的泵速不够；

（2）井眼扩大导致排量降低；

（3）由于页岩坍塌、钻遇松散地层或发生井漏，致使钻井液

or lost circulation;

(4) There may be a rate of penetration that generates cuttings faster than the drilling fluid can carry.

9.1.1.6 Cemented Sticking

Cemented sticking can occur if the cement that is being circulated goes somewhere other than where it was intended to go (Fig.9.7). For example, if a cement plug was being spotted and the cement flowed higher up the string than anticipated, the cement could set before the string could be pulled out of the cement. The string is then stuck. If the cement is not too thick, the string could be jarred loose; otherwise, a washover operation is needed. The causes of cement sticking include

(1) Mechanical failures (e.g., string leaks);

(2) Human error (e.g., miscalculating a displacement or losing track of cement being used to remedy a blowout or a lost-circulation zone);

(3) An oversized hole.

9.1.1.7 Mechanical Sticking

This is a "catch-all" category for sticking problems. Any drilling or completion tool can become mechanically stuck (Fig. 9.8).

（4）机械钻速过快，产生的岩屑大于钻井液的携带能力。

9.1.1.6 水泥卡钻

如果水泥没有顶替到设计位置，则可能会发生水泥卡钻（图9.7）。例如，固井作业过程中，水泥塞水泥面的高度高出设计值，高过了钻柱底部，并且在钻柱提出前水泥已经凝固了，就会发生水泥卡钻。如果卡住钻柱的水泥不太厚，还可以通过震击解卡，否则的话只能通过套铣作业来解除。水泥卡钻的原因包括：

（1）机械故障（如钻柱泄漏）；

（2）人为错误（如顶替量计算错误或漏算了用于处理井涌或井漏的水泥量）；

（3）井眼过大。

9.1.1.7 机械卡钻

机械卡钻是一个"包罗万象"的事故类别，任何钻井或完井工具都可能被机械卡住（图9.8）。

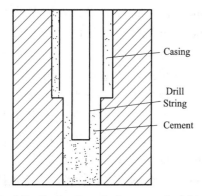

Fig. 9.7　Basic concept of cemented sticking

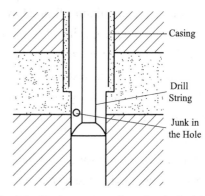

Fig. 9.8　Basic concept of mechanical sticking

(1) Packers. Sometimes the slips on a packer can become wedged so tightly against the casing that they cannot come free. In addition, retrieval failures can happen. In these cases, sometimes a high-force pulse of short duration can knock the packer loose.

(2) Multiple Strings. Multiple strings can jam in a

（1）封隔器：有时，封隔器上的卡瓦会紧紧地楔入套管，使其无法自由移动，也可能无法取回。这时短促的高压脉冲可能会松动封隔器。

（2）多管柱：同一井筒内有

hole. The two, three, or even four strings in the hole can rotate around each other as they are being run into the hole. The strings become intertwined, in which case they are notoriously difficult to retrieve.

(3) Crooked Pipe. If a drillstring is dropped in a mud-filled hole, the string can become permanently bent. This bend can wedge the string against the side of the hole, making it difficult to retrieve.

(4) Junk in the Hole. Junk in the hole is a description for small pieces of man-made materials that either are dropped down the hole or fall off of a downhole tool. Examples of items dropped down the hole include drill collar safety clamps, wrenches, and drillstring tools being made up in the rotary table. Items that can fall off of downhole tools include slips from packers, rubber drillpipe protectors, and (especially prevalent) cones off roller-cone bits. This debris can either fall to the bottom of the hole or can wedge against the side of the drillstring. If debris has wedged the string in the hole, then jarring can possibly knock it loose.

9.1.2 Recognizing the Pipe Sticking Problem

As noted, several types of pipe sticking may occur during drilling. Certain identification markers are peculiar to pipe sticking. Recognition of these markers will help in the decision to select the appropriate procedure to free the pipe.

Often, time is an important factor in determining the severity of a sticking problem. For example, in pressure-differential sticking, after the pipe becomes stuck, filtration continues to deposit solids adjacent to the pipe/mudcake interface and increases the contact area. In addition, filtration continues behind the pipe/mudcake interface. This ongoing deposition decreases the water content of the filter cake and increases its friction coefficient (Fig.9.9).

Two early warning signs of differential sticking are increased torque and drag. Both indicate that an increased frictional force is being encountered while either rotating or moving pipe vertically in the hole.

多管柱可能会发生卡钻，井筒内的两根、三根甚至四根管柱在进入井筒时可能会互相围绕旋转，缠绕在一起，这种情况下，起出这些管柱异常困难。

（3）钻柱弯曲：如果钻柱下入充满钻井液的井筒内，钻柱可能会永久弯曲，这种弯曲会将钻柱楔入井壁，发生卡钻。

（4）井内落物：井内落物是指从井下工具脱落或地面落下的小块人造材料。落物既包括钻铤安全钳、扳手和在转盘面现组装的钻柱工具，也包括从井下工具脱落的落物包括封隔器上的卡瓦、钻杆橡胶保护套和牙轮钻头上的牙轮（最为常见）。这些落物可能会落到井底，也可能会楔入钻柱侧面。如果发生落物卡钻，或许可以通过震击进行解卡。

9.1.2 卡钻的识别

如前所述，钻井过程中可能会发生不同类型的卡钻，每种卡钻都有其特定的识别特征，识别这些特征有助于选择合适的程序来进行解卡。

通常，时间是决定卡钻事故严重程度的一个重要因素。例如，在压差卡钻中，钻柱被卡后，滤失作用会持续让固体颗粒在钻柱与滤饼界面附近沉积，增大二者接触面积。同时，在钻柱与滤饼界面的持续滤失会降低滤饼的含水量，增加滤饼的摩擦系数（图9.9）。

压差卡钻的两个早期预警信号是摩阻和扭矩的增加。这两种情况都表明，钻柱无论是在井筒内旋转还是垂向移动，摩擦力都会增加。

These increases may indicate other drilling problems, but they are most often considered the early warning signs of differential sticking. When pipe is differentially stuck, there is no obstruction in the hole to prevent or retard mud circulation, as opposed to the case of a pipe stuck due to hole bridging or caving (Fig. 9.10).

尽管摩阻和扭矩的增加也可能会表明存在其他钻井复杂情况，但它们更多地被认为是压差卡钻的早期预警信号。当发生压差卡钻时，井筒内没有阻碍或延缓钻井液循环的障碍物，这和砂桥卡钻或坍塌卡钻有所不同（图9.10）。

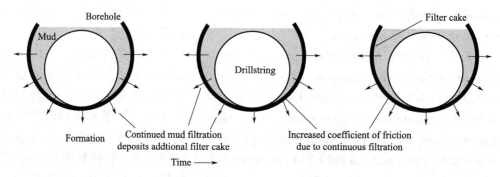

Fig. 9.9　After the pipe is stuck, filtration continues to deposit solids, which build the filter cake and increase the coefficient of friction behind the pipe because of decreased water content

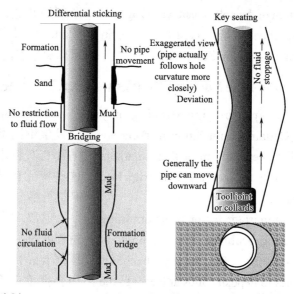

Fig. 9.10　Differential sticking poses no restriction to fluid flow as opposed to bridging; differentially stuck pipe cannot be moved in any direction, as opposed to keyseating

Continued fluid circulation while the pipe is stuck is a basic indicator of differential sticking. Another characteristic is the inability to rotate or move the pipe in either direction. This is the primary distinction from keyseating, in which the pipe becomes stuck as a result of prolonged wear in a dogleg. Although these two types of pipe sticking appear identical in most respects, pipe

压差卡钻时，持续的钻井液循环是一个基本指标，另一个特点是不能向任何方向旋转或移动钻柱。这是与键槽卡钻的主要区别。在键槽卡钻中，由于狗腿的长期磨损，钻柱被卡。尽管这两种类型的卡钻很多方面看起来都

stuck in a keyseat can usually be worked downward, which is not possible if the pipe is differentially stuck. Differential sticking usually occurs when the pipe has remained motionless in the hole for a prolonged period of time. Often pipe becomes stuck while making a joint connection.

Knowledge of the depth intervals at which sticking is more likely to occur is important in evaluating the remedial procedures to be followed. The intervals most likely to create differential sticking are those with high differential pressures. These can be divided into three categories: (1) Drilling through depleted reservoirs; (2) Pressure progressions; (3) Pressure regressions.

A common drilling situation with a high sticking tendency is drilling through a depleted reservoir, one that is not only hydrocarbon-depleted, but also pressure-depleted. Differential pressures in these cases can range as high as several thousand pounds, compared with a differential of only a few hundred pounds before depletion (Fig. 9.11).

一样，但键槽卡钻时钻柱通常可以向下活动，而对于压差卡钻而言这是不可能的。当钻柱在井筒内长时间静止不动时，就会发生压差卡钻。接单根作业，就容易经常发生卡钻。

了解卡钻更容易发生的井段特征对于评估拟采取的卡钻处理过程非常重要。最有可能产生压差卡钻的井段是压差较大的地层，可分为以下三类：(1)枯竭储层；(2)压力逐渐升高地层；(3)压力回落地层。

钻遇油气枯竭和压力枯竭储层时，最容易发生卡钻。这些情况下，压差由地层枯竭前的几百磅可能会上升到几千磅（图9.11）。

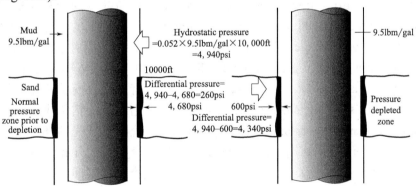

Fig. 9.11 Drilling through a depleted reservoir will create differential pressures greater than drilling the reservoir when it was originally normally pressured

9.1.3 Minimizing Sticking

Using the proper procedures for preventing stuck pipe in intervals that are expected to be troublesome can significantly reduce the number of stuck-pipe occurrences. Low-water-loss muds reduce the initial contact area because they produce a thin, hard filter cake. A thick, soft filter cake is associated with a high-water-loss mud. Pipe cannot be embedded as deeply in a thin

9.1.3 减少卡钻发生的方法

在预期会发生复杂事故的钻井井段，可以通过合理的作业程序预防卡钻，从而显著减少卡钻发生的次数。相比较于高失水泥浆则会产生厚而软的滤饼而言，低失水泥浆则会产生薄而硬的滤饼，从而减少钻柱与滤饼的初始接触面

cake and, therefore, the sticking force is reduced (see Fig. 9.12). Moreover, low-water-loss muds have a reduced filtration rate, which decreases the solids-deposition rate along the pipe/cake interface and minimizes the friction coefficient increase. Oil-based muds offer perhaps the best weapon against stuck pipe. Increased lubricity is most important. These muds develop little or no filter cake, resulting in minimum contact area for the pipe.

积，钻柱不会深埋在薄滤饼中，黏滞力也会降低（图9.12）。此外，低失水泥浆还可以降低滤速，从而降低沿钻柱与滤饼界面的固体沉积速率，降低摩擦系数。油基钻井液可能是防止卡钻的最好武器，可以增加润滑性，几乎不产生或产生很薄的滤饼，会使钻柱与滤饼的接触面积最小。

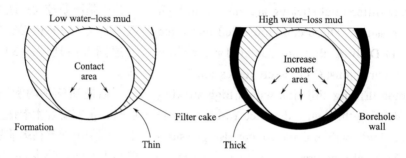

Fig. 9.12 Low-water-loss muds form a thinner, harder filter cake than do high-water-loss muds and are less susceptible to sticking

Drillstring alterations can reduce sticking tendencies by minimizing the pipe area in contact with the borehole wall. Commonly, stabilizers in the bottom hole assembly (BHA) are used to force pipe standoff. Another string alteration uses spirally grooved drill collars or heavyweight pipe instead of conventional or smooth pipe. This pipe has a spiraling, shallow but wide groove cut into the outer diameter over the entire length of the joint. Surface area is reduced by approximately 50%, while weight is reduced only by 4%.

改变钻柱的类型可以减少钻柱与井壁的接触面积，从而降低卡钻发生的可能性。常用的一种方法是用井底钻具组合（BHA）中的稳定器增加钻柱的刚度。还有一种改变钻柱结构的方式是使用螺旋槽钻铤或加重钻杆代替传统钻铤或光钻杆。加重钻杆在整个接头的外径上有一个螺旋形的浅而宽的凹槽，使其表面积减少约50%，而质量仅减少4%。

A field-developed procedure has been used successfully to minimize (temporarily) the mud friction coefficient along the borehole wall. Addition of walnut hulls or similar specialty products has been found to reduce friction by embedding hulls in the filter cake. This material seems to act like ball bearings for pipe. Although the friction reduction—for both drag and torque—is temporary, it usually alleviates the immediate rig site situation. The addition of bentonite to the mud system is another temporary measure for

油田现场通常会通过在钻井液中添加核桃壳或类似的特殊产品来成功地减少（或暂时减少）沿井壁的摩擦系数，嵌入滤饼的核桃壳类似于钻柱的滚珠轴承，可以减少钻柱与井壁的摩擦。尽管这种降摩方式对于钻柱的摩阻和扭矩而言是暂时的，但通常可以缓解钻井现场的紧急情况。在钻井液体系中加入膨润土是降低井壁摩阻系数的一种临

reducing the friction coefficient on the borehole wall. The hydration capabilities of the bentonite reduce sticking tendencies. This relief is also temporary because the water film is eventually lost through further filtration or is replaced with drilling solids with a higher friction coefficient.

9.1.4 Detecting Stuck Sections

This operation involves first determining where the string is stuck in the hole, and then determining the procedure needed to unstick the string. If a pipe string cannot be quickly worked or jarred free, the next step is to determine the free point, or the depth above which the string is free to move.

9.1.4.1 Free Pointing

An initial estimate involves taking a stretch reading to see how far the pipe moves in response to an applied tension. Although not as accurate as the wireline methods that may be used later, stretch readings provide a good first approximation of the free point. Moreover, if the drill string is plugged, stretch readings may be the only way of obtaining this information.

A procedure for obtaining a stretch reading is as follows:

(1) Calculate the buoyant weight of the drill string, including the blocks (or use the hook load recorded immediately before sticking occurred).

(2) Pick up the pipe to this weight and make a mark on the pipe at the rotary table.

(3) Apply a predetermined amount of over pull to the drill string.

(4) Make another mark on the pipe at the rotary table.

The length between these two marks represents the amount of stretch in the pipe.

Once a stretch reading has been recorded, there are two ways of estimating the free point.

The first method uses the relationship

$$L = \frac{\Delta LAE}{F}$$

(9.4)

where, L is the free-pipe length, ΔL is the stretch length, A is the cross-sectional pipe area, E is the modulus of elasticity, and F is the tension load applied.

A simpler approximation of this equation is

$$L = 735{,}294 \frac{\Delta L W_p}{F} \tag{9.5}$$

where, W_p is the unit weight of pipe, lbf/in. $W_p = 2.67$ ($OD^2 - ID^2$), OD is outside diameter in inches, ID is inside diameter in inches, and the pipe stretch constant, K, shown in Table 9.1 can be used with Eq. 9.6

$$L = \frac{K}{F} \tag{9.6}$$

式中，L 是卡点上方钻柱长度，ΔL 是钻柱伸长量，A 是横截面积，E 是弹性模量，F 是施加的拉伸载荷。

更简单的近似计算公式为：

式中，W_p 是钻柱的单位长度重量，lbf/in，$W_p = 2.67$（OD^2-ID^2），OD 是以英寸为单位的外径，ID 是以英寸为单位的内径。表 9.1 中所示的钻柱拉伸常数 K 可与式（9.6）一起使用：

Table 9.1 Constants used to calculation the free point

	Tubing		Drill pipe		
	2 in	2.5 in	2.87 in(10.4 lbm/ft)	3.5 in(13.3 lbm/ft)	4.5 in(16.6 lbm/ft)
K	3,250,000	4,500,000	7,000,000	8,800,000	10,800,000

Although this method is fast, it is not particularly accurate. It can provide an answer to within two or three joints. If the string is to be backed off, a more accurate answer is needed. In addition, if there is more than one type of pipe in the string, the calculations become more complicated. Moreover, if the hole is deviated or doglegged, the drag from the string rubbing against the hole wall may preclude any stretching of the string below that point.

9.1.4.2 Freepoint Tool

The freepoint tool (Fig. 9.13) is far more accurate than the stretch method; however, it requires that a wireline tool be run inside the drillstring. The freepoint tool consists of a set of strain gauges and spring-loaded drag blocks or electromagnets that rub against the inside of the string. As the tool is run into the string, torsion or tension is applied to the string. The degree of pipe movement resulting from the

虽然这个方法很快，但不是特别准确，两个或三个单根内尚可提供参考。如果要提出整个钻柱，则需要更准确的答案。此外，如果钻柱中有多种类型的钻杆，计算会变得更加复杂。如果存在井斜或井眼弯曲，则钻柱与井壁接触摩擦产生的阻力会妨碍接触点以下被卡钻柱的拉伸。

9.1.4.2 卡点测量工具

卡点测量工具（图 9.13）比拉伸法精确得多；然而，它要求在钻柱内部使用电缆。卡点测量工具由一组应变计和弹簧加载的阻力块或电磁铁组成，这些阻力块或电磁铁会与钻柱内部摩擦。当卡点测量工具下入钻柱卡点以上时，对钻柱进行扭转或拉伸，卡

application of this torsion or tension is transmitted to the surface through the wireline. After the tool moves below the stuck point, no movement of the string will be detected.

点测量工具可以检测到因扭转或拉伸迫使钻柱产生移动的程度，并通过电缆传递到地面。当卡点测量工具移动到卡点以下时，将无法检测到钻柱的移动。

Fig. 9.13 Free point indication tool

9.1.5 Procedures to free a differentially stuck drillstring

After a drillstring becomes differentially stuck, three release techniques may be used: (1) Spotting fluids; (2) Hydrostatic reductions; (3) Mechanical methods.

Selection of the appropriate procedure is based on an evaluation of the factors that created the problem in the first place, the time elapsed since initial sticking, mud types and properties, and other issues.

9.1.5.1 Spotting Fluids

A spotting fluid is any type of fluid used to cover a section of a well for any reason. Those used for stuck pipe are usually oil-based products positioned or spotted in an open hole to cover a specified interval (Fig. 9.14).

The oil penetrates the filter cake and invades the seal on the drillpipe. In addition, the oil tends to wet the circumference of the pipe, creating a thin layer between the pipe and the mud cake (Fig. 9.15). This reduces the coefficient of friction and may enable the pipe to be pulled free.

9.1.5 压差卡钻解卡程序

钻柱发生压差卡钻后，可采用以下三种技术解卡：（1）浸泡解卡；（2）降低静液压力；（3）机械方法。

解卡作业程序的选择要基于卡钻致因因素、卡钻时长、钻井液类型和性质及其他影响因素的综合评估。

9.1.5.1 浸泡解卡

浸泡液是指出于某种原因用来浸泡某一井段的流体。用于解卡的浸泡液通常是油基流体，以便在裸眼中浸泡卡钻的井段（图9.14）。

油渗入滤饼，侵入钻杆与滤饼密封区域。同时，油往往会润湿钻柱圆周，在钻柱和滤饼之间形成一层薄薄的油膜（图9.15）。这将会降低摩擦系数，并可能解卡。

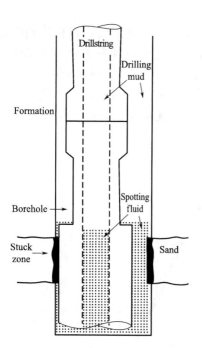
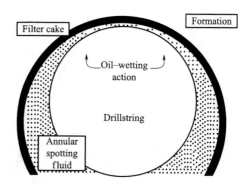

Fig. 9.14 Spotting fluids are usually oil-based muds positioned in the hole to cover a particular interval

Fig. 9.15 Oil has a tendency to wet the circumference of the pipe, reducing friction

Spotting-fluid density is important. Hydrocarbon fluids that are less dense than drilling fluids will migrate or float to the surface. The reverse is true with oil weighted to a density greater than that of the original drilling fluid. To ensure that the spotting fluid will remain where it is placed in the hole, its density should be approximately the same as the mud density. One exception is when pipe is known to be stuck at the bottom of the hole. The spotting- fluid density should then be slightly greater than the mud density to ensure complete encirclement of the pipe in spite of gravity segregation.

One potential problem with weighted spotting fluids is that the mixing time for the chemicals and the barite in oil may appear limited because of the perceived need to spot the fluid as quickly as possible. If the fluid is not mixed properly, the properties necessary to support the barite in suspension may not develop. This can result in barite settling that forms a bridge plug in the pipe or the open hole. Proper fluid mixing is necessary, even if it takes a little longer.

确定浸泡液密度很重要。密度小于钻井液的烃类浸泡液将迁移或浮到地表。反之，当浸泡液的密度大于原始钻井液的密度时，浸泡液将下沉至井底。为确保浸泡液留在井筒内的指定位置，其密度应与钻井液密度大致相同。当卡钻发生在井底时情况有所不同，浸泡液密度应略大于钻井液密度，以确保在重力分离的情况下能完全包围钻柱。

加重浸泡液的一个潜在问题是，添加剂和重晶石在浸泡液中的混合时间可能有限，因为解卡需要尽快浸泡。如果浸泡液未能正确混合配置，可能无法达到支撑悬浮重晶石所需的性能，就会导致重晶石沉降，从而在套管或裸眼中形成桥塞。虽然需要一些配置时间，但是配置满足一定性能的浸泡液配置十分必要。

Spotting-fluid success is directly related to the volume used. Larger volumes cover a longer section of the open hole and are therefore more likely to cover the stuck intervals. It is a mistake to assume that the pipe is stuck only in the drill-collar region and to use only enough fluid volume to cover the collars. This conservative tendency in spotting-fluid use probably contributes to a large number of failures. Minimum spotting-fluid volumes can be determined at the rig site. Using the pipe-stretch calculations described earlier, the uppermost stuck interval can be identified. Once this is done, a sufficient fluid volume should be used to cover this section and all lower open-hole sections.

Pipe sections higher in the hole often become stuck while efforts are being made to free lower sections (Fig. 9.16). It is often beneficial to spot enough fluid volume to cover all exposed permeable zones. Although more expensive initially, this technique may be the most economical procedure overall.

浸泡解卡的成功与否与浸泡液用量直接相关。足够量的浸泡液可以覆盖较长的裸眼段，就有可能覆盖被卡住的层段。如果认为仅卡在钻铤区域，浸泡液用量仅够覆盖钻铤的话，这是错误的做法。这种保守的浸泡液使用方法可能导致更多的问题。最小浸泡液用量可以在井场确定。使用前面描述的钻柱拉伸法计算，可以确定卡段最上端的位置，进而确定足够覆盖该部分和下部裸眼段的浸泡液用量。

当在下部井段竭尽全力解卡时，上部井段也可能会发生卡钻（图9.16）。使用足够的浸泡液浸泡所有暴露的渗透带十分有益。虽然开始比较昂贵，但总体上可能是最经济的方法。

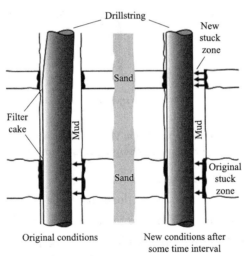

Fig. 9.16 Sections above the original stuck section often become stuck during the time spent trying to free the pipe

After spotting the fluid in the hole, time is required before the pipe can be released. The amount of time depends on a number of factors such as mud properties, mud-displacement efficiencies, hole-to-pipe geometry, and differential pressure. Even though the time cannot be quantified with any degree of precision, raw field data suggest that an average of eight to ten hours is required

浸泡液替入井筒后，需要浸泡一段时间才能解卡。浸泡时间长短取决于钻井液性能、钻井液驱替效率、井眼和钻柱几何形状、压差大小等诸多因素。尽管浸泡时间无法精确量化，但粗略的现场数据表明，平均需要浸泡

for release. Many cases, however, have taken longer. A good field rule is to spot enough fluid to cover the openhole section and wait for at least 12 hours for the fluid to free the pipe.

Correct positioning of the spotting fluid is of critical importance. The volume pumped must be recorded using number of pump strokes or the rig's trip tanks. Fluid should be spotted in the open hole with a volume left in the drillpipe (Fig. 9.17). At specific time intervals, small amounts of fluid must be displaced from the pipe to create annular movement. This may increase fluid effectiveness as well as minimizing potential bridges.

8~10小时才能解卡，甚至更长的时间。一个很好的现场规则就是替入足够的浸泡液覆盖裸眼段，并等待至少12小时，浸泡解卡。

浸泡液位置至关重要，必须利用泵冲数或钻机的计量罐记录浸泡液泵入量。浸泡液需要泵入裸眼中，同时在钻杆中留有一定体积（图9.17），以便按照特定时间间隔，继续从钻杆内将浸泡液少量泵入环空，使得环空内浸泡液产生运动。这既可以提高流动效率，又能减少潜在的桥堵。

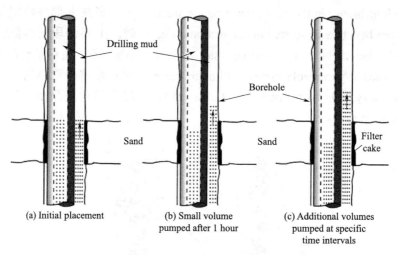

Fig. 9.17 Fluid should be spotted in the open hole with some volume left in the drillpipe for pumping at later intervals

While displacing fluid and waiting for the pipe to release, it is important to maintain a rig hook load that is equal to or slightly less than the load before the pipe stuck. Do not hold extra pull on the pipe while using spotting fluid. Extra surface pull does not increase the chemical effectiveness of the spotting fluid, and it places abnormal stresses on the pipe. Pipe should be raised and lowered as conditions permit and at specific intervals, perhaps hourly, to check whether it can be released. Lowering some weight onto the pipe for a short time can also assist in releasing the stuck pipe from the filter cake. If the pipe is not yet free, release the extra surface pull and allow additional time for the fluid to work.

在置换浸泡液并等待解卡时，很重要的一点是保持钻机大钩载荷等于或略小于卡钻前的载荷。浸泡解卡时，不要对钻柱施加额外的拉力。额外的拉力不会增加浸泡液的化学效力，反而会在钻柱上产生异常应力。在条件允许的情况下，可以按特定的时间间隔（可能是每小时）上下活动钻杆，以检查是否解卡。短时间在钻柱上施加一些力有助于钻柱从滤饼中解卡，如果钻柱尚未自由移动，则可继续释放钻柱重量，并留出一定时间让浸泡液产生效果。

Spotting fluids may be used effectively for preventive as well as remedial purposes. For instance, fluid may be spotted before running casing or tubulars into a well where a sticking potential exists.

9.1.5.2 Hydrostatic Pressure Reduction

Reducing differential pressure is another technique for releasing stuck pipe. Lowering the differential pressure reduces the restraining force on pipe, and it may become possible to pull it free. Reducing hydrostatic pressure, however, may create problems such as kicks or sloughing in other hole sections. Circulating a lower-density fluid reduces hydrostatic pressure. Another common procedure is a localized pressure reduction to reduce hydrostatic pressure on and below the stuck interval while maintaining full hydrostatic pressure above.

Localized pressure reduction uses stuck-zone detection procedures, conventional backoff procedures, and drill- stem-testing (DST) tools (Fig. 9.18). After the stuck interval has been identified, the pipe above that section is unscrewed and pulled from the hole, and a DST tool is attached. The pipe is rerun and screwed into the fish. Water is displaced into the drill pipe to reduce hydrostatic pressure to a precalculated value. The test tool is opened and the pipe is pulled free.

This technique is not universally applicable in all drilling environments. The pressure reduction below the DST tool can create potentially large negative differential pressures. Kicks can occur, or the hole can collapse. The technique is better suited for hard, low-permeability rocks.

浸泡液可有效用于预防和处理卡钻。例如，在将套管或油管下入存在卡钻风险的油井之前，可先注入一些浸泡液。

9.1.5.2 降低静液压力

另一种解卡技术是减小静液压力。通过降低压差减小钻柱上的上提下放阻力，可能会实现解卡。然而，降低静液压力也可能会在其他井段造成井涌或井壁坍塌等复杂情况。通过降低循环流体密度可以降低静液压力，还有一种常见的方法是局部降压，即降低卡段及卡段以下井段的静液压力，同时在卡段以上维持正常静液压力。

局部降压法通过卡段识别、常规倒扣和钻柱测试（DST）工具来实现（图9.18）。确定卡段后，倒扣并提出卡段上方的钻柱，然后连接DST工具，重新下入并对扣连接到井底的落鱼中。用水顶替DST工具上方钻杆内的钻井液，将静液压力降低到预先计算值，打开DST测试工具并连同底部钻柱一起提出。

这种方法并不适用于所有钻井环境，DST工具以下井段的压力降低可能会产生较大的负压差，从而导致井涌或井塌。所以这项技术更适合用于坚硬、低渗透地层。

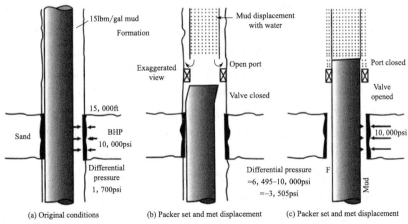

Fig. 9.18 Localized pressure reduction is possible using conventional test tools

9.1.5.3 Mechanical Methods

Mechanical methods physically destroy the bond between the pipe and the filter cake. They are based either on impact loading using jarring devices or on destruction of the cake by grinding procedures. Another method is to wash over the pipe—in essence, coring over the stuck pipe.

With the drillstring out of the hole, the fishing tools are made up. A fishing string with a jar is often called a jarring string. A typical jarring string consists of an overshot or screw-in sub, drill collars, a jar, perhaps an accelerator, perhaps a bumper sub, and drillpipe. The makeup of jarring strings varies considerably and depends on the fish and the amount of jarring force needed. There are no hard-and-fast rules for making up a jarring string. The amount of impulse, force, and energy developed and applied by the jar to the string is highly dependent on the makeup of the jarring string.

1. Jarring

Jarring is simply the process of impacting the fish with a large force impulse. This is not unlike hitting a stuck item with a hammer. The same phenomenon is true when using jarring to fish for stuck tools. In this case, the hammer is called a jar. The jar is placed in the drillstring in a position where it can apply a hammer blow to the fish. This is accomplished in the following manner. The string is stretched, putting strain energy into the string above and below the jar. The amount of tension put into the string over and above the weight of the string above the jar is called the overpull. At some predetermined load value, the jar is triggered. The top and bottom parts of the jar disconnect from each other and are free to travel up for the top part (called the hammer) and down for the bottom part (called the anvil). Both parts of the string contract at what is known as the free contraction velocity and build kinetic energy. Eventually, after the anvil and hammer have traveled a certain distance (called the stroke), the hammer and anvil will impact. Most of the kinetic energy is converted back into strain energy that then propagates up and down the string. Some of this energy will propagate to the stuck point and hopefully jar the fish loose. The magnitudes of the force, energy, and impulse involved depend on the initial strain energy, the stroke length, and the wave-propagation characteristics of the jarring string. With each hammer blow, this wave propagates to the stuck point to provide a short-duration (milliseconds) pulse of force. Eventually the fish will come loose. The bad news is that this may take days or weeks. At some point, it may be more economical to abandon the hole and drill a new one.

2. Washover Pipe

Wash pipe is perhaps the most widely used procedure to free severely stuck pipe. A wash pipe is a large-diameter, thin-walled tubular with a grinding shoe on the bottom. The shoe is used to destroy the bond between the pipe and the cake as the wash pipe is rotated and lowered. After a specific length has been freed in this manner, the previously stuck section can be mechanically unscrewed and retrieved. This procedure is repeated until the entire drill string has been recovered.

A common occurrence when using wash pipe is that it also becomes stuck. Its large OD then prevents further retrieval operations. Sidetracking around the fish (or fishes) is the only remaining viable option. Retrieval presents several problems. Because the borehole is known to have sticking tendencies, the danger of stuck wash pipe is especially great because of the large outer surface area of the pipe. Occasionally, it is impossible to free and remove all pipe in the hole completely. When this occurs, it becomes necessary to sidetrack.

3. Backoff Procedures

After the stuck point has been found, the method of recovery must be determined. Often, the string is broken just above the stuck point, and a jarring string is run into the hole. The backoff procedure, as this is called, involves unscrewing or cutting the string above the stuck point. Unscrewing the string is the preferred method

分动能被转换成应变能，然后沿钻柱上下传播。有一些能量会传播到卡点，可能会把落鱼震松。所涉及的力、能量和冲量的大小取决于震击管串的初始应变能、冲程长度和震动波传播特性。每次震击时，震动波传播到卡点，提供持续时间较短（毫秒）的脉冲力，最终落鱼会被震击解卡。但是这可能需要几天或几周的时间，在某种程度上，放弃原井眼，侧钻一个新井眼可能更经济。

2. 铣管

对于比较严重的卡钻，使用铣管进行解卡是最常见的方式。铣管是一种底部带有磨鞋的大直径薄壁管。磨鞋的作用是在铣管旋转和下行时，破坏钻柱和滤饼之间的黏结。按照这种方式套铣一定长度后，可以通过机械方式松扣并取回套铣段以上的被卡钻柱，重复此步骤，直到整个钻柱解卡。

常见的情况是，套铣解卡时，铣管也可能会被卡住。铣管外径较大，令进一步打捞作业变得困难。在落鱼附近侧钻是唯一可行的选择。打捞存在几个问题，由于井眼存在卡钻风险，而铣管的外表面面积较大，卡钻的危险性就更大。有时，不可能完全解卡井筒中的所有钻杆，出现这种情况，就有必要侧钻。

3. 倒扣程序

找到卡点后，必须确定解卡方法。通常情况下，钻柱恰好在卡点上方断开，将震击管串下入井眼。倒扣是指松扣或切断卡点以上的钻柱。松扣是优先考虑的方法，因为这样可以保持钻柱完

because it leaves the string intact. Breaking the string involves explosive, chemical, or mechanical cutting of the metal.

To unscrew a string that is stuck, a string shot is run into the hole. A string shot is a small amount of explosive. The tool joint that is to be unscrewed is found using a collar locator. Then the string shot is run into the middle of the inside of the tool joint. The driller then applies torque and tension to the string. The amount of torque should be sufficient to unscrew the string after the shot, but not before. The string shot is then exploded. The resulting torque in the string should unscrew the string at the explosion point. The approach is similar to hammering a reluctant screw. If all goes well, which it often does not, the string should come loose at that point. The string is then pulled out of the hole, leaving the fish stuck in the hole.

9.1.6 Economics of Avoiding or Freeing Stuck Pipe

Previous discussions have covered various efficient procedures to avoid or free stuck pipe. The economics of each option should be a primary factor when considering various preventive and remedial procedures. From an economic viewpoint, it is better to prevent differential sticking than to be forced to use remedial measures. However, when remedial efforts are necessary, the cost of large volumes of spotting fluids is small compared to rig costs. The use of wash over techniques should be based on economics. Drill pipe recovery "at any cost" should be replaced with sound economic judgment. Some confusion exists as to which techniques are best to retrieve pipe when spotting fluid fails. Economics are the controlling factor at this point, and the drilling supervisor must decide whether it is less expensive to fish the stuck pipe or simply to sidetrack and redrill the interval.

好无损。切断方式包括爆炸、化学方法或机械切割。

如果要对被卡钻柱进行松扣，就需要下入松扣炸药包，松扣炸药包含有少量的炸药。首先使用接箍定位器找到需要进行松扣的工具接头位置，其次将松扣炸药包下入工具接头内的中间位置，然后司钻在钻柱施加扭矩和拉力，扭矩应足够可以在松扣炸药包爆炸后（而不是爆炸前）进行松扣，接下来引发松扣炸药包，施加在钻柱上的扭矩应在爆炸点处使钻柱松扣。这种方法类就像用力钉一个螺钉一样，往往很难成功。如果一切顺利，那么此时钻柱就应该在卡点位置松扣，然后把松扣的钻柱从井筒内提上来，落鱼留在井底，但事实往往并非如此。

9.1.6 避免卡钻或解卡的经济评价

前面已经讨论了各种预防和处理卡钻的有效措施。在考虑这些措施时，每种措施的经济性应该是一个主要因素。从经济角度看，防止压差卡钻要比采取处理措施进行解卡要好。然而，当需要采取处理措施时，与钻机成本相比，大量解卡液的成本就显得小很多。使用套铣技术解卡应以经济性为前提，要以合理的经济判断取代"不惜一切代价"地打捞钻杆。当泡解卡液失败时，哪种技术最适合打捞钻杆难以抉择。在这点上，经济性是决定因素，钻井监督必须要做的选择是以较低的成本来打捞被卡钻柱，还是简单地侧钻并重新钻进一段。

Altering the drillstring configuration is an easy and inexpensive technique to avoid stuck pipe. Minimizing the contact area between the pipe and the wellbore can be accomplished using spirally grooved collars and heavyweight drillpipe. Fig. 9.19 shows several example BHAs which have the same buoyed weight. The assemblies in Figs. 9.19b and 9.19c effectively reduce contact area without sacrificing available bit weight.

改变钻柱结构是避免卡钻的一种简单而廉价的技术。可以通过使用螺旋槽钻铤和加重钻杆减少钻柱与井壁之间的接触面积。图9.19给出了几个具有相同浮力的BHA示例。图9.19（b）和图9.19（c）所示的钻具组合在不牺牲钻头有效重量的情况下有效地减少了接触面积。

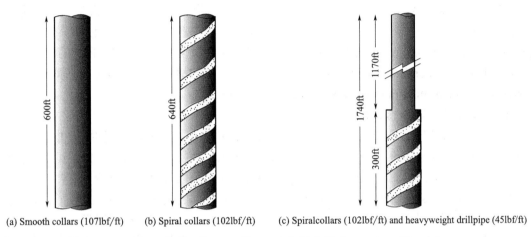

(a) Smooth collars (107lbf/ft)　　(b) Spiral collars (102lbf/ft)　　(c) Spiralcollars (102lbf/ft) and heavyweight drillpipe (45lbf/ft)

Fig. 9.19 Combining spiral collars and heavyweight drillpipe exposes the minimum area to sticking while maintaining the equivalent buoyed weight

Drilling fluids play an important role in relation to stuck pipe. Water-based muds with low filtration properties, low friction coefficients, thin filter cakes, and low rates of filter-cake buildup can reduce the severity of sticking if it does occur.

High-density muds exhibit little variation in cost because of the relative ease of decreasing the filtration properties of muds with high percentages of barite. Lower-density muds exhibit greater variation in the cost to reduce filtration properties, but the difference is small compared to total rig costs. Oil-based muds provide maximum protection against stuck pipe. Oil-based mud costs (as well as environmental concerns) have historically been considered prohibitive, when in fact these muds may be cheaper compared with water-based muds. Oil-based and synthetic-oil-based muds are becoming standard in many wells to avoid hole problems.

钻井液对卡钻起着重要的作用。对于低滤失性、低摩擦系数、薄滤饼和低滤饼堆积率的水基钻井液而言，即便发生卡钻，也可以将卡钻严重程度降低。

高密度钻井液降低滤失性能的成本变化不大，因为重晶石含量高的钻井液相对容易降低滤失性能。低密度钻井液降低滤失性能的成本变化较大，但与钻机总成本相比差异仍然较小。油基钻井液可最大限度地防止卡钻，但油基钻井液成本（及环保问题）历来被认为很高，而事实上，考虑到卡钻等事故影响，与水基钻井液相比，油基钻井液可能更便宜。油基和合成油基钻井液已成为许多油井避免卡钻的标准类型。

9.2 Loss of Circulation

9.2.1 Description

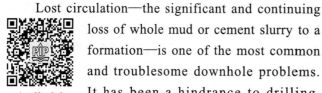
Audio 9.2

Lost circulation—the significant and continuing loss of whole mud or cement slurry to a formation—is one of the most common and troublesome downhole problems. It has been a hindrance to drilling, completion, and workover operations ever since rotary rigs first came into use, and it continues to have a profound negative impact on well economics. Estimates of the direct and indirect costs of lost-circulation problems in the drilling industry worldwide run into the hundreds of millions of US dollars annually.

Although drilling ahead and primary cementing pose particular risks, lost circulation can occur during any well procedure that involves pumping fluid down the hole. Indications of lost circulation may range from a gradual drop in pit level to a partial or complete loss of returns. In extreme cases, the fluid level in the annulus may drop rapidly, sometimes by hundreds of meters.

Lost circulation invariably results in higher costs for materials, services, and additional rig time. Depending on the timing and severity of its occurrence, it can lead to the loss of formation-evaluation data because the information normally obtained from mud returns and drilled cuttings is no longer available. Lost circulation can also result in reduced well productivity if the loss zone is also a potential pay interval. If the wellbore-fluid level drops far enough and fast enough, the drop can allow fluid to enter the wellbore from a higher-pressure formation. When this influx or kick does occur, it makes well control all the more difficult because of the inability to circulate kill fluid.

The loss can be partial or total, as shown in Fig. 9.20. In partial lost circulation, mud continues to flow to surface with some loss to the formation. Total loss circulation, however, is when all the mud flows into a formation with

9.2 井漏

9.2.1 概述

井漏是整个钻井液或水泥浆在地层中持续大量地漏失，是最常见和最麻烦的井下复杂情况之一。自旋转钻机首次投入使用以来，它一直是钻井、完井和修井作业的障碍，并继续对油井经济产生深远的负面影响。全世界钻井界因井漏问题造成的直接和间接损失估计每年达数亿美元。

尽管钻进和固井作业会发生一些特定的风险，但井漏更常见，它可以发生在任何涉及向井下泵送液体的钻井过程中。井漏可以通过钻井液池液位逐渐下降、钻井液部分或全部失返等现象来判定。在极端情况下，环空中的液面可能迅速下降，有时下降数百米。

井漏会导致材料花费、服务费用和额外钻井时间增加。在特定的时间和一定的严重程度下，井漏可能会导致地层评价数据丢失，因为井漏通常意味着返回的钻井液和钻屑中所含的信息不再可用。如果井漏区也是一个潜在的产层段，井漏也会导致油井产能降低。如果井筒液面下降得足够长，速度足够快，高压地层内的地层流体就会侵入井眼，发生井侵或井涌，由于压井液无法循环，使得井控更加困难。

钻井液漏失可以是部分漏失，也可以是全部失返，如图9.20所示。在部分漏失中，钻井液继续流向地面，同时有一部流入地层。然而，

no return to surface. In total loss circulation, if drilling is allowed to continue, it is referred to as blind drilling. This is not a common practice in the field, unless the formation above the thief zone is mechanically stable (i.e., there is no production and the fluid is clear water) and it is feasible, in terms of economics and safety, to go ahead with the drilling operation.

9.2.2 Lost circulation zones and causes

For lost circulation to occur, there must be (1) a formation with flow channels that allow passage of hole fluid from the wellbore and (2) an overbalance or positive pressure differential between the wellbore and the formation. Both of these conditions must be present, although one or the other may predominate. For example, a very small overbalance may be sufficient to drive fluid into a highly porous and permeable rock, while even a relatively nonporous, impermeable rock can accept considerable amounts of fluid if the overbalance is large enough to induce hydraulic fracturing.

(1) Permeable zones. Some types of rocks, because of their high primary porosity and permeability, almost seem to be designed to cause lost-circulation problems. Unconsolidated formations, gravel beds, loose conglomerates, and shallow or highly depleted sandstones have long been recognized as having natural lost-circulation tendencies. Lost circulation in these rocks most often manifests itself as a gradual drop in pit level, although continued drilling time and additional exposure to the wellbore may result in partial or complete mud losses (see Fig. 9.21, the sections marked with an "A").

(2) Natural fractures. Secondary porosity and permeability—such as occur in naturally fractured sandstones, shales, and carbonates—are also conducive to lost circulation (Fig. 9.21, the section marked with a "C"). Natural fractures may be either horizontal or vertical depending on a rock's depth, mechanical characteristics, and stress environment. In a horizontal fracture network, lost circulation may first manifest itself

9.2.2 井漏发生的区域及原因

井漏的发生必须具备以下条件：(1) 一个具备流通通道的地层，该流通通道允许井内流体从中通过；(2) 井筒和地层之间存在压力失衡或正压差。发生井漏，这两种情况都必须存在，可能是其中一种占主导地位。例如，一个非常小的压力失衡可能足以将井筒内的流体驱入一个高孔高渗的地层；对于低孔低渗地层而言，如果压力失衡大到足以引起水力压裂，也可以向地层内渗入相当数量的流体。

(1) 渗透带。某些类型的岩石，由于其高原生孔隙度和渗透率，发生井漏是显而易见的问题。松散地层、砾石层、松散砾岩和浅层或压力衰竭砂岩长期以来被认为具有自然井漏倾向。这些岩石中的井漏通常表现为钻井液池液位逐渐下降，持续钻井或过多暴露可能会导致部分漏失或全部失返（图9.21中标有"A"的部分）。

(2) 天然裂缝。天然裂缝性砂岩、页岩和碳酸盐岩中出现的次生孔隙度和渗透率也容易产生井漏（图9.21中标有"C"的部分）。天然裂缝可以是水平的，也可以是垂直的，这取决于岩石的深度、力学特性和应力环境。在水平裂缝网中，井漏可能首先表

as a gradual lowering of the pit level, with a complete loss of returns occurring as additional fractures are encountered. Vertical fractures, on the other hand, will take progressively increasing amounts of mud as drilling progresses and more of the fractures are exposed.

(3) Induced fractures. If lost returns occur in an area where offset wells have not experienced lost circulation, then the problem is likely the result of fracturing that is induced during well operations, rather than the result of a natural fracture network. Most induced fractures are related in some way to drilling-fluid or cementing programs, although sometimes the well architecture may itself be a contributing factor as, for example, when a surface or intermediate casing string is set too high. (Fig. 9.21, the section marked with a "D").

现为钻井液池液位逐渐降低，当遇到其他裂缝时，会发生钻井液全部失返。另外，随着钻井的进行，钻遇越来越多的垂直裂缝也将逐渐增加钻井液漏失量。

（3）人工裂缝。如果井漏发生在邻井未经历井漏的区域，则问题可能是井作业期间引发的压裂造成的，而不是天然裂缝造成的。大多数人工裂缝在某种程度上与钻井液或固井程序有关，当然也会受井身结构本身的影响，例如，表层或中间套管设计下深不够（图9.21中标有"D"的部分）。

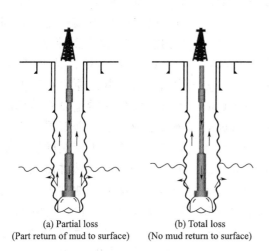

Fig. 9.20 Lost circulation

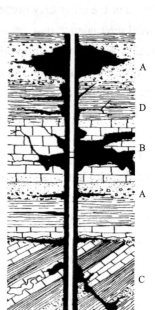

Fig. 9.21 Types of lost circulation

(4) Caverns. The most severe lost-circulation problems occur in cavernous or extremely vugular formations (Fig.9.21, the section marked with a "B"). These are typically limestones that have been leached by water. The void spaces in these formations can be large enough that when they are encountered, the drillstring may actually drop by as much as several meters preceding a sudden, complete loss of returns.

（4）溶洞。最严重的井漏问题出现在洞穴状或大溶洞地层中（图9.21中标有"B"的部分）。这些洞穴通常出现在被水淋溶的石灰岩中。这些地层中的洞穴空间可能很大，当遇到这些溶洞时，钻柱可能会突然下降几米，钻井液完全失返。在钻头遇到溶洞带

Rough drilling may occur just before a bit encounters a cavernous zone.

9.2.3 Prevention of Lost circulation

To prevent or at least minimize lost-circulation problems, it is necessary to address the conditions that cause lost circulation to occur, either by sealing off the problem formation or by reducing the wellbore pressure differential.

9.2.3.1 Mud System

Prevention of lost circulation starts with the mud system. The best way to seal off a potential loss interval is to keep filtrate losses to a workable minimum and to maintain a thin, firm, impermeable filter cake along the borehole wall. The mud specific density should be as low as possible, but high enough to control the formation pressure.

In an area where high porous, permeable zones are a known problem, and a low-weight, low-solids mud is being used, it sometimes is a good idea to pretreat the mud with solid LCM. This material should be fine enough to pass through the shale shaker with the other mud components and sized so as to plug small openings in the formation. If mud losses are fracture related, however, such pretreatment will not be effective, especially in weighted-mud systems.

The mud-weight schedule is perhaps the single most important factor in preventing lost circulation. The closer that the hydrostatic pressure of the mud column gets to the formation-fracture pressure, the more likely lost circulation becomes. Local drilling conditions and well parameters will determine how much overbalance is required to optimize drilling performance, control formation pressures, and allow for abnormal or unexpected conditions.

If the well cannot be safely drilled using a conventional mud system, then a rig equipped for underbalanced drilling should be considered.

9.2.3 井漏的预防

为了防止或至少减少井漏的发生，有必要通过封堵漏失地层或减小井筒压差来解决问题。

9.2.3.1 选择合理的钻井液体系

防止井漏从钻井液体系开始。封堵潜在漏失层段的最佳方法是将钻井液滤失保持在可操作的最小值，并沿井壁保持一个薄的、牢固的、不透水的滤饼。钻井液密度应尽可能低，但至少要达到足以平衡地层压力。

在高孔高渗地层，井漏是显而易见的问题，使用低密度、低固体含量的钻井液，有时最好使用固体堵漏材料（LCM）对钻井液进行预处理。LCM应足够细，能够与其他钻井液成分一起通过振动筛，并且尺寸应能塞住地层中的小开口。然而，如果钻井液漏失与裂缝有关，这种预处理将不起作用，特别是在加重钻井液系统中。

合理的钻井液设计密度是防止井漏的最重要因素。钻井液的静液柱压力越接近地层破裂压力，井漏的可能性越大。钻井现场条件和油井参数将决定优化钻井性能、控制地层压力及考虑异常或意外情况所需的过平衡压力附加量。

如果使用常规钻井液体系无法安全钻井，则应考虑欠平衡钻井。

9.2.3.2 Equivalent Circulating Density (ECD)

Even when the mud weight is far less than that required to fracture the formation, lost circulation can still result from a high ECD caused by excessive pump pressure and poor hydraulics practices. The mud's rheological properties (viscosity, yield point, and gel strength) should be specified to maintain its desired cuttings suspension and transport properties, but at the same time should enable the well to be circulated at an optimal pump pressure.

High surge pressure is a major contributor to lost circulation. Surge effects can be minimized by avoiding excessive speed when tripping in the hole, breaking circulation gradually, and maintaining circulation at the minimum pump rate needed to ensure adequate hole cleaning.

9.2.3.3 Casing Setting Depth

Selection of casing setting depths is crucial to preventing lost circulation and is closely related to the design of the mud program. In many wells, it is necessary to set one or more strings of intermediate casing to protect low-pressure zones from the higher mud weights required for deeper intervals. In selecting these casing points, the well planner should ensure that they are not themselves located in potential loss zones.

Complete prevention of lost circulation is not possible. This is because some formations— such as inherently fractured, low-pressure, cavernous, or high-permeability zones—are not avoidable when encountered during the drilling operation if the target zone is to be reached. However, mitigating the problem of lost circulation is possible if certain precautions, especially those related to induced fractures, are taken:

(1) Maintain proper mud weight;

(2) Minimize annular friction pressure losses during drilling and tripping in;

(3) Maintain adequate hole cleaning and avoid restrictions in the annular space;

(4) Set casing to protect upper weaker formations during a transition zone;

9.2.3.2 控制循环当量密度（ECD）

即使当钻井液密度远小于压裂地层所需的密度时，泵压过高和操作不当导致的高 ECD 仍然会导致井漏。应选择合适的钻井液流变特性（黏度、动切力和静切力），来维持岩屑悬浮和确保携岩能力，同时能够保证钻井液在最佳泵压下循环。

高激动压力是造成井漏的主要原因。通过避免下钻速度过快、逐渐停止循环、在确保井眼清洁最小泵速下循环等操作，可以将激动压力降至最低。

9.2.3.3 确定合理的套管下深

套管下入深度的选择是防止井漏的关键，这与钻井液设计密切相关。在许多油气井中，有必要设计一个或多个中间套管，以保护低压区不受较深层段所需较高钻井液密度的影响。在选择这些套管下入点时，钻井设计人员应确保下入点本身不位于潜在的漏失区。

完全防止井漏是不可能的。这是因为，如果要达到目的层，在钻井作业期间钻遇一些复杂地层（如裂缝、低压、洞穴状或高渗透层）是无法避免的。但是，如果采取某些预防措施，特别是与人工裂缝相关的预防措施，则可以缓解井漏问题：

（1）保持适当的钻井液密度；

（2）将钻进和起下钻过程中的环空压耗降至最低；

（3）保持足够的井眼清洁，避免环空受限；

（4）下套管，在过渡区保护上部较弱地层；

(5) Update formation pore pressure and fracture gradients for more accurate log and drilling data;

(6) If lost circulation zones are anticipated, work out preventive measures prior to drilling that zone by treating the mud with lost circulation materials.

9.2.4　Diagnosis of Lost Circulation

There are a number of methods for combating lost circulation, each of which is effective when properly used. Selecting the best method for a particular situation involves three diagnostic steps:

(1) Determining at what depth the loss is occurring

(2) Describing the type of loss zone

(3) Evaluating the severity of the loss

9.2.4.1　Depth

Intuitively, one might expect lost circulation to occur at or near the bottom of a well, where the ECD is at its highest. It is far more common, however, for the loss zone to be farther up the hole—typically near the casing shoe—where fractures may have been opened, resealed, and then reopened as the well was drilled deeper with increasing mud weights.

9.2.4.2　Methods of Locating Lost-Circulation Zones

The usual way to combat lost circulation during drilling is to monitor the possible presence of LCM across the suspected zone of loss. At shallow depth, the location of the losses into naturally permeable zones need not be known exactly. At greater depths (more than 1,500m) or when severe losses are occurring, the exact location of the "thief" zone must be determined before efficiently sealing the hole and continuing to drill. A number of methods have been developed for this purpose and are discussed below.

(1) Temperature survey. A temperature-recording device is run twice on wire and records the temperature at various depths (Fig. 9.22). First, the device is run under static conditions—when the mud temperature is in

（5）通过更准确的测井和钻井数据，更新地层孔隙压力和破裂压力值；

（6）如果预测会出现井漏，则在钻开漏失层前，在钻井液中加入堵漏材料，确定预防措施。

9.2.4　井漏诊断

防止井漏的方法有很多种，只要使用得当，每种方法都是有效的，通常需要以下三个诊断步骤：

（1）确定漏失发生的深度；

（2）描述漏失层的类型；

（3）评估漏失的严重程度。

9.2.4.1　深度

直观地说，人们可能认为井漏发生在井底或井底附近，此处的当量循环密度最高。然而，更常见的情况是，漏失层位于井眼上方更远的位置，通常位于上一层套管鞋附近。套管鞋处地层可能已经产生裂缝、重新闭合，然后随着钻井深度的增加和钻井液密度的增加而重新打开。

9.2.4.2　确定漏失层

在钻井过程中防止井漏的常用方法是使用LCM监测疑似漏失层的存在。在浅处，漏失到自然渗透带的位置不需要确切知道。在更深的井段（超过1500m）或发生严重漏失时，必须在有效封堵和继续钻进之前确定漏失层的准确位置。有很多方法用来确定漏失层的位置，下面将对它们逐一进行讨论。

（1）井温测定法。电缆温度记录仪下入两次，记录不同深度的温度（图9.22）。当钻井液温度与地层温度平衡时，在钻井液静

equilibrium with the formation—to provide a base log. Enough fresh, cool mud is then pumped into the hole so that the change in temperature can be recorded by a second survey. The temperature above the loss zone will be lower than that recorded in the first run. Below the thief zone, the mud remains static and its temperature will be higher than that of the mud flowing into the formation. The new temperature survey will show an anomaly across the zone where the losses are occurring, and their location can be determined by the depth where the recorded line changes its gradient. This method gives good results in areas where the temperature gradient is of the order of 1.8°C/100 m. One benefit of this method is that it can be used with drilling fluids containing large amounts of LCM.

态条件下下入电缆温度记录仪，提供基础记录。然后将足够的新配置冷却钻井液泵入井筒，进行第二次测量记录温度变化。漏失层上方的温度将低于第一次运行时记录的温度。而在漏失层以下，由于钻井液保持静止，其温度将高于流入地层的钻井液温度。第二次温度测量结果会显示漏失发生区域测量值的异常，漏失层位置可由第二次测量曲线温度梯度变化的深度所确定。这种方法成功地应用在温度梯度约为1.8°C/100m的区域，它的一个优点是可以用于含有大量LCM的钻井液。

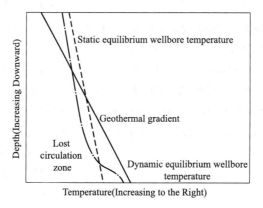

Fig. 9.22　Temperature survey

(2) Radioactive-tracer survey. Two gamma ray logs are run to determine the exact position of the thief zone. The first is recorded to establish the normal radioactivity of the downhole formation as a basis for comparison. Then, a small amount of radioactive material is displaced around the hole where losses are suspected to occur. A second gamma ray log is run and compared with the base log. At the thief zone, a steep change of radioactivity can be seen. The precise point of loss can be determined with this method, although it requires special equipment and is expensive.

(3) Hot-wire survey. As in the temperature survey,

（2）放射性示踪测量法。通过两次伽马测井来确定漏失层的确切位置。第一次记录是为了确定井下地层的正常放射性，作为对比的基础。然后将加了少量放射性示踪物质的钻井液替入可能会发生漏失的井段周围，进行第二次伽马测井，并与基础测井进行比较。在漏失层，可以看到放射性的急剧变化。虽然这种方法需要特殊的设备，而且价格昂贵，但它可以确定精确的漏失点。

（3）热敏电阻测量法。热敏电

the change in mud temperature is monitored. A calibrated resistance wire that is sensitive to changes in temperature is run to a certain depth, and then fresh, cool mud is pumped into the hole. If a change of temperature at the tool is observed, then the tool is placed above the point of loss. If no change is recorded, then it is placed below the thief zone. This method can be used in any mud system, but a large amount of mud is required to find the exact location of the loss.

(4) Spinner survey. A small spinner attached to the end of a cable is run in the hole to the location where the losses are suspected to occur. The spinner either spins or turns in response to mud movement, and the revolutions per minute of this response are recorded on film. Near the thief zone, acceleration can be observed as mud flows into the formation. This method delivers the best results when there are no sealing agents in the mud, but it requires large volumes of mud.

9.2.4.3 Type and Severity of Losses

Once the loss zone is located, it can be described in terms of its lithology and the type of loss that is occurring. For example, if there is a slow but steady decrease in pit level and if mud logs or other data indicate that the loss zone is composed of sandstone, then high permeability and porosity are likely the causes of the problem. On the other hand, if the loss of returns is sudden, induced fracturing is the most likely cause.

The severity of the problem can be expressed in terms of the amount of mud lost and the static fluid-level drop. Seepage causes a gradual lowering of the pit level (generally from 0.15 to 1.5 m^3/hr). Losses in the 1.5 to 15 m^3/hr range are considered partial. Complete losses involve fluid-level drops ranging from 60 to 150 m, while severe complete losses involve drops of more than 150 m where there is evidence of vugs or caverns. In the worst case of lost circulation, an underground blowout, the loss zone is taking not only drilling mud, but also formation fluid from a higher-pressure interval.

阻测量和温度测量一样主要监测钻井液的温度变化。把一根对温度变化敏感的校准电阻丝下入井筒，再将新配置的冷却钻井液泵入，如果测量工具温度变化，就意味着工具放在了漏失点上方。如果温度没有任何更改，则意味着将工具在漏失层下方。这种方法适用于任何钻井液，但需要大量钻井液才能找到准确的漏失位置。

（4）螺旋流量计法。通过电缆将一个小的螺旋流量计下入疑似发生井漏的位置。螺旋流量计根据钻井液运动旋转或转动，每分钟的旋转速度记录在胶片上。在漏失层附近，钻井液流入地层时可以观察到加速度。这种方法在钻井液中没有堵漏剂时效果最好，但需要大量钻井液。

9.2.4.3 确定井漏的类型和严重程度

一旦确定了漏失层位置，就可以通过漏失层岩性和漏失类型来描述井漏。例如，如果钻井液池水位缓慢但稳定地下降，钻井液记录或其他数据表明漏失层是砂岩，则高渗透性和孔隙度可能是问题的原因。另外，如果钻井液失返是突然的，则诱导裂缝是最可能的原因。

井漏的严重程度可以用钻井液漏失量和静态面下降程度来表示。渗透性漏失导致钻井液池液面逐渐降低（通常0.15～1.5m^3/h）。1.5～15m^3/h范围内的漏失视为部分失返。完全失返会有60～150m的液位下降，而对于存在岩穴或溶洞的严重失返会有150m以上的液面下降。井漏最严重就是地下井喷，钻井液和高压地层的流体都会漏失到漏失层。

9.2.5 Controlling Lost Circulation.

Techniques for controlling lost circulation are designed to seal off the loss interval. They may entail

(1) Allowing the formation to heal itself by removing the conditions that caused the lost circulation;

(2) Using LCM or drilled solids to bridge off the interval;

(3) Spotting a high-viscosity plug across the interval;

(4) Squeezing the interval with cement;

(5) Setting pipe across the interval;

(6) Abandoning or sidetracking the loss interval.

Depending on the location, type, and severity of the problem, remedial measures may involve a combination of these techniques. No one method is applicable to all types of lost circulation.

When lost circulation results from induced fracturing, a pause in operations or a change in drilling practices may help to eliminate the original cause of the fracture.

In some cases, stopping circulation and allowing solids to build up against the borehole wall may heal an induced fracture. One such procedure involves pulling the pipe into a protective casing or a secure portion of open hole, shutting down the mud pumps for a minimum of six to eight hours, attempting to fill the hole with water, and then gradually resuming circulation in stages.

Reducing the mud weight is an effective way of reducing the hydrostatic pressure of the mud column and, thus, the pressure differential with the formation. This is only feasible, of course, if there is no danger of a kick. Another step would be to adjust the mud viscosity and gel strength based on hole conditions—either increasing the viscosity and gel strength to help slow the flow of mud into permeable zones, or decreasing the viscosity and gel strength to reduce the pump pressure required for circulation, thereby lowering the ECD and reducing losses from induced fractures.

When lost circulation occurs, sealing the zone is necessary, unless the geological conditions allow blind

9.2.5 井漏控制

通过井漏控制技术封堵漏失层段，主要包括：

（1）通过消除造成井漏的条件，使地层自行愈合；

（2）使用堵漏剂或钻屑桥堵漏失层段；

（3）在漏失段打入高黏度浆液塞；

（4）在漏失层段挤水泥；

（5）漏失层段下入管柱封隔；

（6）放弃或侧钻漏失层。

根据井漏的位置、类型和严重程度不同，可能会综合使用上述方法，没有一种方法适用于所有类型的井漏。

当诱导裂缝造成井漏时，暂停作业或改变钻井方法可能有助于消除裂缝的影响。

在某些情况下，停止循环并允许岩屑堆积在井壁上，可能会治愈诱导裂缝。其中一个步骤是将钻柱提到套管内或裸眼的安全部分，关闭钻井泵至少6～8小时，尝试向井眼注水，然后分阶段逐步恢复循环。

降低钻井液密度是降低静液压力，从而降低压差的有效方法。当然，这只有在没有溢流危险的情况下才可行。还有就是根据井眼条件调整钻井液黏度和静切力，既包括通过增加黏度和静切力来减缓钻井液流入渗透层，也包括通过降低黏度和静切力以降低循环所需的泵压，从而降低ECD，减少诱导裂缝的漏失。

发生井漏时，必须封堵漏失层，除非地质条件允许盲钻（在

drilling (which is unlikely in most cases). The common lost circulation materials that are mixed with the mud to seal loss zones may be grouped as fibrous, flaked, granular, or a combination of the three.

These materials are available in course, medium, and fine grades for an attempt to seal low to moderate loss circulation zones. In the case of severe loss circulation zones, the use of various plugs to seal the zone becomes mandatory. It is important, however, to know the location of the lost circulation zone prior to setting a plug. Various types of plugs used throughout the industry include bentonite–diesel oil squeeze, cement–bentonite–diesel oil squeeze, cement, and barite.

大多数情况下不太可能）。加入钻井液中用以封堵漏失层的常见堵漏材料可分为纤维状、片状、粒状或三者的组合。

这些材料有粗、中、细三种等级，用于封堵小漏到中漏的漏失层。对于严重井漏的情况，必须使用各种段塞来封堵漏失段。然而，在打段塞之前，知道井漏的位置十分重要。整个行业使用的各种段塞类型包括挤压膨润土—柴油、水泥—膨润土—柴油、水泥和重晶石。

9.3 Hole Deviation

9.3 井斜

Hole deviation is the unintentional departure of the drill bit from drilling along a preselected borehole trajectory. Whether drilling a straight or a curved hole section, the tendency of the bit to walk away from the desired path can lead to a higher drilling cost and legal problems with regard to the lease boundary (Fig. 9.23).

Audio 9.3

井斜是指钻头偏离设计井眼轨道的现象。无论是直井段还是斜井段，钻头偏离设计轨道的趋势都会导致钻井成本的升高和有关租赁边界的法律问题（图9.23）。

Fig. 9.23 Wellbore deviation

It is not exactly known what causes a drill bit to deviate from its intended path. It is, however, generally agreed that one or a combination of several of the following factors may be responsible for hole deviation:

(1) Heterogeneous nature of formation and dip angle;

目前还不清楚是什么导致钻头偏离其设计轨道。然而，人们普遍认为，造成井斜的原因有：

（1）地层及地层倾角的非均质性；

(2) Drill string characteristics and dynamic behavior;

(3) Applied WOB;

(4) Hole inclination angle from vertical;

(5) Drill bit type and basic mechanical and hydraulic design;

(6) Hydraulics at the bit;

(7) Improper hole cleaning.

A resultant force acting on a drill bit causes hole deviation to occur. The mechanics of this resultant force is complex and is mainly governed by the mechanics of the BHA, the rock-bit interaction, bit operating conditions, and to a lesser extent, the drilling-fluid hydraulics. The forces imparted to the drill bit owing to the BHA are directly related to the makeup of the BHA, (stiffness, stabilizers, reamers, etc.).

The BHA is a flexible elastic structural member that can buckle under compressive loads. The buckled shape of a given design of BHA depends on the amount of applied WOB. The significance of the BHA buckling is that it causes the axis of the drill bit to misalign with the axis of the intended hole path, causing the deviation. Pipe stiffness and length and number of stabilizers (their location and clearances from the wall of the wellbore) are two major parameters that govern the BHA-buckling behavior. The buckling tendency of the BHA is minimized by reduction of WOB and use of stabilizers with outside diameters that are almost in gauge with the wall of the borehole.

The contribution of the rock-bit interaction to bit deviating forces is governed by the following factors:

(1) Rock properties (cohesive strength, bedding or dip angle, and internal friction angle)

(2) Drill bit design features (tooth angle, bit size, bit type, bit offset [in the case of roller cone bits], teeth location and number, bit profile, and bit hydraulic features)

(3) Drilling parameters (tooth penetration into the rock and its cutting mechanism)

The mechanics of the rock-bit interaction is a very complex subject and is the least understood problem

（2）钻柱特性和动态特征；

（3）钻压；

（4）井斜角；

（5）钻头类型及其机械设计和水力设计；

（6）钻头水力学；

（7）井眼清洁不当。

作用在钻头上的合力会引起井斜。这种合力是复杂的，主要受 BHA 力学特性、钻头与岩石的相互作用、钻头工作条件以及钻井液水力学的影响。BHA 施加在钻头上的作用力与 BHA 的组成（刚度、稳定器、扩眼器等）直接相关。

BHA 是柔性弹性结构，可在压缩载荷下发生屈曲。给定设计的 BHA 屈曲形状取决于施加钻压的大小。BHA 屈曲的意义在于，它导致钻头轴线与设计井眼轨道轴线不一致，从而导致偏差。钻杆刚度和长度、稳定器数量（它们的位置和与井壁的间隙）是控制 BHA 屈曲行为的两个主要参数。通过减少钻压和使用外径几乎与井壁相等的稳定器，可以将 BHA 的屈曲趋势最小化。

钻头与岩石的相互作用对钻头造斜力的影响由以下因素决定：

（1）岩石性质（黏结强度、层理或倾角及内摩擦角）；

（2）钻头设计特征 [齿角、尺寸、类型、移轴量（牙轮钻头）、齿位置和数量、钻头轮廓和钻头水力特征]；

（3）钻进参数（牙齿破碎岩石及切削机理）；

钻头与岩石相互作用的力学机理是一个非常复杂的问题，也

contributing to hole deviation. Fortunately, the advent of downhole measuring-while drilling (MWD) tools that allow the monitoring of the advance of the drill bit along the desired path has made our lack of understanding with regard to the mechanics of hole deviation more acceptable.

9.4 Borehole Instability

Borehole instability is an undesirable condition of an open hole interval that does not maintain its gauge size and shape and its structural integrity. The causes may grouped into the following categories:

Audio 9.4

(1) Mechanical—due to in situ stresses;
(2) Erosion—due to fluid circulation;
(3) Chemical—due to interaction of borehole fluid with the formation.

There are four different types of borehole instabilities, as shown in Fig. 9.24, including: (1) Hole closure or narrowing; (2) Hole enlargement or washouts; (3) Fracturing; (4) Collapse.

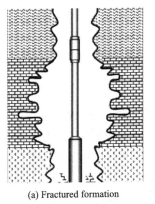
(a) Fractured formation

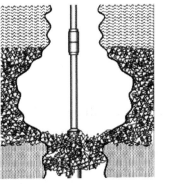
(b) Unconsolidated formation

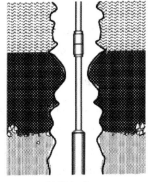
(c) Hole closure

Fig. 9.24 Borehole instability–associated problems

The problems associated with each type are addressed in the following sections.

9.4.1 Hole Closure

Hole closure is a time-dependent process of borehole narrowing, sometimes referred to as creep, under the overburden pressure. It generally occurs in plastic flowing shale and salt sections. The associated problems are

(1) Increase in torque and drag;
(2) Increase in potential for pipe sticking;
(3) Increase in difficulty of casings landing.

9.4.2 Hole Enlargement

Hole enlargements are commonly termed washouts; that is, the hole becomes undesirably larger than intended. Hole enlargements are generally caused by hydraulic erosion and mechanical abrasion owing to the drill string and in inherently sloughing shale. The associated problems are

(1) Increase in difficulty of cementing;
(2) Increase in potential for hole deviation;
(3) Increase in hydraulics requirements for effective hole cleaning;
(4) Increase in potential for problems during logging.

9.4.3 Fracturing

Fracturing occurs when the wellbore drilling-fluid pressure exceeds the formation fracture pressure. The associated problems are

(1) Lost circulation;
(2) Possible kick occurrence.

9.4.4 Collapse

Borehole collapse occurs when the drilling-fluid pressure is too low to maintain the structural integrity of the drilled hole. The associated problems are

(1) Pipe sticking;
(2) Possible loss of well.

9.5 Exercise

(1) Calculate the pull-out force, given the following well data:

① Depth of the stuck point = 10,000 ft;
② Drill collar OD = 6.25 in;
③ Hole size = 8.5 in;

④ Mud cake thickness = ⅛ in;
⑤ Coefficient of friction = 0.20;
⑥ Mud weight = 10 lbm/gal;
⑦ Length of the embedded portion = 50 ft;
⑧ Formation pressure at the stuck point = 4,950 psi.

(2) Determine the volume and pressures during spotting, given the following well data:

① Drill collar OD = 6.25 in;
② Hole size = 8.5 in;
③ Hole depth = 10,000 ft;
④ Mud weight = 10 lbm/gal;
⑤ Length of embedded drill collar = 100 ft;
⑥ Spotting fluid weight = 7.5 lbm/gal;
⑦ Assume 3° hole inclination, 5% washout, and inside diameter of the string uniform with 3 in.

(3) Calculate the allowable mud weight at 5,000 ft, given the following data:

① Cake thickness = ⅛ in;
② Diameter of the hole = 12¼ in;
③ Pipe diameter = 6¼ in;
④ Coefficient of friction = 0.25;
⑤ Pore pressure at 5,000 ft = 2,340 psi;
⑥ MOP = 75 kip.

(4) An 8½ in. hole is drilled through a depleted sand section of 100 ft. The wellbore pressure is 1,500 psi greater than the formation pressure. If a 6¼ in. drill collar is stuck over the entire section of the sand, calculate and plot the pull-out forces for cake thicknesses of ¹⁄₃₂ in, ¹⁄₁₆ in, ³⁄₃₂ in, and ⅛ in, using coefficients of friction 0.1, 0.15, 0.2, 0.25, and 0.3. State your conclusions.

(5) Calculate the optimum fishing time, given the following data:

① Well depth = 3,000m
② Estimated stuck depth = 2,500m
③ Daily operating rate of the rig = $1,500/day
④ Value of drill string below stuck point = $100,000
⑤ Time taken to drill original hole below stuck point = 30h

④ 泥饼厚度 =⅛in；
⑤ 摩擦系数 =0.20；
⑥ 钻井液密度 =10 lbm/gal；
⑦ 嵌入部分的长度 =50ft；
⑧ 卡点地层压力 =4950psi。

(2) 根据以下条件，确定解卡剂的用量和注入压力：

① 钻铤外径 =6.25in；
② 井眼尺寸 =8.5in；
③ 井深 =10000 ft；
④ 钻井液密度 =10lbm/gal；
⑤ 嵌入钻铤的长度 =100ft；
⑥ 解卡剂密度 =7.5lbm/gal；
⑦ 假设井斜为3°，冲蚀率为5%，钻柱内径统一为3in。

（3）根据以下数据，计算5000ft处的最大允许钻井液密度：

① 滤饼厚度 =⅛in；
② 井眼直径 =12¼in；
③ 钻杆外径 =6¼in；
④ 摩擦系数 =0.25；
⑤ 5000ft处的孔隙压力=2340psi；
⑥ MOP=75kip。

（4）钻穿100 ft枯竭砂岩段的8½in井眼，井筒压力比地层压力高1500psi。如果6¼in的钻铤卡在整个砂岩段 摩擦系数分别取0.1、0.15、0.2、0.25和0.3，请计算并绘制¹⁄₃₂ in、¹⁄₁₆ in、³⁄₃₂ in 和 ⅛ in 滤饼厚度的上提力。

（5）根据以下数据计算最佳打捞时间：

① 井深3000m；
② 估计卡点深度为2500m；
③ 钻机的日运行费率为1500美元/天；
④ 卡点以下钻柱价值为100000美元；
⑤ 在卡点下方钻原始井眼所用的时间为30小时。

References/ 参考文献

[1] Bourgoyne A T, Millheim K K, Chenevert M E, et al. Applied Drilling Engineering (revised printing) [M], Textbook Series, SPE, Richardson, Texas 2.1991.

[2] Robert F Mitchell, Stefan Z Miska, Bernt S Aadnoy. Fundamentals of Drilling Engineering [M], Richardson, TX 75080-2040 USA. 2011.

[3] Bernt S Aadnoy, Iain Cooper, Stefan Z Miska, et al.. Advanced Drilling and Well Technology [M], Richardson, TX 75080-2040 USA. 2009.

[4] John Ford. Drilling Engineering [M]. Heriot-Watt University.

[5] Cunha J C. Effective Prevention and Mitigation of Drilling Problems [J]. World Oil & Gas Technologies, 2002, Volume 2 (September): 28-34.

[6] DeLuca M. Surface BOP Maintaining Its Niche [J]. Offshore Engineer, 2005, 30 (2): 19-22.

[7] Derrick Engineering Company. For Sale: Oil Field Masts and Substructures, Drill Rigs Available for Immediate Sale [M], http://www.derrickengineering.com/classifi eds.htm (accessed 16 December 2010). 2010.

[8] Lake L W. Petroleum Engineering Handbook [M], Vol. II, Chap. 14, II-607. Richardson, Texas: SPE. 2006.

[9] Aadnoy B S. Modern Well Design: Second Edition [M]. Leiden: Francis and Taylor. 2010.

[10] Aadnoy B S. and Chenevert, M.E. Stability of Highly Inclined Boreholes [M]. SPE Drill Eng 2 (4): 364-374. 1987.

[11] Hareland G, Hoberock L L. Use of Drilling Parameters to Predict In-Situ Stress Bounds [J]. Paper SPE/IADC 25727 presented at the SPE/IADC Drilling Conference, Amsterdam, 22-25 February. DOI:10.2118/25727-MS. 1993.

[12] Jorden J R, Shirley O J.Application of Drilling Performance Data to Overpressure Detection [J]. J. Pet.Tech. 18 (11): 1387-1394. SPE-1407-PA. DOI: 10.2118/1407-PA. 1966.

[13] Annis M R. High-Temperature Flow Properties of Water-Base Drilling Fluids [J]. JPT 19 (8): 1074-1080;Trans., AIME, 240. DOI: 10.2118/1698-PA. 1967.

[14] Azar J J. and Sanchez, R.A. Important Issues in Cuttings Transport for Drilling Directional Wells [J]. Paper SPE39020 presented at the Latin American and Caribbean Petroleum Engineering Conference and Exhibition, Riode Janeiro, 30 August-3 September. DOI: 10.2118/39020-MS. 1997.

[15] Bailey W J, Peden J M.A Generalized and Consistent Pressure Drop and Flow Regime Transition Model for Drilling Hydraulics [J]. SPEDC 15 (1): 44-56. SPE-62167-PA. DOI: 10.2118/62167-PA.2000.

[16] Brand F, Peixinho J, NouarC. A Quantitative Investigation of the Laminar-to-Turbulent Transition: Application to Efficient Mud Cleaning [J]. Paper SPE 71375 presented at the SPE Annual Technical Conference and Exhibition, New Orleans, 30 September-3 October. DOI: 10.2118/71375-MS.2001.

[17] Chabra R P, Richardson J F. Non-Newtonian Flow in Process Industries [M].Oxford: Butterworth-Heinemann. 1999.

[18] Escudier M P, Oliveira P J, Pinho F T. Fully Developed Laminar Flow of Purely Viscous Non-

Newtonian Liquids Through Annuli, Including Effects of Eccentricity and Inner Cylinder Rotation [J]. International J.of Heat and Fluid Flow 23 (1): 52-73. 2002.

[19] Kelessidis V C, Mpandelis G. Flow Patterns and Minimum Suspension Velocity for Effi cient Cuttings Transport in Horizontal and Deviated Wells in Coiled-Tubing Drilling [J]. Paper SPE 81746 presented at the SPE/ICoTA Coiled Tubing Conference and Exhibition, Houston, 8-9 April. DOI: 10.2118/81746-MS. 2003.

[20] Charlez P A, Easton M, Morrice G. Validation of Advanced Hydraulic Modeling Using PWD Data [J]. Paper 8804 presented at the Offshore Technology Conference, Houston, 4-7 May. DOI: 10.4043/8804-MS. 1998.

[21] Clark R K, Bickham K L. A Mechanistic Model for Cuttings Transport [J]. Paper SPE 28306 presented at the SPE Annual Technical Conference and Exhibition, New Orleans, 25-28 September. DOI: 10.2118/28306-MS.1994.

[22] Doan Q T, Oguztoreli M, Masuda Y, et al. Modeling of Transient Cuttings Transport in Underbalanced Drilling [J]. SPEJ 8 (2): 160-170. SPE-85061-PA. DOI: 10.2118/85061-PA.2003.

[23] Leising L J, Walton I C. Cuttings Transport Problems and Solutions in Coiled Tubing Drilling [J]. SPEDC17 (1): 54-66. SPE-77261-PA. DOI: 10.2118/77261-PA.2002.

[24] Li J, Walker S. Sensitivity Analysis of Hole Cleaning Parameters in Directional Wells [J]. SPEJ 6 (4):356-363. SPE-74710-PA. DOI: 10.2118/74710-PA.2001.

[25] Lubinski A, Hsu F H, Nolte K G. Transient Pressure Surges Due to Pipe Movement in an Oil Well [J]. Oil & Gas Science and Technology - Rev, 1977, IFP 32 (3): 307-348.

[26] Masuda Y, Doan Q, Oguztoreli M, et al. Critical Cuttings Transport Velocity in Inclined Annulus: Experimental Studies and Numerical Simulation [J]. Paper SPE 65502 presented at the SPE/CIM International Conference on Horizontal Well Technology, Calgary, 6-8 November. DOI: 10.2118/65502-MS.2000.

[27] Mitchell R F. Surge Pressures in Low-Clearance Liners [J]. Paper SPE 87181 presented at the IADC/SPEDrilling Conference, Dallas, 2-4 March. DOI: 10.2118/87181-MS.2004.

[28] Zamora M, Roy S. The Top 10 Reasons To Rethink Hydraulics and Rheology [J]. Paper SPE 62731 presented at the 2000 IADC/SPE Asia Pacifi c Drilling Technology Conference, Kuala Lumpur, 11-13 September.2000.

[29] Zhu H, Kim Y D, Kee D. Non-Newtonian Fluids With a Yield Stress [J]. J. of Non-Newtonian Fluid Mechanics, 2005, 129 (3): 177-181.

[30] Duman O B, Miska S, Kuru E. Effect of Tool Joints on Contact Force and Axial-Force Transfer in Horizontal Wellbores [J]. SPE Drill & Compl 18 (3): 267-274. SPE-85775-PA. DOI: 10.2118/85775-PA.2003.

[31] Guo B, Miska S, Lee R L.Constant Curvature Method for Planning a 3-D Directional Well [J]. Paper SPE 24381 presented at the SPE Rocky Mountain Regional Meeting, Casper, Wyoming, 18-21 May. DOI:10.2118/24381-MS. 1992.

[32] Inglis T A. Directional Drilling. London [M]: Graham and Trotman.1987.

[33] Lubinski A. Developments in Petroleum Engineering [M], Vol. 1: Stability of Tubulars, Deviation Control. Ed. Stefan Miska. Houston: Gulf Publishing Company.1987.

[34] Miska S, Miska W. Modeling of Complex Bottom Hole Assemblies in Curved and Straight Holes [J]. Archives of Mining Sciences 51 (1): 35-54.2006.

[35] Sampaio J H B Jr. Planning 3D Well Trajectories Using Spline-in-Tension Function [J]. J. of Energy Resources Technology 129 (4): 289-300.2007.

[36] Sawaryn S J, Thorogood J L. A Compendium of Directional Calculations Based on the Minimum Curvature Method [J]. Paper SPE 84246 presented at the SPE Annual Technical Conference and Exhibition, Denver, 5-8 October. DOI: 10.2118/84246.2003.

[37] Warren T.Steerable Motors Hold Their Own Against Rotary Steerable Systems [J]. Paper SPE 104268 presented at the SPE Annual Technical Conference and Exhibition, San Antonio, Texas, USA, 24-27 September. DOI: 10.2118/104268-MS. 2006.

[38] Paul W Ellis. Blowout Prevention and Well Control [M]. Editions Technip, 27 Rue Ginoux 75737, Paris Cedex 15.1981.

[39] Goins W C, Riley Sheffielf. Blowout Prevention [M], 2nd ed., Gulf Publishing Co., Book Division. Houston, London, Paris.

[40] Mitchell Bill.Oil Well Drilling Engineering [M], 8th ed. 1974: Golden, CO.

[41] Baker Oil Tools. Tech Facts: Engineering Handbook [M]. Houston: Baker Hughes, 1993.

[42] Boresi A P, Schmidt R J. Advanced Mechanics of Materials [M], sixth edition. New York City: John Wiley& Sons.2002.

[43] Mitchell R F. Lateral Buckling of Pipe With Connectors in Horizontal Wells [J]. SPE J. 8 (2): 124-137. SPE-84950-PA. DOI: 10.2118/84950-PA.2003.

[44] Mitchell R F. Tubing Buckling—The State of the Art [J]. SPE Dril & Compl 23 (4): 361-370. SPE-104267-PA. DOI: 10.2118 104267-PA.2008.

[45] Mitchell R F. Fluid Momentum Balance Defines the Effective Force [J]. Paper SPE 119954 presented at theIADC/SPE Drilling Conference and Exhibition, Amsterdam, 17-19 March. DOI: 10.2118/11 954-M.2009.

[46] API RP 65-Part 1, Cementing Shallow Water Flow Zones in Deep Water Wells [M], first edition. Washington DC: API.2002.

[47] API RP 65-Part 2, Isolating Potential Flow Zones During Well Construction [M], first edition. Washington DC: API.2010.

[48] Halliburton. The Red Book: Halliburton Cementing Tables [M]. Duncan, Oklahoma: Halliburton Company.2001.

[49] Kosmatka S H, Wilson M L. Design and Control of Concrete Mixtures [M], 15th edition. Skokie, Illinois: Portland Cement Association.2011.